Advances in Smart Grid Power System

Advances in Smart Grid Power System

Network, Control and Security

Edited by

Anuradha Tomar
Electrical Engineering Department
Eindhoven University of Technology
Eindhoven, The Netherlands

Electrical Engineering Department
JSS Academy of Technical Education
Noida, India

Ritu Kandari
Department of Electronics and Communication Engineering
Indira Gandhi Delhi Technical University for Women
Delhi, India

ELSEVIER

ACADEMIC PRESS
An imprint of Elsevier

Academic Press is an imprint of Elsevier
125 London Wall, London EC2Y 5AS, United Kingdom
525 B Street, Suite 1650, San Diego, CA 92101, United States
50 Hampshire Street, 5th Floor, Cambridge, MA 02139, United States
The Boulevard, Langford Lane, Kidlington, Oxford OX5 1GB, United Kingdom

Library of Congress Cataloging-in-Publication Data
A catalog record for this book is available from the Library of Congress

British Library Cataloguing-in-Publication Data
A catalogue record for this book is available from the British Library

ISBN: 978-0-12-824337-4

For information on all Academic Press publications visit our website at https://www.elsevier.com/books-and-journals

Publisher: Brian Romer
Acquisitions Editor: Lisa Reading
Editorial Project Manager: Alice Grant
Production Project Manager: Poulouse Joseph
Cover Designer: Vicky Pearson

Typeset by TNQ Technologies

Contents

Contributors

Rashmi Agarwal, Department of Electrical Engineering, J.C.Bose University of Science & Technology, YMCA, Faridabad, Haryana, India

A.V. Ajay, Department of Computer Science and Engineering, NIE Institute of Technology, Mysuru, Karnataka, India

Divya Asija, Department of Electrical & Electronics Engineering, Amity School of Engineering and Technology, Amity University Uttar Pradesh, Noida, India

Chittibabu B, Department of Electronics and Communication Engineering, Indian Institute of Information Technology Design and Manufacturing Kancheepuram, Chennai, Tamil Nadu, India

Narendra Babu P, Department of Electrical Engineering, National Institute of Technology Meghalaya, Shillong, Meghalaya, India

Manoj Badoni, Electrical and Instrumentation Engineering Department, Thapar Institute of Engineering and Technology, Patiala, Punjab, India

Y.S. Brar, Department of Electrical Engineering, IKGPTU, Kapurthala, Punjab, India

Sumana Chattaraj, Indian Institute of Technology Jodhpur, Rajasthan, India

M. Gowtham, Department of Computer Science and Engineering, NIE Institute of Technology, Mysuru, Karnataka, India

Neeraj Gupta, Department of Electrical Engineering, National Institute of Technology, Srinagar, Jammu and Kashmir, India

Pankaj Gupta, Department of Electronics and Communication Engineering, Indira Gandhi Delhi Technical University for Women, Delhi, India

Sanjay K. Jain, Department of Electrical and Instrumentation Engineering, Thapar Institute of Engineering and Technology, Patiala, Punjab, India

Karan Singh Joshal, Department of Electrical Engineering, National Institute of Technology, Srinagar, Jammu and Kashmir, India

Ritu Kandari, Department of Electronics and Communication Engineering, Indira Gandhi Delhi Technical University for Women, Delhi, India

Navdeep Kaur, Department of Electrical and Instrumentation Engineering, Thapar Institute of Engineering and Technology, Patiala, Punjab, India

Md. Irfan Khan, Supreme & Co. Pvt. Ltd., Kolkata, West Bengal, India

Bharti Koul, Department of Electrical Engineering, NIT Hamirpur, Hamirpur, Himachal Pradesh, India

Y. Bhuvan Kumar, Department of Computer Science and Engineering, NIE Institute of Technology, Mysuru, Karnataka, India

Ashwani Kumar, Department of Electronics and Communication Engineering, Indira Gandhi Delhi Technical University for Women, Delhi, India

Rajender Kumar Beniwal, Electrical Engineering Department, Sobhasaria Group of Institutions, Sikar, Rajasthan, India

P. Manoj, Department of Computer Science and Engineering, NIE Institute of Technology, Mysuru, Karnataka, India

Sushri Mukherjee, Indian Institute of Technology Delhi, Hauz Khas, New Delhi, India

Rangababu P, Department of Electronics and Communication Engineering, National Institute of Technology Meghalaya, Shillong, Meghalaya, India

Kaibalya Prasad Panda, Department of Electrical Engineering, National Institute of Technology Meghalaya, Shillong, Meghalaya, India

Gayadhar Panda, Department of Electrical Engineering, National Institute of Technology Meghalaya, Shillong, Meghalaya, India

Dharmbir Prasad, Asansol Engineering College, Asansol, West Bengal, India

B. Rajanarayan Prusty, School of Electrical Engineering, Vellore Institute of Technology, Vellore, India

Pushpendra, Department of Electrical Engineering, J.C.Bose University of Science & Technology, YMCA, Faridabad, Haryana, India

Mamata Rath, School of Management (Information Technology), Birla Global University, Bhubaneswar, Odisha, India

Koushik Sarkar, Supreme & Co. Pvt. Ltd., Kolkata, West Bengal, India

Alka Singh, Electrical Engineering Department, Delhi Technological University, Delhi, India

Kanwardeep Singh, Department of Electrical Engineering, Guru Nanak Dev Engineering College, Ludhiana, Punjab, India

Rudra Pratap Singh, Asansol Engineering College, Asansol, West Bengal, India

Karan Singh Joshal, Department of Electrical Engineering, National Institute of Technology, Srinagar, Jammu and Kashmir, India

Anuradha Tomar, Electrical Engineering Department, Eindhoven University of Technology, Eindhoven, The Netherlands; Electrical Engineering Department, JSS Academy of Technical Education, Noida, India

Debesh Shankar Tripathy, National Institute of Science and Technology, Berhampur, Odisha, India

Rajkumar Viral, Department of Electrical & Electronics Engineering, Amity School of Engineering and Technology, Amity University Uttar Pradesh, Noida, India

D.B. Vishwas, Department of Computer Science and Engineering, NIE Institute of Technology, Mysuru, Karnataka, India

Chapter 1

An introduction to the smart grid-I

Pankaj Gupta, Ritu Kandari, Ashwani Kumar
Department of Electronics and Communication Engineering, Indira Gandhi Delhi Technical University for Women, Delhi, India

1. Introduction

Traditional power systems are designed considering power flow to be unidirectional from generating station to consumers. The process involves step-up of voltage at the generating station and thereby transmission of electricity through transmission lines, which further steps down the voltage at the distribution substation for feeding to consumers. Generators, transformers, transmission lines, feeders, electromechanical metering, control, and protection units are the main components of a traditional power system. This system is a centralized generation system where the generating stations are centralized with no direct link between the generating unit and end user. Most thermal power plants today are becoming old, their efficiencies are coming down day by day, and concerns have been raised about their depletion of coal [1,2]. The potential for technological advancement in new business opportunities, aging assets, environmental concerns, and a lack of circuit capacity create the need for rapid replacement of these facilities, which is the biggest challenge [3]. Security of supply requires a reliable electricity supply as a growing number of critical loads are connected to grid. However, over time, power systems continue upgrading to provide more information, real-time monitoring, and control by making use of communication architecture [4]. Promoting renewable energy may be one possible solution to tackle the challenges posed by conventional energy resources [5]. Adding renewable energy sources (RESs) to the maximum extent possible can lead to excess generation from installed capacity and may reduce the gap between the peak energy requirement and available energy. These generations are locally available and location-specific, so they are being embedded into the low-voltage distribution grid. The distribution system has seen major changes because of distributed generation (DG) being embedded into the distribution system, making the power flow

Advances in Smart Grid Power System. https://doi.org/10.1016/B978-0-12-824337-4.00001-1

bidirectional [6]. Generation is most commonly done from RESs: solar, photovoltaic (PV), wind, etc. Liberalization of the energy market now allows private operators and consumers to participate in buying and selling energy along with utilities [7–9]. This upgrade has transformed the conventional power system into a smart grid, which applies extensive communication to every component of the power system for online monitoring, control, protection, and metering purposes [10,11]. Two-way interactions between customers and utilities, along with extensive use of sensors throughout the transmission and distribution lines, make the grid smarter [12]. From a customer's perspective, customers can organize their consumption patterns based on energy availability, cost, and time of use; they may reduce their monthly energy bills by the maximum amount possible [13]. The smart grid helps the customer to manage their schedule for usage of electricity and choose the favorable time to sell or purchase electricity [14]. One can participate and save on energy bills, by generating power from a roof top solar PV generation [15] and by using electric vehicle [16]. One of the major factors that put us in risk today is the huge demand of electricity. Shortfalls of electricity means there is an energy gap between generation and consumption. The distribution loss, demand response management, energy management during peak hours, peak shortage [17], global warming, reduced gas emission [18], competitive bidding [19], electric vehicles [16], cybersecurity, Internet of Things [20], and advanced communication [10] are some areas of research in the smart grid scenario. Technically, control [21] and protection [22] of the distribution system in a smart grid scenario are a big challenge. A broad comparison between traditional and smart grids is given in Table 1.1.

This chapter presents an introduction to the smart grid with a view toward the smart grid framework and road maps issued by the National Institute of Standards and Technology (NIST) for future smart grids [23–25]. The rest of the paper is organized as follows. Section 2 discusses the traditional power system in brief. Section 3 presents the NIST concept of the smart grid framework. Section 4 discusses smart grid infrastructure and technologies. Energy storage technologies are presented in Section 5. Benefits of the smart grid as well as various challenges are discussed in Sections 6 and 7, respectively. Section 8 concludes the chapter.

2. Traditional power system model

An electric power system (EPS) is a network of energy providers and consumers interconnected with the help of transmission and distribution lines. Fig. 1.1 is a pictorial view of a traditional power system model consisting of generating stations, step-up transformer, transmission lines, step-down distribution transformer, and distribution lines feeding various loads. Besides this, there are control centers that help in operations and control of the EPS. The power system protection unit protects the EPS during various types of system

TABLE 1.1 A comparison of smart and traditional grids.

Feature	Traditional grid	Smart grid
Power flow	Unidirectional	Bidirectional
Communication	One-way	Two-way
Distribution system	Passive	Active
Monitoring	Manual	Self
Restoration	Manual	Self-healing
Complexity	Less complex	More complex
Customer involvement	No	Yes
Resiliency	Not resilient	Better resiliency
Sensors	Fewer sensors	Extensive use of sensors
Control	Centralized	Centralized and decentralized
Metering system	Traditional (electromechanical)	Advanced metering (electronic)

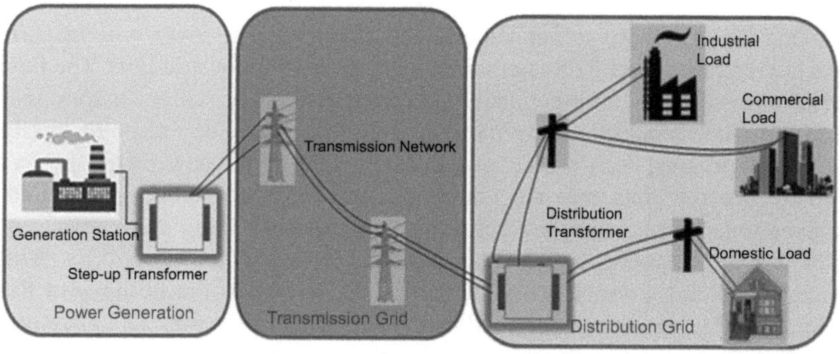

FIGURE 1.1 Traditional power system model.

faults. The transmission of extra-high-voltage AC lines comprises 800/765, 400, 220–132 kV, whereas the voltage of commonly used high-voltage DC lines is as high as 500 kV.

Power generation in a traditional EPS is typically from thermal power stations and hydropower stations, with a small proportion from nuclear power stations, nonconventional energy resources, etc. Subtransmission customers are connected to a 33 kV line and primary customers to an 11 kV line, while

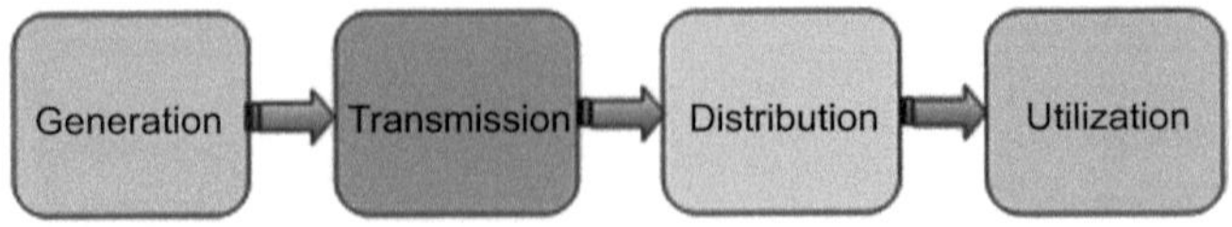

FIGURE 1.2 Power flow in a traditional electric power system.

secondary customers—i.e., residential loads—are connected to a 400 V line, which may further be stepped down to 220 V for various domestic purposes. Customers may be industrial, commercial, or residential domestic types of loads. The EPS structure is historically vertical in nature—that is, energy is produced at the generation station and then transmitted through transmission lines at high voltage before being further stepped down at the distribution transformer to feed various types of loads through distribution lines and feeders. This means that power flow is unidirectional in a traditional EPS—i.e., from generating stations to consumers—as shown in Fig. 1.2. Hence, the control and protection of traditional power systems are designed assuming unidirectional power flow [26]. The customers are only consumers of electricity and have no role in electricity pricing, nor can they contribute to the grid even if they own a captive power plant.

3. Smart grid framework

Probable smart grid structures and expected features have been extensively researched during the last decade [27,28]. The overall focus is on continuously making the grid more automated and computerized. The smart grid supports the interconnection of RESs in the low-voltage distribution grid [29]. The flow of electricity is now bidirectional—i.e., from generating station to consumer end and vice versa [22]. The generating stations generate necessary electricity according to total load requirements. All sectors of the power system have remote control units that are connected to one central unit. As required, conventional technologies are being modified and becoming more advanced, whereas those technologies that become obsolete are removed. With innovations and advancements in existing technologies, the utility grid has become more resilient and intelligent [30]. The integration of smart and intelligent devices for efficient, sustainable, and secure electric power flow from generation to consumption makes the system smart. It enables smart and automatic applications like energy storage, advance metering infrastructure [31], smart distribution management [32], and demand response. Real-time measurement and monitoring, big data, and intelligent computing are the key features [33] that make the grid smart. Using advanced automated control mechanisms, energy management schemes [32], smart sensors, and smart meters [31] improves system stability, making it more reliable. Advancements in technology have made integration of RESs possible [34], hence making the flow of electricity bidirectional and allowing for active consumer participation.

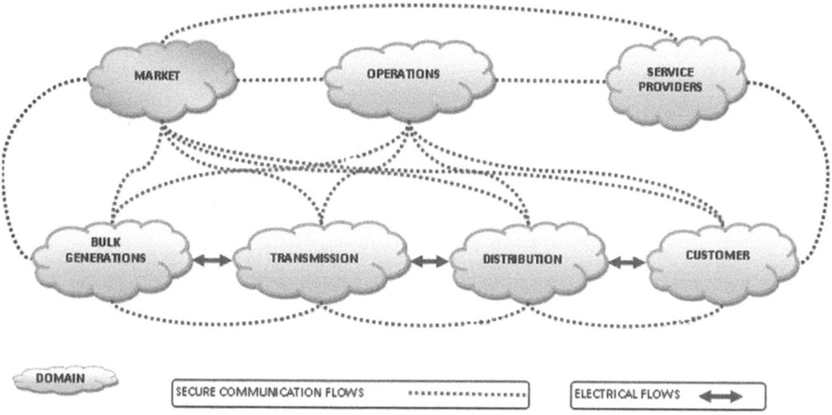

FIGURE 1.3 Smart grid conceptual framework by the National Institute of Standards and Technology.

Customers may have the pleasure of "easy plug and play" [35], which means they now can export and import energy according to their choices and costs.

Fig. 1.3 is a pictorial view of the smart grid concept developed by the NIST [23,24]. It consists of seven domains, namely service providers, operations, markets, generation, transmission, distribution, and customer domains. The smart grid conceptually takes cares of secure communication between the various domains. The smart grid operates between customers, bulk generation, service providers, and the energy market with the help of an operating center. There is a secure communication interface between all the domains. The smart grid conceptually takes care of electrical energy flows between bulk generation to customers and vice versa. Together, the energy and communication interface between various domains conceptualizes the smart grid framework. The bulk-generation domain includes conventional and nonconventional energy resources. The customer domain includes industrial, commercial, and domestic customers. The role of each domain as described in Refs. [23,24] is summarized in Table 1.2 and discussed in the next subsections.

3.1 Bulk-generation domain

The bulk-generation domain includes all possible types of electricity generators that may participate in providing power to grid [23,24]. The bulk-generation domain helps to generate power. Bulk generation may be further classified into three classes, namely (1) renewable sources with variable generation, (2) renewable sources with nonvariable generation, and (3) nonrenewable sources with nonvariable generation, as shown in Fig. 1.4. There is an electrical connection between the bulk-generation and transmission domains. The bulk-generation domain shares information with the operation, market, and transmission domains via a secure communication network.

TABLE 1.2 Participants and services in the domains of the smart grid framework.

Domain	Participants and roles
Customers	Comprises industrial, commercial, and residential customers; may produce, store, or manage electricity usage; may also actively participate in energy saving and pricing
Bulk generation	Includes all possible types of electricity generation that may participate in providing power to grid; promotes renewable energy, maintains stability and reliability of system; may also store energy
Transmission	Transmission of bulk power over long distances; may store and generate electricity
Distribution	Distribution of energy generated by utilities and customers
Markets	Includes operators and contributors of energy markets; competitive bidding, selling, and purchase of power; also enables small utilities to participate in power markets
Service providers	Includes companies facilitating services to consumers and utilities; helps in processing business of all smart grid domains: billing, customer account management, management of energy, etc.
Operations	Supervisors ensure the flow of electricity

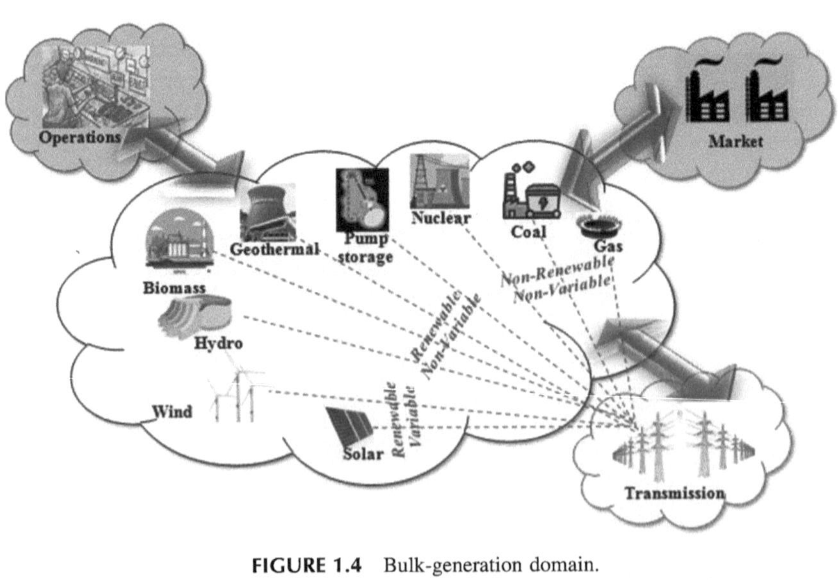

FIGURE 1.4 Bulk-generation domain.

The main requirements for the bulk-generation domain include the promotion of renewable sources and private utilities, greenhouse gases emissions control, the provision of energy storage, wide area monitoring (WAM), interoperability among various domains, and maintenance of system stability and reliability.

3.2 Transmission domain

The transmission domain transmits the bulk power generated at the generating station to the distribution system and vice versa. It transfers electrical power from generation sources to distribution through multiple substations; the generating station voltage is from 20 to 23 kV, which is stepped up to high voltage of 800 or 400 kV. The high voltage is then stepped down to 230 kV, then to 132 kV, 66 kV, 11 kV, and finally 440 V at the customer end. The transmission network is operated by two operators—one is the regional transmission operator and the other is an independent system operator (ISO) [36] whose primary responsibility is to maintain the stability of the electric grid. Stability is maintained by balancing generation with load. If the two are balanced, the system will be stable, and frequency and voltage will be within limits. The other services are energy and support ancillary services. The ancillary services provide frequency support, voltage support, and spinning reserve.

The transmission domain has three parts, namely substation storage, measurement, and control. The substation helps maintain all equipment, such as the transformers, measuring equipment, protection system, and metering equipment, used for the transmission system. The measurement and control section measures, records, and controls with the intent of optimizing grid operation. The storage part provides control by charging and discharging the energy storage unit in accordance with requirements. If generation is more than consumption or load demand, the excess energy is stored in the storage device. This energy is utilized when generated power is in deficit.

3.3 Distribution domain

The distribution domain is the important interface between the transmission and customer domains. This domain interacts with the transmission, customer, operations, and market domains for smooth operation of the power system as shown in Fig. 1.5. Smart metering systems are also well connected with the help of the distribution domain. The electrical distribution system may be arranged in a variety of structures including radial, mesh, and loop. For improved reliability, a radial system may also have a loop or mesh system. The distribution system has undergone significant changes in order to facilitate the interconnection of RESs, electric vehicles, and energy storage devices.

Smart grid architecture provides well-connected electrical access for residences, electric vehicles, and RESs through an efficient energy management system [37]. The protection of the distribution system has undergone a complete

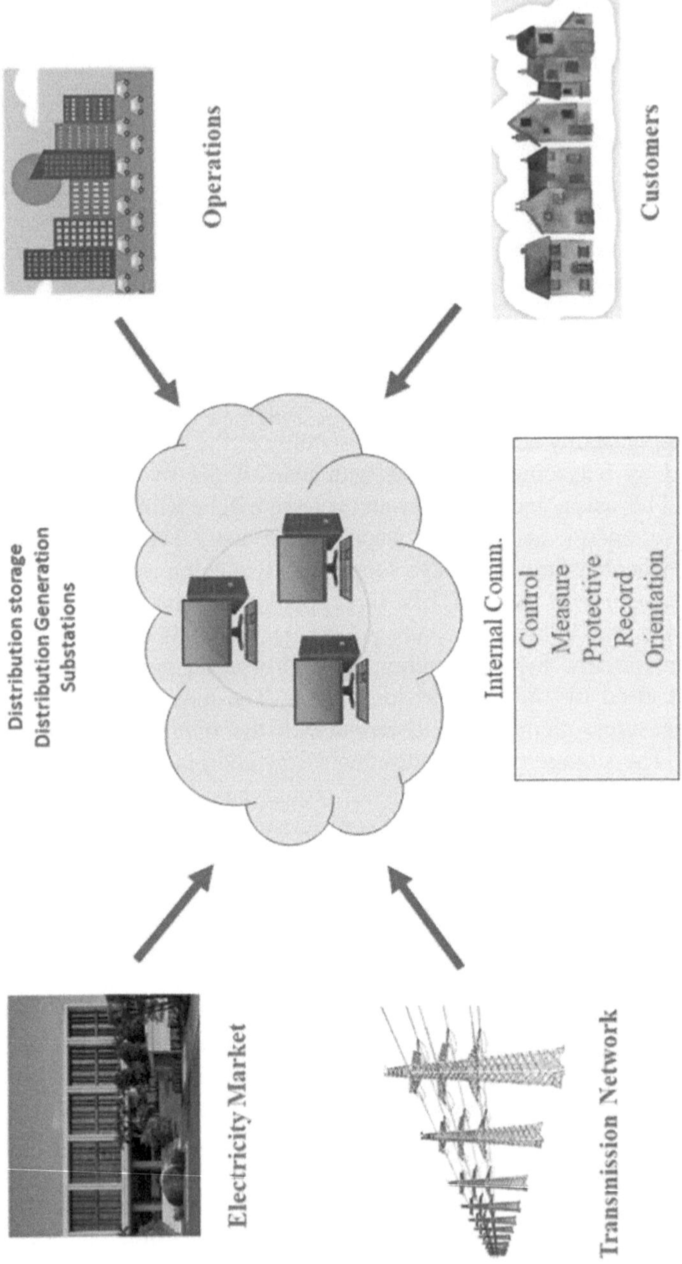

FIGURE 1.5 Distribution domain.

change because it was designed under the assumption of unidirectional power flow; however, power flow is now bidirectional. The major components of the distribution domain are the smart infrastructure, communication, management, and protection systems. The smart infrastructure system is further divided into two parts, a smart energy system and smart information systems. The smart distribution of power is possible because of the smart distribution domain, which is now supported by the two-way flow of power and information.

3.4 The customer domain

In the customer domain, the customer is the end user who consumes the power. These customers are basically of three types, namely (1) domestic home customers who use power for home equipment, (2) commercial or building consumers who consume power in commercial buildings or sectors, and (3) industrial power sectors or consumers. The power range for home users is up to about 20 kW; for commercial buildings, it is 20–200 kW, and industrial users, it should be greater than or equal to 200 kW. The energy services interface (ESI) [38] provides a secure end interface for utility-to-customer interactions. A dedicated communication system is essential for the flow of data from one customer end to another customer end or from one to another part of the smart grid network. The ESI communicates with the advanced metering infrastructure through the Internet. The ESI communication system also communicates with the devices present inside the system through the home area network (HAN) or local area network. Building and home automation [39] using microgeneration, which includes all types of DGs, is also part of the customer domain. The major function of this building and home automation is to control various functions inside buildings in the smart grid. Electric vehicles also play a vital role in the customer domain by balancing energy during peak and off-peak hours [40].

3.5 Market domain

Market means where grid assets are bought or sold. There are some high-priority challenges as far as the market domain is concerned. The first is extension of price, and for that, the distributed energy resource (DER) power for each customer should be closely monitored to determine the price. The second challenge is the expansion of aggregator capabilities, which may help the DER and small utilities to participate in the exchange of power in big power markets. Next is interoperability—that is, the ability to operate the smart grid concept in such a manner that all domains are able to speak with one another. Growth management and regulation of energy retailing and wholesaling are also important aspects of the market domain. The important components of the market domain are DER aggregation, market management, market operations, wholesale trading, ancillary services, and retailing. The DER aggregator combines small participants in the market to sell their power to the energy market—i.e., to enable them to actively participate in the energy businesses. Market management helps manage the ISOs for wholesale markets. The ancillary operations provide frequency, voltage, and spinning reserve support. The retailing section of the market domain sells power to end consumers.

3.6 Service provider domain

The service provider domain consists of companies or organizations that provide services to customers and utilities. It supports the business processes of power system. The service provider domain helps in processing the business of all domains. Business processes include traditional utility services: billing of customers, customer account management, management of energy used, etc. It is difficult to develop the key interfaces and standards that enable a dynamic market-driven ecosystem, so things provided by the service provider ecosystem should be economical, reliable, user-friendly, and environmentally friendly; as well, the critical power infrastructure should be intact. The components of the service provider domain are installation and maintenance, building management, home management, emerging services, account management, customer management, and finally, the billing system. This domain speaks to the market's customers and operations through a dedicated communication system by exchanging data between them. The installation management component of the service provider domain helps in installing and maintaining premises equipment. The building management section monitors and controls all building energy. As far as the service provider is concerned, the main part is billing. Customer information should be properly accessed for billing, and the bill should be prepared on time. Emerging services are basically innovation services of the smart grid operation—innovations in future smart grid operations that can make our smart grid operations more accurate, reliable, and secure. The last business process is customer management. The customer management section maintains relationships with customers so that customer relationships are properly maintained in a fair manner.

3.7 Operations domain

The next domain of smart grid architecture is the operations domain. It comprises mainly the supervisors of electricity flow management and is responsible for smooth operation of the EPS. The main target of the operations domain is to maintain, operate, monitor, and control the smart grid in a secure, dependable, and reliable manner. The main components of the operations domain are maintenance and construction, finance, supply chain logistics, records and assets, security management, communication networks, meter reading and control, operations planning, and extension planning.

All components of the operations domain communicate with each other through a secure communication channel. The priority is uninterrupted power supply and the quality of power to customers. The monitoring part of the operations domain consists of the phasor measurement unit (PMU) or WAM systems. The supervisory control and data acquisition system, PMU, or WAMS provides monitoring support. In the monitoring system, this equipment supervises network connectivity, power, voltage, frequency, etc. Besides monitoring, control is the other objective of this domain. The control section supervises and provides automatic control support. The protection and fault management system helps to identify the fault, its elimination, and service restoration. Major disturbances are taken care of by this system. The disturbances are detected, maintained, and repaired, and service to the customer is then restored. Next is analysis—system operation, control, monitoring, and fault detection—every part is analyzed to determine whether it is operating properly relative to expectations. This analysis is a must in the smart grid system, and reporting statistics calculations are a part of it. The next section is training and real-time network calculation; this section provides records and assets, the facility for dispatches that stimulate the system and provide tracking reports on network equipment inventory. The other sections are maintenance, extension planning, and customer support.

4. Smart grid—infrastructure and technologies

4.1 Smart information system

The smart grid expects two-way communication between various domains, especially in the distribution system where the end users—i.e., customers—are present. Their metering, billing, and feedback are important data for designing a self-sustained smart information system [41]. In the smart information system, smart meters and smart routers are connected through a HAN with the help of low-cost Zigbee metering communication [42] devices placed for the exchange of information within the smart grid [43]. A smart meter records customer information like consumption, load, and balance, which is then sent to the utility for various decision-making activities like monitoring, billing, and schedule planning. The automatic meter reading (AMR) system automatically collects information like diagnostics, utilization, and customer status from energy meters and sends that information to a centrally located energy management system for processing, billing, analysis, and maintenance. The smart meter plays a vital role in sending all customer information to the utility so that troubleshooting analysis may be performed at a given time.

4.2 Automatic meter infrastructure

Automatic meter infrastructure (AMI) differs from traditional AMR in that it may have two-way communication with the meters. AMR is certainly a helpful instrument in smart grid infrastructure, but AMI has the edge by enabling two-way communication that helps in the availability of real-time information [44].

AMI consists of three components. The first is the meter data management system (MDMS), which is present inside the utility data and control center. The second is the smart meters, which are located basically in the consumer's periphery. The third is the data control centers, which are present at the periphery or premises of the utility. The AMR system allows meters to be read by utility personnel over wireless links. At the customer end, data are collected from on-site energy meters. The AMR system allows a customer to read meter data wirelessly at the customer's own premises. The wireless communication system helps reduce the cost associated with meter reading and also supports the billing process. The basic block diagram of AMI is shown in Fig. 1.6.

4.3 Phasor measurement unit and sensors

PMUs measure the electrical quantities of an EPS and thus help maintain the health of the EPS. The major functions of PMUs are islanding detection, fault detection, locating disturbances, online monitoring, and devising experimental setups for monitoring the active distribution system [45]. Fig. 1.7 shows the block diagram of a PMU.

The second important feature of smart grid infrastructure is the extensive use of sensors. Sensor networks are used as a monitoring and measuring unit. A smart grid uses sensors because of their quality of service and resource constraints. The best part about using sensors in the network is that they can be maintained remotely.

4.4 Smart grid communication

Smart grid communication is responsible for the flow of information among all domains. Reliable information exchange and security are important for effective implementation of the smart grid. The basic functional requirement is the quality of service of critical data. Maintaining the reliability of such a big system is not a trivial task. One of the major smart communication methods is wireless communication. It has a wireless mesh network, cellular communications, cognitive radio. Of all these, wireless communication is quite important for the successful operation of a smart grid [46]. Wired communication could be either fiber optic or power line communication (PLC). Wireless communication offers significant advantages over wired technology: low installation cost, remote applications, mobility, etc. Cognitive radio has been proposed for smart grids based on the IEEE 802.22 standard for use as a secondary or backup radio to deal with noncritical information [47]. Wired communication certainly includes fiber optic communication. It has high bandwidth capacity and immunity characteristics. Although the installation cost of optical fiber is high, it is still cost-effective because of its high-speed communication, which is the backbone of the future smart grid. Along with fiber optics, PLC is also used even though it is a very old technology. This is

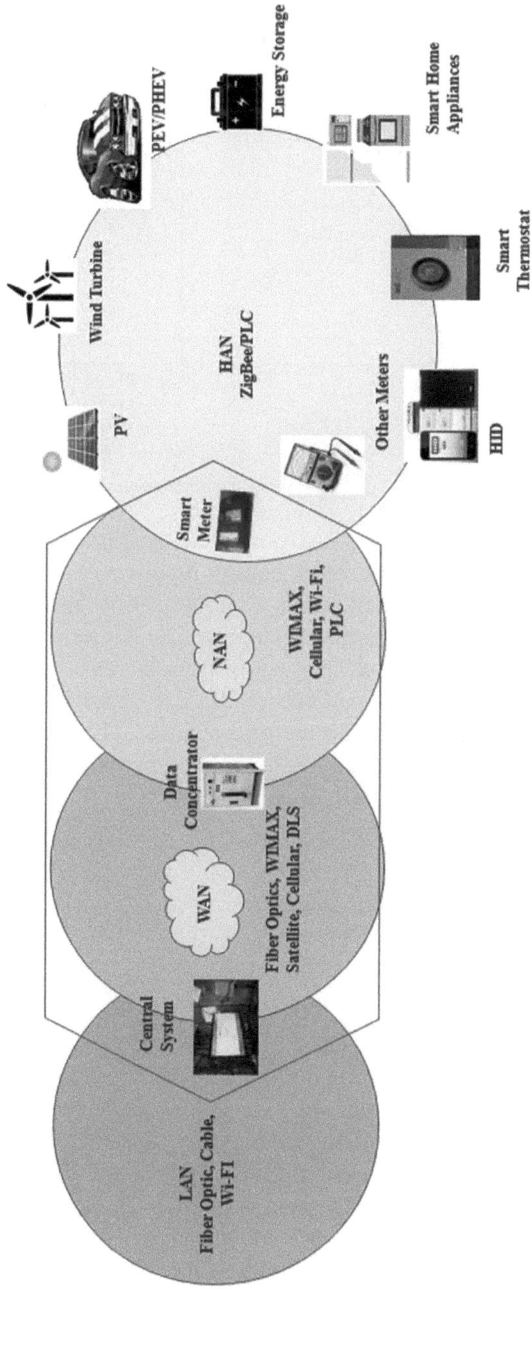

FIGURE 1.6 Automatic meter infrastructure.

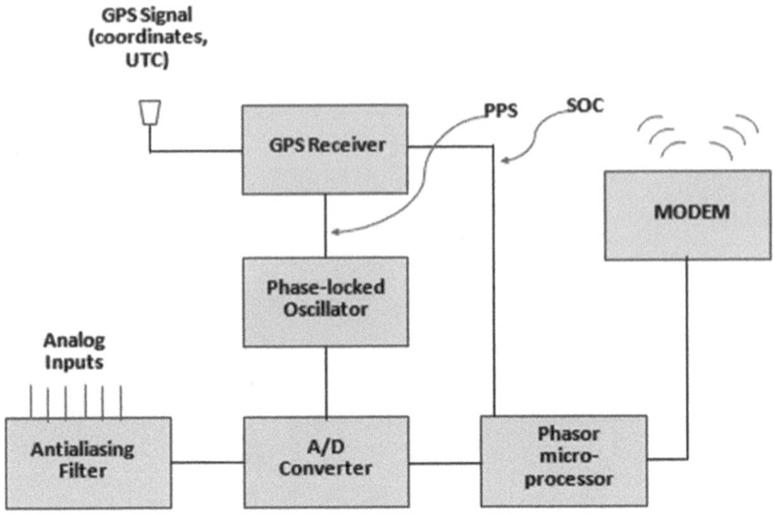

FIGURE 1.7 Phasor measurement units.

because PLC carries information and power on the same conductor. PLC is used for remote metering and load control applications. Smart metering infrastructure is well suited for wired communication based on PLC, and certainly a huge PLC infrastructure is expected in future smart grids. Also, broadband PLC can transfer data seamlessly from the smart grid controller to load and vice versa. The users and source can use communication as a platform for their interactions to make the grid smarter. The smart grid communication network concept, expressed with the help of a hierarchical multilayer architecture as shown in Ref. [48], is as follows:

- *customer area network that may be an industrial area network, a building area network, or a HAN*
- *neighborhood and field area networks*
- *wide area network*

4.5 Information management

Information management is carried out in different steps. The large amount of information generated from metering, sensing, and monitoring is collected from the grid. Advanced information management systems and techniques are needed for information analysis, stochastic analysis, and optimization. Information analysis is conducted on the data to support processing interpretations and correlations of new observations [49]. And finally, the information is integrated from different sources having different kinds of representations.

The smart meter is connected directly to the data control center or MDMS using a dedicated communication network. A second one is the vendor proprietary

AMI solution-based connection. In this, a greater number of smart meters, which are connected to the meter concentrator through the communication network, connected to the MDMS of the utility section. Here each meter communicates with the MDMS over a communication network. That is the first requirement. The second is that this interface connects to the utilities through a wired or wireless communication network, or a broadband network, to provide connectivity to the MDMS.

4.6 Smart grid protection system

Faster and more accurate protection is demanded by the emerging smart grid in the 21st century. Reliability, fault identification, prediction and prevention of fault occurrence, diagnosis and recovery, microgrid protection, and security are the important requirements. The smart protection system tends to make all the components perform their respective functions under defined conditions for defined intervals of time. Local power generation can reduce the likelihood of cascading failure. An ideal mix of smart grid resources like renewable and storage lead to a flattened net demand response curve and eventually increase system reliability. Major blackouts can be prevented by proper predictions. Self-healing can be effective if the power grid is divided into small sections. Postfault decision-making should occur at the substation level for immediate action.

Distribution system protection has seen major changes due to the introduction of DG, energy storage systems (ESSs), and electric vehicles that make the distribution system active [22]. Adaptive protection may be a solution to the active distribution system or microgrid. Anti-islanding protection of microgrids is quite challenging. Overcurrent grading is no longer a reliable method for protection of the distribution system because bidirectional power flow has challenges such as blinding of protection and false tripping [22].

5. Energy storage technologies

Although power from renewable energy resources is intermittent in nature, advanced and efficient energy storage devices for the smart grid can solve this problem to some extent. Storage technology has advanced greatly with the passage of time [50]. Hydro and pumped storage was the conventional way of storing energy. But much variability is present due to differences in generation and demand due to the integration of RESs with intermittent outputs. However, storage technologies are advancing to better enable effective implementation of renewable sources. Various ESSs are emerging to fill this gap. These storage technologies make the demand profile flat and hence make the system more reliable. Research is ongoing to provide storage systems in the microgrid using flywheels, supercapacitors, batteries etc. Battery ESSs are one such technology undergoing much advancement. The presence of a storage system

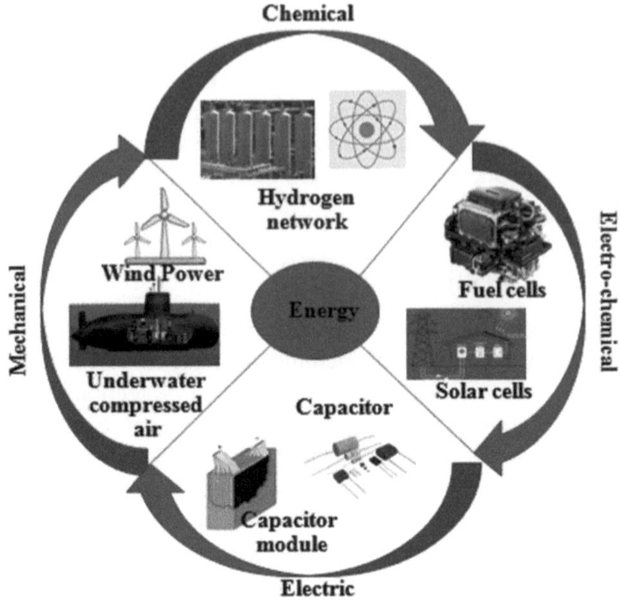

FIGURE 1.8 Types of energy storage systems.

in the microgrid helps improve the flexibility of the system. The types of ESSs may be classified as shown in Fig. 1.8.

Generation, transmission, and utilization of power is optimized using storage technologies that hence enable better efficiency across the system. Energy storage devices store excess energy during off-peak times and discharge it to support the grid when power is in deficit. ESSs charge them from RESs and usually are not allowed to draw energy from the grid. Energy storage devices play a vital role in the smart grid by supporting the system by keeping the voltage profile flat and improving the power quality. Different storage devices have their respective applications. Dynamic storage technologies are also under research. Energy storage devices let the consumer store energy during off-peak hours and utilize it during peak hours. This is economical for the consumer as well. For proper charging and discharging of the device, control techniques are applied to them so they perform well and on time, which helps make the system stable.

Energy storage helps in maintaining the power balance between generation and demand, thereby maintaining grid voltage and frequency. Based on the application, any storage device can be used, but usually battery storage devices and supercapacitors are used as storage devices in smart grids. Batteries, having high energy density, remain active for long durations and support slow transient power demand. Supercapacitors are used in systems with fast transient demands.

TABLE 1.3 Types of energy storage systems (ESS).

ESS type	Advantages	Disadvantages
Pumped hydro energy storage	High energy capacity	High environmental impact
Compressed-air energy storage	Low capital costs	Needs auxiliary fuel input
Lithium-ion battery	Long life, high efficiency	High initial cost
Supercapacitor	Long life, high efficiency	Low energy density, high initial cost
Lead–acid battery	Recyclable	Limited charge depth
Flywheel	High power density	Low energy density
Hydrogen energy storage	High energy density	Low efficiency

To support both slow and fast transients, hybrid storage devices are now being used. The merits and demerits of various types of ESSs are tabulated in Table 1.3. The benefits of using storage technology are as follows:

- *Energy management:* Storage devices are a vital part of smart grid energy generation. The management of total energy can be done using storage devices. Due to the intermittent nature of renewable energy resources in a smart grid, it is often difficult to manage the power flow. When power generation is high, demand may be low, while during peak-load hours, the RES may not be able to provide sufficient energy to the system. This can be managed using energy storage devices. The excess energy can be stored in storage devices during off-load times and further utilized during peak-load times.
- *Peak shaving:* By relying on storage devices, consumption can be reduced. During peak-load times, the energy storage device is discharged to serve the priority load. During off-peak hours, they are recharged and excess power is stored.
- *Voltage support:* An energy storage device supports the grid by instantly charging or discharging according to load demand. This also helps maintain the voltage level, which may vary due to generated power and load power mismatches within the tolerable range set by the IEEE standard.
- *Power quality:* By providing support to the microgrid, storage devices help the grid to maintain the system power quality. Whenever there is a mismatch between generation and consumption, power quality issues arise. Energy storage devices support the grid instantly by charging or discharging and hence help to maintain power quality.

The most commonly used types of batteries are explained in the next section.

5.1 Chemical or electrochemical storage system

Batteries are the most commonly used storage devices. They convert chemical energy to electrical form and play a vital role in stand-alone microgrid operations that have no support from the utility grid. The excess power generated during off-peak hours is stored in them. During peak-load hours, this stored energy is provided to the load. Batteries are manufactured in different sizes depending on their applications. Their efficiency lies between 60% and 90%.

5.1.1 Sodium sulfur battery

This battery contains sodium and sulfur, which act as the anode and cathode, respectively. This battery has high energy density, requires minimal maintenance, and is economical. It is widely used in power system applications. Its fast response makes it suitable for power regulation applications. But it is not reliable at high temperatures. Also, it may produce combustible products with water, so there is a risk of hazard. Fig. 1.9 shows a sodium sulfur battery.

5.1.2 Sodium nickel chloride battery

The sodium nickel chloride battery stores energy through charging and discharging reactions between nickel and sodium chloride.

This battery is mainly used in electrical vehicles and uninterruptable power supply as well as for backup power. Its temperature range is quite high. Its efficiency is better than that of sodium sulfur batteries. It is highly flexible and temperature-resistant. Fig. 1.10 shows a sodium nickel chloride battery.

5.1.3 Vanadium redox flow battery

The vanadium redox flow battery uses the properties of vanadium in different oxidation states. Vanadium has the property that it may exist in four different

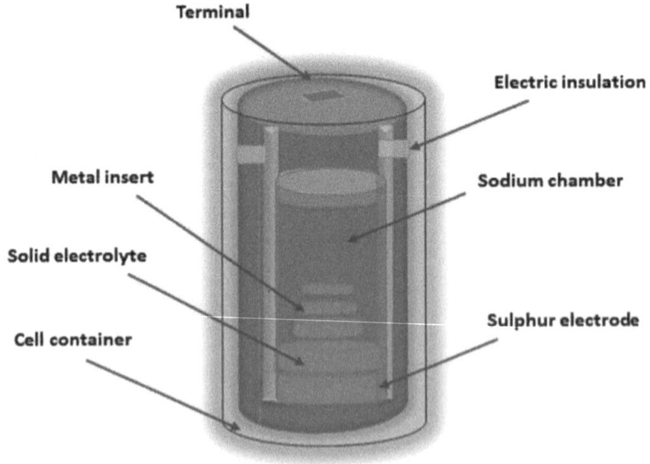

FIGURE 1.9 Sodium sulfur battery.

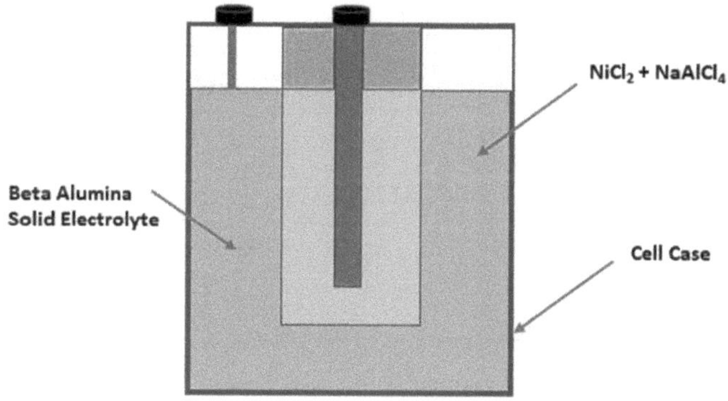

FIGURE 1.10 Sodium nickel chloride battery.

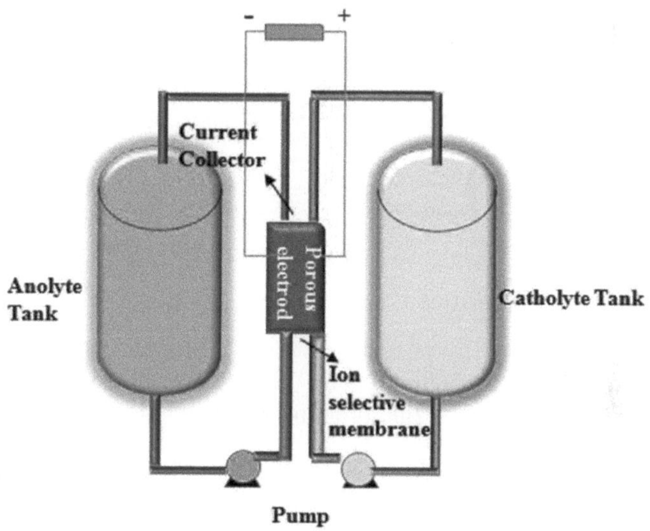

FIGURE 1.11 Vanadium redox flow battery.

oxidation states in solution. This property of vanadium is used to make the battery. The benefit of this battery is that it is rechargeable.

The operating temperature of these batteries is low. They are very bulky but have a lower relative cost. They are used in power management, pumps, grid energy storage, etc. Fig. 1.11 shows a vanadium redox flow battery.

5.1.4 Zinc bromine battery

The zinc bromine battery consists of zinc and bromide. During charging, zinc is reduced to a thick film, and bromide is converted to bromine. Conversely, during discharging, zinc is oxidized, and bromine is reduced to bromide.

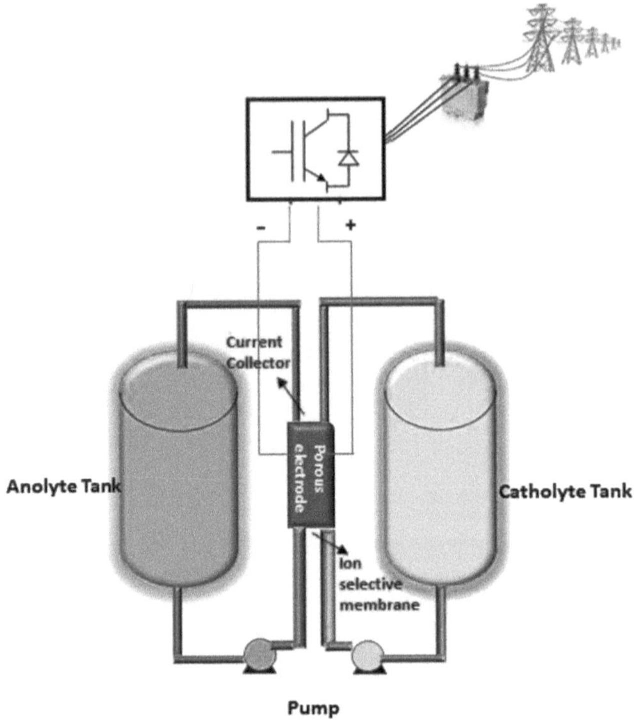

FIGURE 1.12 Zinc bromine battery.

To maintain the stability of the compound, cooling systems are a must at high temperature. These batteries have high energy density but are quite expensive. Fig. 1.12 shows a zinc bromine battery.

5.1.5 Lithium-ion battery

Lithium-ion batteries are the most commonly used battery storage device. They have high energy and high power density.

Lithium-ion batteries consist of carbon compounds on the positive electrode with an oxide layer at the negative electrode. Their efficiency is high compared with that of other batteries, and they have good battery life. They are temperature dependent. Their main drawback is their high cost.

5.1.6 Lead—acid batteries

Lead—acid batteries are the conventionally and most widely used batteries due to their cost-effectiveness.

They have high efficiency and a low rate of discharge. A lead—acid battery consists of lead oxide at the anode and metal at the cathode. It does not have high energy density and is temperature dependent. Its water level needs to be maintained from time to time. These batteries are not suitable for high-power applications. Fig. 1.13 shows a lead—acid battery.

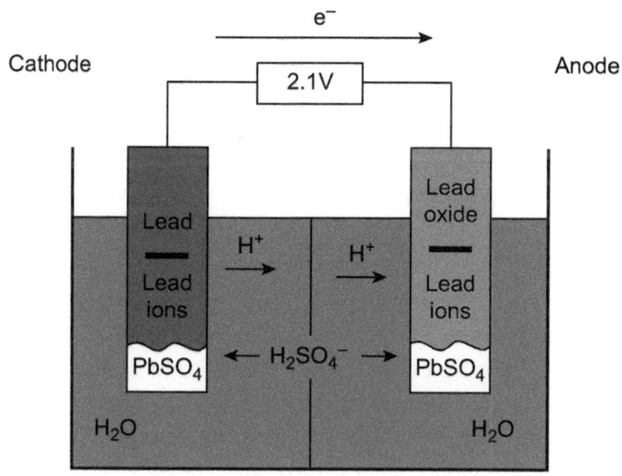

FIGURE 1.13 Lead–acid battery.

5.2 Supercapacitors

Supercapacitors have gained much attention in the last decade as a storage device. Supercapacitors have a good lifetime and very high efficiency. They have even replaced batteries in some applications. The working of a supercapacitor is the same as for normal capacitor.

They have very high energy because they contain an electrolytic solution. They are mostly used in portable devices because of their immense energy and power density. The main drawback of supercapacitors is their high cost. Fig. 1.14 shows a supercapacitor.

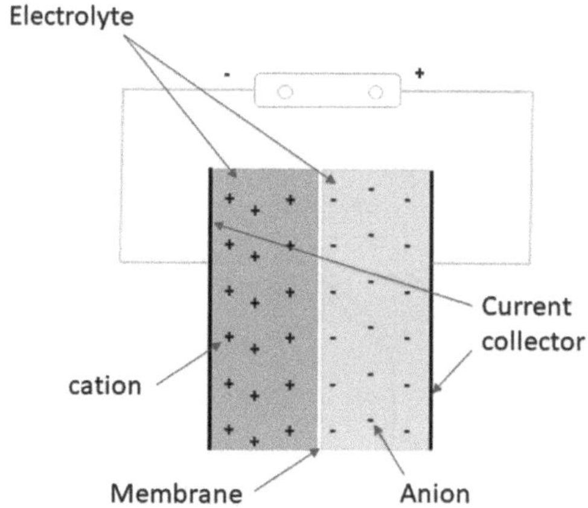

FIGURE 1.14 Supercapacitors.

5.3 Mechanical storage systems

In mechanical storage systems, mechanical energy is converted into electrical. During off-peak durations, surplus energy is consumed and mechanically stored. Mechanical energy storage devices are very flexible. The most commonly used mechanical ESSs are the flywheel ESS, pumped hydro storage system, gravity ESS, and compressed-air energy storage system (CAES).

5.3.1 Pumped hydro storage system

Hydro storage devices store electrical energy by pumping water from a lower level to a higher level of the reservoir in the form of potential energy. It is a conventional way of storing energy, but it has certain disadvantages like large capital investment, size of the hydro storage system, site selection, and long construction duration. Fig. 1.15 shows a pumped hydro storage system.

5.3.2 Flywheel storage system

In a flywheel ESS, the mass of the flywheel is used to store energy in the form of kinetic energy. Its energy and power density are quite high. A flywheel ESS has five components: a flywheel, an electrical machine, a power electronic device, a set of bearings, and a containment unit. The energy stored in a flywheel depends on the moment of inertia of its rotor and its speed. Fig. 1.16 shows a flywheel storage system.

5.3.3 Compressed-air energy storage system

The CAES system stores energy in the form of compressed air pressure in the reservoir. This compressed air rotates the turbine, which is coupled with a generator to produce electricity. Fig. 1.17 shows a CAES.

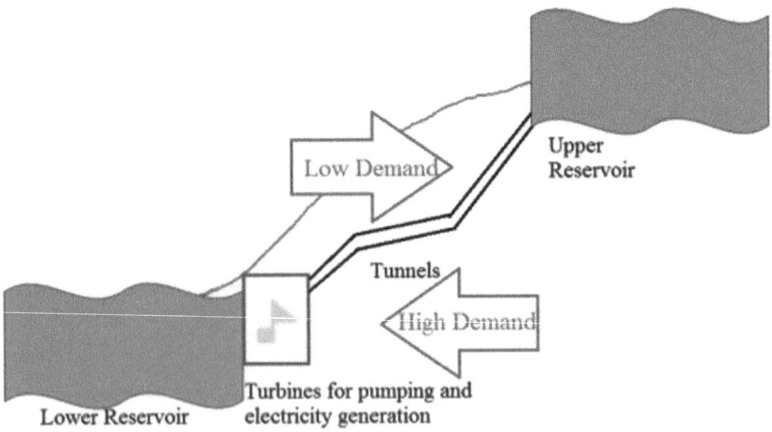

FIGURE 1.15 Pumped hydro storage system.

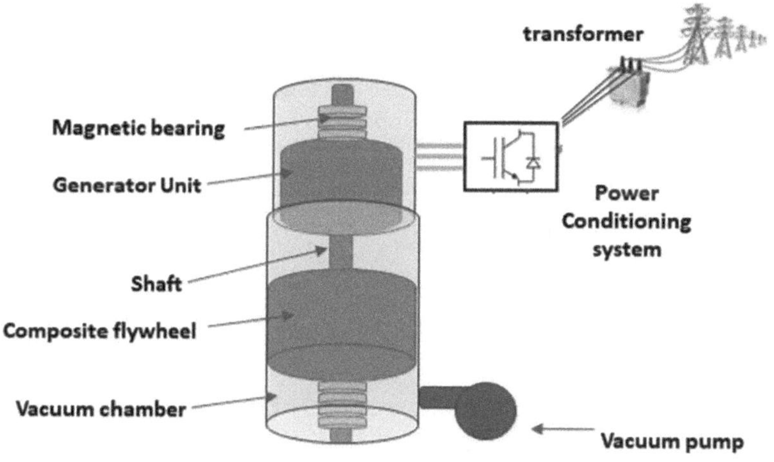

FIGURE 1.16 Flywheel storage system.

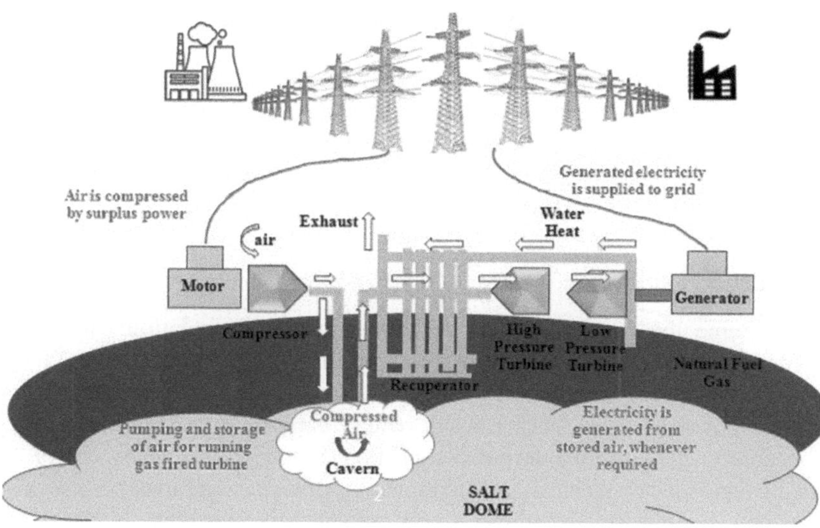

FIGURE 1.17 Compressed-air energy storage system.

5.4 Hybrid energy storage

As the name suggests, this is the combination of two or more energy storage devices. The idea is to use two or more storage devices in a system so they may complement each other. If one has high power density, the other should have high energy density so that system output can be optimized. Mostly, batteries and supercapacitors are used for the hybrid energy storage device. Batteries typically have high energy density and low power density, while

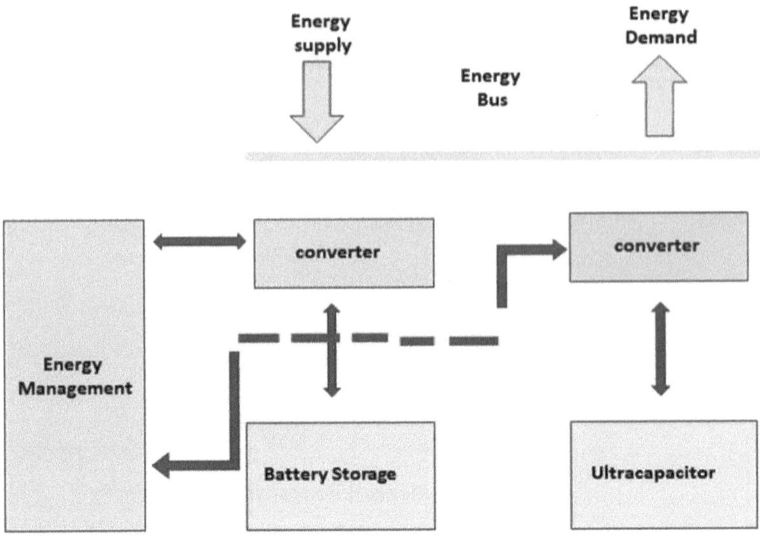

FIGURE 1.18 Hybrid energy storage system.

supercapacitors have high power density. These storage devices may or may not be directly connected to the bus. They are usually connected through power electronic devices depending on the type of bus. Overall system efficiency, storage capacity, and lifetime may be increased using a hybrid ESS. Fig. 1.18 shows a hybrid ESS.

5.5 Electric vehicles

To overcome the adverse effects of environmental degradation caused by conventional transportation systems running on internal combustion engines, electric vehicles have been introduced. These electric vehicles can run on alternate power sources. Only a heavy battery can propel the vehicle, so it is often combined with an internal combustion engine. Hence, they are often called hybrid electric vehicle, providing better efficiency and lower emissions. They can be charge-sustaining, where regenerative braking is used for charging. To overcome problems of charging speed and battery size, plug-in hybrid electric vehicles (PHEVs) having external charging facilities were introduced. Uncontrolled charging of such vehicles can lead to problems like unavailability of power to priority loads. But if proper charging approaches are applied, they can be considered distributed storage sources. Electric vehicles should be charged during no- or low-load hours so that their charging does not affect any priority load.

Electric vehicles can be considered part of a smart grid power system. They can be treated as loads that take power from the grid. However, their charging patterns can be scheduled to correspond with load demand. An advanced way

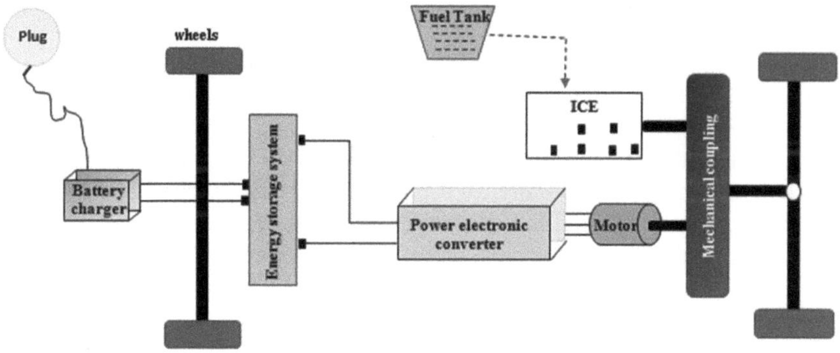

FIGURE 1.19 Electric vehicle.

of connecting an electric vehicle as part of a grid is by controlling its charging and discharging behavior. This is also known as vehicle-to-grid (V2G) charging [51]. In V2G mode, the electric vehicle is charged during off-peak hours, thus balancing demand and supplying the grid with additional energy during peak hours. Traders sell and buy electricity from the market. Balance group managers maintain and provide accounting to the transmission system operator (TSO). The TSO utilizes this information to assess system operation. The distribution system operator maintains the distribution network to provide all charging facilities for electric vehicles, and the TSO does so for the transmission line. The PHEV exchanges information with the connected units. The PHEV is shown in Fig. 1.19.

6. Smart grid—advantages

6.1 Environmental concerns and reduced emission

The smart grid allows for integration of distributed energy resources, generally RESs, to the utility grid and hence results in reduced emission. It helps improve air quality and reduces air pollution and carbon footprints.

6.2 Energy conservation

Nonconventional energy sources like solar and wind generation are used for DG. Being renewable, they also can produce clean, emission-free electricity. If used judiciously along with storage devices, the issue of their intermittent nature may also be resolved. With the help of smart grid architecture, their integration with the utility grid helps in shifting the load as required. Also, by using the smart meters, consumers are given participation in the system and hence can moderate their usage during high energy-usage hours. Hence, conserving energy in the system is achieved.

6.3 Competitive price of energy

Using AMI, the information is collected, analyzed, and sent to the utility for billing and forecasting purposes. Real-time pricing helps both the consumer and supplier to judiciously manage their demand and supply. The smart grid helps in the bidirectional flow of electricity and information, hence allowing the customer to buy or sell any excess amount of energy during peak hours. Dynamic pricing may lead to benefits for both. The customer gets options to buy energy at the best price, and system losses are reduced as well.

7. Smart grid technical issues and challenges

7.1 Interconnection issues and bidirectional power flow

With smart grid architecture, involvement of the consumer is built into the system, so the distributed energy sources produce energy along with the main utility grid. This integration of DG to the system may cause technical issues such as unsynchronized reclosing, voltage gradient at the point of common coupling, induction of harmonics, and power quality degradation and failure of protection. So proper synchronization of DG with the utility grid must be achieved. Private operators must follow stringent grid codes. IEEE standard 1547 [34], which gives the specifications for DER interconnection and interoperability with the utility grid, needs to be followed. The interconnection of DG in the distribution grid has a negative impact on the power quality and protection system. These DGs have power electronic converter interfaces, which generally introduce harmonics in the system and cause power quality problems. Bidirectional power flow may cause failure of overcurrent protection, blinding of protection, and false tripping [52]. Unintentional islanding of DG is another major concern of the distribution system.

7.2 Cybersecurity and privacy issues

Cybersecurity is one of the major challenges in the smart grid scenario. Vulnerability may allow an attacker to enter the system to penetrate the privacy of customers, may access and hack various software, and later manipulate data to make the grid unstable or create unforeseeable situations. Smart meters are a preferred choice of hackers so that the vulnerability may be monetized. Wide deployment of monitoring and measurement devices could also lead to system vulnerability. In the smart grid network, many points of vulnerability are present.

- *Denial-of-service attack:* When the attacker blocks the consumer from availing the facilities they are eligible for, the attack is known as a denial-of-service attack. Man-in-the-middle attacks may access the information between the sender or receiver or may modify the data as well. The latter version is considered an active attack, which is more harmful than the former.

- *Replay attack:* In this attack, the attacker reads the message sent by the sender and sends it multiple times to the receiver. This can be detrimental, as it may also lead to blackout in the worst-case scenario.

Security requirements [24] for a smart grid may include the following:

- *Confidentiality:* There must be proper means to protect the privacy of consumers and their information. Any unauthorized intervention should be avoided. If confidentiality is not maintained, it can have bad impacts on the consumer. To prevent breaches of confidentiality, authorized restrictions on information should be preserved.
- *Integrity:* Unauthorized modification of consumer information can be considered a loss of integrity. This modified information can hamper the smart grid system. The smart grid should be designed so that such losses can be avoided.
- *Availability:* The smart grid should ensure timely access to information. Unavailability of resources may make the system unreliable and inefficient.
- Privacy preservation in the smart grid.

Privacy [24] concerns may be expressed as follows:

- *Personal information:* The personal information of the consumer comes under this section. This could be anything that may disclose the identity of the consumer: consumer financial status, social status, physical status, etc.
- *Personal privacy:* This is related to the consumer's physical details, includes medical status.
- *Behavioral privacy:* Information regarding the energy consumption history of any customer also reflects their behavioral patterns. Knowing the behavioral pattern, one may be able to ascertain a customer's day-to-day activities and predict future activities.
- *Personal communications privacy:* Consumer information regarding communication with others comes under this. If this information is leaked, it is considered an invasion of privacy.

7.3 Information technology and database management

A cloud server is used to store and process information in the smart grid. This cloud is subject to different possible cyberthreats. This information needs to be encrypted to prevent unauthorized access. Data management and preprocessing this information is a big task in itself.

7.4 Competitive pricing

A demand-response program allows a consumer to vary usage patterns to contribute to the smart grid system. Real-time pricing requires implementation of smart grid infrastructure throughout the system. This dynamic pricing is done by varying the price of energy during the day. Due to the intermittent nature of DERs, prediction technologies are used for hour-ahead or day-ahead prediction, which may or may not be accurate. This is because consumption patterns of consumers may change with time. Further, sellers may not know exact consumption patterns because of privacy issues. So this needs a well-optimized and precise prediction in order to get the benefit of competitive pricing.

7.5 Regulatory issues

There are issues with the regulatory framework. It is not clear from electricity bills how trading can be done from the domestic consumer's end. Multiple suppliers are allowed in an area so consumers can get electricity at the best price, but it is unclear how this switching between suppliers can be done by the consumer. Pricing strategies, mechanisms for incentives, and customer awareness strategies need to be addressed by policy-makers. Strategies should be developed for installing advanced storage technologies of higher capacity to serve during peak-load hours. Policy regimes cannot sufficiently address the issues of advanced development support, implementation of advanced technologies, involvement of incentives, etc.

8. Conclusion

This chapter introduces the smart grid from the perspective of the smart grid framework issued by the NIST. It also presents basic knowledge about the smart grid infrastructure and various ESSs. Various technical issues and challenges related to effective implementation of the smart grid also are covered.

References

[1] T. Hammons, Impact of electric power generation on green house gas emissions in europe: Russia, greece, italy and views of the eu power plant supply industry—a critical analysis, Int. J. Electr. Power Energy Syst. 28 (2006) 548–564.
[2] R. Heinberg, D. Fridley, The end of cheap coal, Nature 468 (2010) 367–369.
[3] R.W. Bentley, Global oil & gas depletion: an overview, Energy Pol. 30 (2002) 189–205.
[4] F.F. Wu, K. Moslehi, A. Bose, Power system control centers: past, present, and future, Proc. IEEE 93 (2005) 1890–1908.
[5] P. Menanteau, D. Finon, M.-L. Lamy, Prices versus quantities: choosing policies for promoting the development of renewable energy, Energy Pol. 31 (2003) 799–812.
[6] K. Balamurugan, D. Srinivasan, T. Reindl, Impact of distributed generation on power distribution systems, Energy Procedia 25 (2012) 93–100.
[7] T. Jamasb, M. Pollitt, Electricity market reform in the european union: review of progress toward liberalization & integration, Energy J. 26 (2005).

[8] I. Pavić, M. Beus, H. Pandžić, T. Capuder, I. Štritof, Electricity markets overviewmarket participation possibilities for renewable and distributed energy resources, in: 14th International Conference on the European Energy Market (EEM), IEEE, 2017, 2017, pp. 1–5.
[9] F. Iovino, N. Tsitsianis, The european energy markets: The liberalization processes, in: Changes in European Energy Markets, Emerald Publishing Limited, 2020.
[10] S.M. Amin, B.F. Wollenberg, Toward a smart grid: power delivery for the 21st century, IEEE Power Energy Mag. 3 (2005) 34–41.
[11] G. Dileep, A survey on smart grid technologies and applications, Renew. Energy 146 (2020) 2589–2625.
[12] Y. Kabalci, A survey on smart metering and smart grid communication, Renew. Sustain. Energy Rev. 57 (2016) 302–318.
[13] S. Löbbe, A. Hackbarth, T. Stillahn, L. Pfeiffer, G. Rohbogner, Customer participation in p2p trading: a german energy community case study, in: Behind and Beyond the Meter, Elsevier, 2020, pp. 83–104.
[14] M. Radenković, Z. Bogdanović, M. Despotović-Zrakić, A. Labus, S. Lazarević, Assessing consumer readiness for participation in iot-based demand response business models, Technol. Forecast. Social Change 150 (2020) 119715.
[15] G.L. Barbose, S. Forrester, N.R. Darghouth, B. Hoen, Income Trends among US Residential Rooftop Solar Adopters, Technical Report, Lawrence Berkeley National Lab.(LBNL), Berkeley, CA (United States), 2020.
[16] X. Li, Y. Tan, X. Liu, Q. Liao, B. Sun, G. Cao, C. Li, X. Yang, Z. Wang, A cost-benefit analysis of v2g electric vehicles supporting peak shaving in shanghai, Electr. Power Syst. Res. 179 (2020) 106058.
[17] L. Ju, H. Li, J. Zhao, K. Chen, Q. Tan, Z. Tan, Multi-objective stochastic scheduling optimization model for connecting a virtual power plant to wind-photovoltaic-electric vehicles considering uncertainties and demand response, Energy Convers. Manag. 128 (2016) 160–177.
[18] M. Attia, N. Haidar, S.M. Senouci, E.-H. Aglzim, Towards an efficient energy management to reduce Co_2 emissions and billing cost in smart buildings, in: 15th IEEE Annual Consumer Communications & Networking Conference (CCNC), IEEE, 2018, 2018, pp. 1–6.
[19] J. Xiao, X. Kong, Q. Jin, H. You, K. Cui, Y. Zhang, Demand-responsive virtual power plant optimization scheduling method based on competitive bidding equilibrium, Energy Procedia 152 (2018) 1158–1163.
[20] J. Sakhnini, H. Karimipour, A. Dehghantanha, R.M. Parizi, G. Srivastava, Security aspects of internet of things aided smart grids: a bibliometric survey, Internet of Things, 2019, 100111.
[21] Y. Yoldaş, A. Önen, S. Muyeen, A.V. Vasilakos, İ. Alan, Enhancing smart grid with microgrids: challenges and opportunities, Renew. Sustain. Energy Rev. 72 (2017) 205–214.
[22] P. Gupta, R.S. Bhatia, D.K. Jain, Adaptive protection schemes for the microgrid in a smart grid scenario: technical challenges, in: IEEE Innovative Smart Grid Technologies-Asia (ISGT Asia), IEEE, 2013, 2013, pp. 1–5.
[23] G.W. Arnold, D.A. Wollman, G.J. FitzPatrick, D. Prochaska, D.G. Holmberg, D.H. Su, A.R. Hefner Jr., N.T. Golmie, T.L. Brewer, M. Bello, et al., NIST Framework and Roadmap for Smart Grid Interoperability Standards, Release 1.0, 2010. Technical Report.
[24] C. Greer, D.A. Wollman, D.E. Prochaska, P.A. Boynton, J.A. Mazer, C.T. Nguyen, G.J. FitzPatrick, T.L. Nelson, G.H. Koepke, A.R. Hefner Jr., et al., Nist framework and roadmap for smart grid interoperability standards, release 3.0, 2014. Technical Report.

[25] G.J. FitzPatrick, D.A. Wollman, Nist interoperability framework and action plans, in: IEEE PES General Meeting, IEEE, 2010, pp. 1–4.
[26] P. Gupta, R. Bhatia, D. Jain, Average absolute frequency deviation value based active islanding detection technique, IEEE Trans. Smart Grid 6 (2014) 26–35.
[27] B.M. Buchholz, Z.A. Styczynski, Smart grids worldwide, in: Smart Grids, Springer, 2020, pp. 367–408.
[28] C. Lima, Smart grids IEEE p2030, in: ETSI Workshop-Standards: An Architecture for the Smart Grid, volume 5, 2011.
[29] T.S. Basso, R. DeBlasio, Advancing Smart Grid Interoperability and Implementing NIST's Interoperability Roadmap: IEEE P2030TM Initiative and IEEE 1547TM Interconnection Standards, National Renewable Energy Laboratory, 2010.
[30] S.M. Amin, Smart grid security, privacy, and resilient architectures: opportunities and challenges, in: IEEE Power and Energy Society General Meeting, IEEE, 2012, 2012, pp. 1–2.
[31] J. Zheng, D.W. Gao, L. Lin, Smart meters in smart grid: an overview, in: IEEE Green Technologies Conference (GreenTech), IEEE, 2013, 2013, pp. 57–64.
[32] P. Samadi, H. Mohsenian-Rad, R. Schober, V.W. Wong, Advanced demand side management for the future smart grid using mechanism design, IEEE Trans. Smart Grid 3 (2012) 1170–1180.
[33] C. Tu, X. He, Z. Shuai, F. Jiang, Big data issues in smart grid–a review, Renew. Sustain. Energy Rev. 79 (2017) 1099–1107.
[34] I.S. Board, IEEE Standard for Interconnecting Distributed Resources with Electric Power Systems: 1547-2003, IEEE, 2003.
[35] F. Salvadori, C.S. Gehrke, A.C. de Oliveira, M. de Campos, P.S. Sausen, Smart grid infrastructure using a hybrid network architecture, IEEE Trans. Smart Grid 4 (2013) 1630–1639.
[36] T. Sueyoshi, M. Goto, Environmental assessment by DEA radial measurement: US coal-fired power plants in ISO (independent system operator) and RTO (regional transmission organization), Energy Econ. 34 (2012) 663–676.
[37] T. Lv, Q. Ai, Interactive energy management of networked microgrids-based active distribution system considering large-scale integration of renewable energy resources, Appl. Energy 163 (2016) 408–422.
[38] E.-K. Lee, R. Gadh, M. Gerla, Energy service interface: accessing to customer energy resources for smart grid interoperation, IEEE J. Sel. Area. Commun. 31 (2013) 1195–1204.
[39] A. Demeure, S. Caffiau, E. Elias, C. Roux, Building and using home automation systems: a field study, in: International Symposium on End User Development, Springer, 2015, pp. 125–140.
[40] S. Mal, A. Chattopadhyay, A. Yang, R. Gadh, Electric vehicle smart charging and vehicle-to-grid operation, Int. J. Parallel Emergent Distr. Syst. 28 (2013) 249–265.
[41] X. Ji, M. Swoboda, P.-L.P. Rau, N. Sun, Designing a smart information system: the influence of feedback on energy conservation persuasion, Enterprise Inf. Syst. 14 (2020) 480–495.
[42] S.-W. Luan, J.-H. Teng, S.-Y. Chan, L.-C. Hwang, Development of a smart power meter for AMI based on zigbee communication, in: International Conference on Power Electronics and Drive Systems (PEDS), IEEE, 2009, 2009, pp. 661–665.
[43] V. Namboodiri, V. Aravinthan, S.N. Mohapatra, B. Karimi, W. Jewell, Toward a secure wireless-based home area network for metering in smart grids, IEEE Syst. J. 8 (2013) 509–520.

[44] D. Colonna, W.L. Hopper, F. Cezar, S.V. Udin, A. Zhelezov, Configurable Secure Mobile Tool For AMI and AMR Metering Systems, 2019. US Patent 10,182,273.

[45] A. Monti, C. Muscas, F. Ponci, Phasor measurement units and wide area monitoring systems, Academic Press, 2016.

[46] Z. Fan, P. Kulkarni, S. Gormus, C. Efthymiou, G. Kalogridis, M. Sooriyabandara, Z. Zhu, S. Lambotharan, W.H. Chin, Smart grid communications: overview of research challenges, solutions, and standardization activities, IEEE Commun. Surv. Tutorials 15 (2012) 21–38.

[47] R. Yu, Y. Zhang, S. Gjessing, C. Yuen, S. Xie, M. Guizani, Cognitive radio based hierarchical communications infrastructure for smart grid, IEEE Net. 25 (2011) 6–14.

[48] M. Kuzlu, M. Pipattanasomporn, S. Rahman, Communication network requirements for major smart grid applications in HAN, NAN and WAN, Comput. Net. 67 (2014) 74–88.

[49] H. Liang, A.K. Tamang, W. Zhuang, X.S. Shen, Stochastic information management in smart grid, IEEE Commun. Surv. Tutorials 16 (2014) 1746–1770.

[50] X. Tan, Q. Li, H. Wang, Advances and trends of energy storage technology in microgrid, Int. J. Electr. Power Energy Syst. 44 (2013) 179–191.

[51] Y. Zhang, M. Lu, S. Shen, On the Values of Vehicle-to-Grid Electricity Selling in Electric Vehicle Sharing, Manufacturing & Service Operations Management, 2020.

[52] P. Gupta, R.S. Bhatia, D.K. Jain, Islanding detection of active distribution system with parallel inverters, in: IEEE International Conference on Signal Processing, Informatics, Communication and Energy Systems (SPICES), IEEE, 2017, 2017, pp. 1–5.

Chapter 2

An introduction to the smart grid-II

Rajender Kumar Beniwal
Electrical Engineering Department, Sobhasaria Group of Institutions, Sikar, Rajasthan, India

1. Introduction

The traditional power system was designed to generate electrical energy at central plants, step up the voltage to transmit over transmission lines, and step down at the distribution level. This results in huge power losses; in India these losses are around 30% [1]. It was originally designed as a hierarchical model—i.e., a top-down approach—so there is no involvement of customers and no provision of space for small power producers. With reforms in the power sector, this model has been restructured and now includes small independent power producers. This helps to fill the gap between demand and supply by greater electricity generation but is unable to completely meet increased demand and also results in increased power system complexity. Revolutionary technologies, communication techniques, and Internet and cloud computing open up new research areas for the power grid. This leads to evaluation of the smart grid. The smart grid is an intelligent grid that can make decisions. There is no specific definition of a smart grid, as each working group has its own view about the definition. The IEC definition of a smart grid is, "The Smart Grid is a developing network of transmission lines, equipment, controls and new technologies working together to respond immediately to our 21st Century demand for electricity."

The IEEE definition of a smart grid is, "The smart grid is a revolutionary undertaking-entailing new communications-and control capabilities, energy sources, generation models and adherence to cross jurisdictional regulatory structures."

The smart grid supports bidirectional flow of electricity and information/data communication with its associated systems. With the help of two-way communication, if a device fails to operate or any other fault occurs, the smart grid self-heals the fault before it becomes a major issue for the system [2]. Smart grid technologies are defined as self-sustained framework that can

Advances in Smart Grid Power System. https://doi.org/10.1016/B978-0-12-824337-4.00002-3

self-heal from any abnormal system conditions. These technologies work with minimal human intervention and ensure a reliable, sustained and continuous supply to the end user [3]. These technologies provide long-term benefits to power companies. Smart grid technologies such as smart information systems, automatic meter infrastructure, phasor measurement units and sensors, smart grid communication, information management, protection systems, and energy storage are presented in Chapter 1. Here, we focus on the energy management system and load/demand-side management in the next sections.

2. Energy management system

Energy management is directly concerned with how effectively energy is utilized. The electrical power system is a huge network from which industrial, commercial, and domestic customers consume energy. It is practically very difficult to meet the energy demand for every consumer. To deal with such a situation, energy management is especially necessary. This can be done with the help of a smart grid to control the production and consumption of energy. This can be achieved by two-way communication between consumers and energy producers as well as devices connected to the smart grid. Here, energy management is achieved not only by lowering consumer demand but also by optimal use of generation through conventional power plants and renewable sources [4].

In older times, people thought that energy was an infinite resource, and this illusion raised their standard of living. In view of this, many types of energy resources and demand-side management approaches were needed to maintain the high living standard and also pointed to concerns about economic security and global warming.

Following are the main reasons for the importance of energy management:

- ➢ The consumption of energy has increased every year due to modernization and population growth.
- ➢ Conventional or fossil fuels are limited in nature and responsible for the production of greenhouse gases.
- ➢ Many countries do not have rich sources of fossil fuels.
- ➢ The environment is suffering from the extinction of sources.

An energy management system (Fig. 2.1) requires dedicated policies and effective programs at national and international levels. Technological development is also required. Effort should be made to maximize the potential of energy management programs.

There is always a discussion at electricity boards or companies about energy efficiency at the customer end and to avail energy sources near the load center or user end so that losses will be minimized. With the evolution of the smart grid, this discussion ended. The power industry, due to conventional techniques, limited capacity, increasing cost, new directives/regulations, and global warming, is now considering load/demand management. Now, companies are trying to change the way customers consume energy in a planned

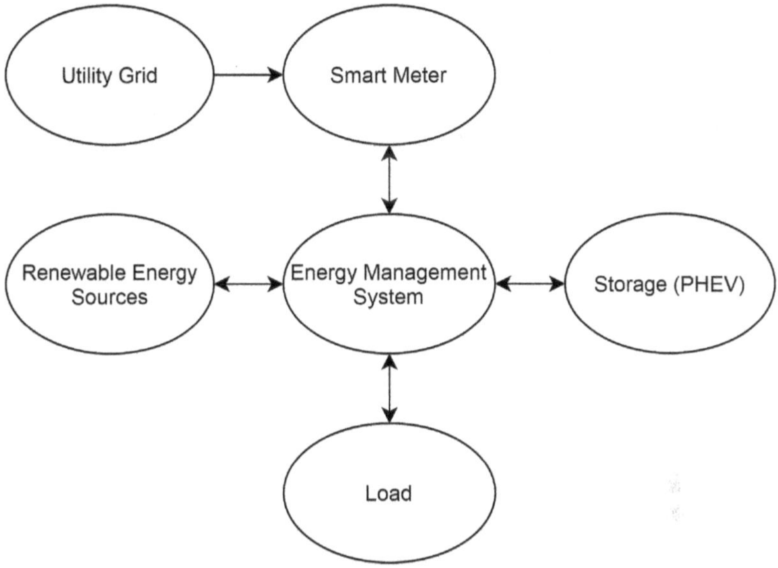

FIGURE 2.1 Generalized block diagram of energy management system [5].

manner so that the company can produce a significant desired change in load shape—i.e., change in the routine and in magnitude, which is called demand-side planning or management [6]. This can be done by using energy-efficient applications and a source of energy as near as possible to the point of use.

2.1 Components of energy management system

The energy management system can be divided in different components:

A brief introduction about these components is as follows:

I. Communication Network: The smart energy management system helps in the generation, transmission, and distribution phases. But it is more effective at the distribution end or the power-consumption side. In this system, we are able to maintain power quality and consumer demand and have control over the energy distribution. This can be done with a wireless network like WiMAX, WLAN, Z-wave, DASh-7, ZigBee (Xbee), Bluetooth, Global System for Mobile Communication, Wi-Fi or wired broadband systems, digital subscriber line, coaxial cable, or optical fiber [7].

II. Smart Meter: The smart meter establishes an information exchange medium between the service provider and customer. It records energy consumption data and sends it to the utilities and provides tariff-related information from the utility to the customer. This helps customers schedule optimal energy usage.

III. Renewable Energy Resources: EMS helps in the integration of available renewable energy resources at the customer level. Many consumers have solar, wind, biomass, or other plants that reduce the load demand of customers and reduce electricity bills.

IV. Plug-in Hybrid Vehicles (PHEVs) or Storage: PHEVs or plug-in electric vehicles can be used as a dynamic energy storage medium on the basis of their daily travel patterns. Although a dedicated battery bank is also used for storage, the batteries can be charged at the time of lowest energy rates or surplus energy from renewable sources and can be used as source of energy when required [8].

V. Load Management: Load management is a very important part of EMS. The performance of EMS is mainly based on load management. Load management is directly associated with consumer requirements and level of comfort. Proper scheduling and usage of load is very helpful in lowering energy consumption and reducing electricity bills. The load can be divided into three categories based on their use:

- ➢ Adjustable load: load or appliances that can be managed for comfort and pricing. This kind of load contains lighting load, HVAC, refrigerator load, or any thermostat-based load whose energy consumption rate can be controlled and used.
- ➢ Shiftable load: the load or appliances whose use timing can be shifted or changed. This type of load is off when the price of electricity is high and can be used at times of low pricing. These loads include dishwashers, washers, dryers, water pumps, electric irons, and PHEVs.
- ➢ Fixed load: appliances that must be used when required. This type of load cannot be shifted or altered: TV, microwave, laptop/computer, printer, etc.

Further load profiles can be improved by using energy-efficient appliances:

VI. Monitoring and Control: This module is responsible for the monitoring, recording, scheduling, and control of energy consumption. Energy consumption is continuously monitored, and the daily load pattern of the consumer is recorded so that a user profile can be created on the basis of the record. On the basis of current load usage, the customer's predefined load schedule, and per-unit cost, optimal scheduling is done.

There are many methods to implement these techniques—e.g., the Xbee network can be interfaced with different types of microcontroller and field programmable grid array boards or computers. Bluetooth can be interfaced with Android software. All these systems are installed with advanced sensors that help in energy management.

3. Demand/load-side management

Demand-side management (DSM) is getting more attention due to stable and reliable operation of the power system network combined with environmental concerns. Due to the revolution in information and communication technology, DSM techniques are easier to implement on the customer end. DSM works on the present state of energy generation and its consumption by managing the customer's load.

The consumption rate of electrical energy is uncertain. The need for required energy is predicted on the history of generation and demand data so that a balance is maintained [9,10]. DSM results in a reduced cost of energy consumption. It increases efficiency by reducing peak-load demand, and a balance is maintained between electricity demand and supply.

DSM consists of load curve pattern-change technologies that include distributed generation, energy storage, curtailment of loads, and end user load control and efficiency. Some techniques used for changing the shape of load curves are listed in Table 2.1. Energy management consists of hardware (appliances and sensors) as well as nonhardware (software) techniques to alter customer load curves.

DSM is defined as the change in a consumer's regular rate of consumption of electrical energy for a given change in the cost of electricity. In the demand-response (DR) program, a consumer reduces electricity usage at times of high prices to get economic benefits. To do this, energy consumption is reduced during peak-load hours. In a DR program, consumers intentionally adjust their normal consumption by changing their timing of use, instantaneous demand, or whole consumption of electricity [11].

DSM can be classified (Fig. 2.2) [12] as

1. Price based: This is a market-based or an economic-based method. In this case, the customer directly receives benefits in billing based on time of use, dynamic pricing, and real-time pricing. In this method, the electricity tariff is not fixed but continuously changes with demand for electricity.
2. Incentive based: This is a system-based or security-based method. This can be further classified as
   - **(a)** Classical incentive method: in this method, a bill credit or discount is given to the participating customer.
   - **(b)** Market-based incentive method: in this method, money is given as a reward on the basis of load reduction during peak conditions.

4. Case study

An energy management system has been presented by Ganji and Shahidehpour, using a microgrid [14]. This study was carried out in a Galvin Center smart home model at the Illinois Institute of Technology. In this model, the

TABLE 2.1 Load-shaping techniques used in demand-response management.

Sr. No.	Name of technique	Shape	Effects
1.	Peak clipping	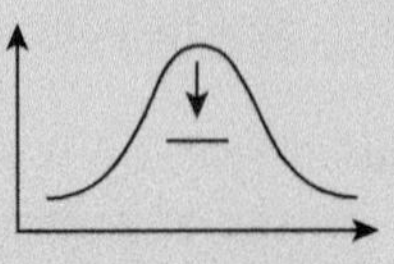	Load reduction at the time of peak load time from the grid and flattening of the load curve.
2.	Valley filling	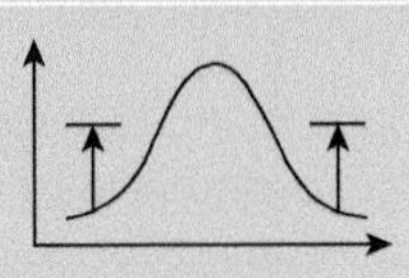	Increased load during off-peak hours to flatten the load curve and improve the load factor and efficiency.
3.	Load shifting	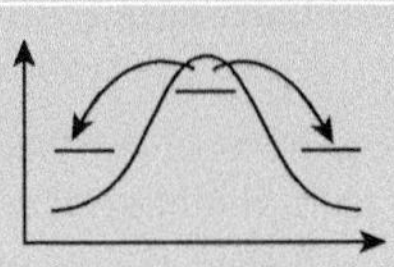	Increase in load during off-peak hours and decreased load during peak hours to flatten the load curve.
4.	Strategic conservation	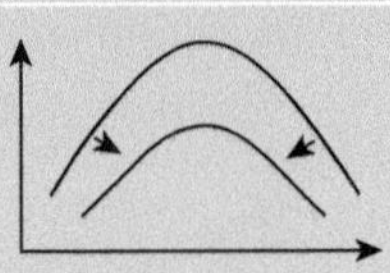	Full-day decrease in load using smart energy-efficient equipment.
5.	Strategic load growth	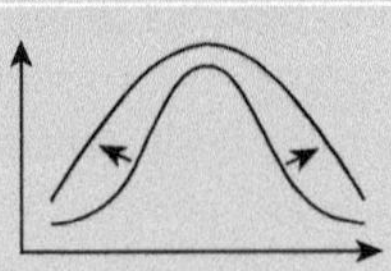	Step-by-step increase in the sale of electricity for financial growth.
6.	Flexible load shape		Provide reliable supply to customer of low power quality.

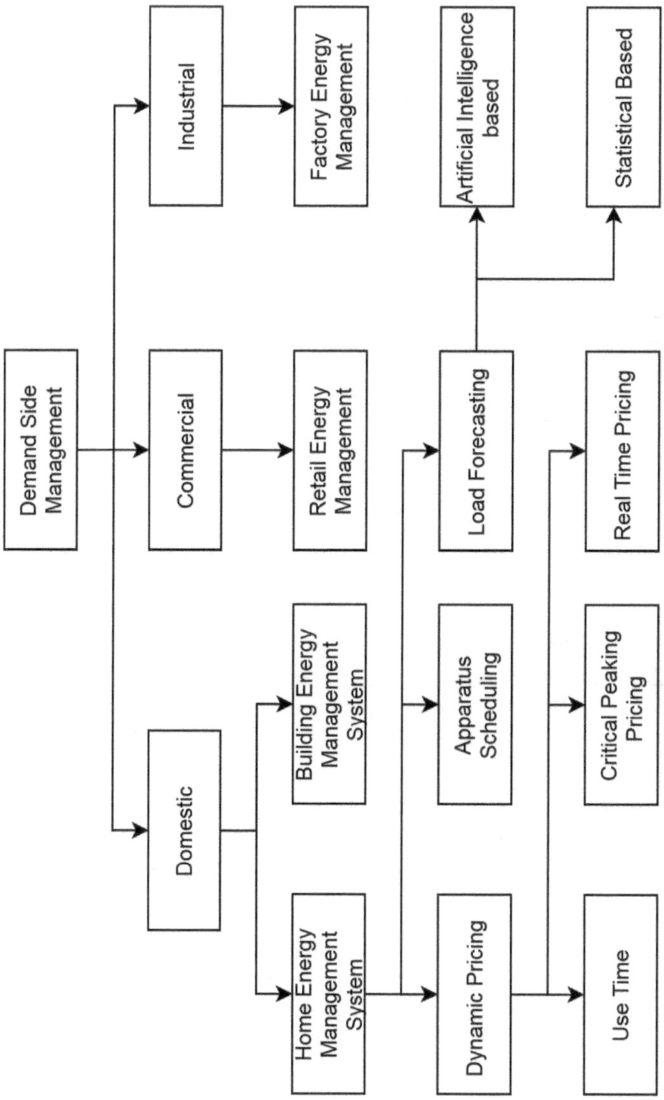

FIGURE 2.2 Various techniques of demand management [13].

entire house is divided into different modules, and each module is equipped with a local controller and a master control system that manages the local controllers. This setup consists of a Zigbee module with different sensors (current sensor, relay, microcontroller, etc.) for wireless data transmission.

The load is divided into the following three categories:

➢ An adjustable load that contains lighting, HVAC, and refrigerator loads that can be managed for comfort and pricing.
➢ Shiftable loads like dishwashers, washers, dryers, PHEVs, etc. This type of load is off when the price of electricity is high and can be used at times of low pricing.
➢ Fixed loads that cannot be shifted or altered like TV, microwave, and laptop/computer.

In this study, a 24 h simulation was scheduled on the basis of electricity pricing and availability of energy from solar and wind plants. In this home energy management system (HEMS), the consumer's load requirement is monitored, and load profile scheduling is done for optimal use of energy based on market cost and available power of 1.5 kW from wind and 2 kW from solar plant.

On the basis of the three categories of load profiles mentioned earlier, different cases are simulated with constraints like per-hour price, customer comfort, and availability of generation and storage:

a. on the basis of flat-rate price
b. optimal scheduling at an hourly rate excluding PHEV
c. optimal scheduling at an hourly rate including PHEV
d. optimal scheduling at an hourly rate including PHEV and a scheduled load limit
e. using reduction in load (price/kWh)
f. consumer's preset shiftable load use timing
g. change in consumer's shiftable load use timing

In case (a), the HEMS and shiftable load are considered, and this case reduces peak-load demand by 18% with a 0% savings on bills. Case (b) is with HEMS and shiftable load when PHEV storage is not used. This gives 11% savings on bills and an 18% reduction in peak load. The PHEV battery storage included in case (c) results in a lowering of peak demand of 10% and a 14% savings on electricity charges with the condition that the battery is charged at 17 h. In case (d), the load is managed by the aggregator, which curtails the load if it is beyond the customer's load limit. This gives a 14% reduction in billing and again lowers the peak load by 18%. When the load is per the consumer's comfort and at an affordable limit of price (rate/kWh), as in case (e), peak load is reduced by 30% with a 34% cut in utility charges. In case (f), the peak is reduced by 18% as in case (a) and produces savings of 13% in billing amount, but it can be further improved by using adjustable load in off-peak hours up to 24% in case (g). All the data considered for the different cases are given in Table 2.2.

TABLE 2.2 Hourly base load (kW) data per day for a house [14].

Hour	Fix load	Adjustable load	Adjustable load (minimum)	Case a	Case b	Case c	Case d	Case e	Case f	Case g	Price/ kWh
1	2.5	3.1	2.2	5.6	4.34	2.94	2.94	2.94	2.7	2.7	1.7
2	2.3	2.8	2	5.1	6.24	6.24	6.24	6.24	6	6	1.6
3	1.9	2.4	1.7	4.3	8.4	8.4	8.4	8.4	6.7	6.7	1.5
4	1.7	2.2	1.5	3.9	8	8	8	8	6.3	6.3	1.5
5	1.7	2.1	1.5	4.8	4.7	4.7	4.7	4.7	4.7	4.7	1.6
6	1.9	2.2	1.5	3.9	3.9	3.9	3.9	3.9	3.9	3.9	1.9
7	2.2	2.7	1.9	6	4.6	4.6	4.6	4.6	4.6	4.6	2.1
8	1.2	3.1	2.2	5.7	4.3	4.3	4.3	4.3	4.3	4.3	2.6
9	1.5	3.5	2.5	6.9	5	5	5	5	5	5.24	3.2
10	0.8	4	2.8	6.2	4.8	4.8	4.8	4.8	4.8	5.04	3.3
11	1.4	4.7	3.3	6.1	3.1	3.1	3.1	3.1	3.1	4.1	5.1
12	2.1	5.4	3.8	7.5	6.5	6.5	6.5	6.5	6.74	7.5	4.3
13	3.4	6	4.2	9.4	6.4	6.4	6.4	6.4	6.64	6.4	4.7
14	3.8	6.2	4.3	10.5	9	9	9	**9**	10	**9.7**	4.1
15	3.9	6.5	4.6	11.4	8.4	8.4	8.4	8.4	9.4	9.1	4.5

Continued

TABLE 2.2 Hourly base load (kW) data per day for a house [14].—cont'd

Hour	Fix load	Adjustable load	Adjustable load (minimum)	Case a	Case b	Case c	Case d	Case e	Case f	Case g	Price/kWh
16	4.1	6.7	4.7	10.8	8.8	8.8	8.8	8.8	8.8	8.8	4.8
17	5.6	6.9	4.8	12.5	**10.5**	**11.9**	**10.5**	8.43	**10.5**	8.43	6
18	5.8	7	4.9	**12.8**	9.8	8.4	8.8	7.7	8.8	7.7	8.9
19	5.3	6.9	4.8	12.2	9.2	7.8	7.8	5.73	7.8	5.73	10.2
20	4.4	6.2	4.3	11.3	9.6	8.2	8.2	6.34	8.2	6.34	9.5
21	4.1	5.9	4.1	10	9	9	10	7.23	10	7.23	8.5
22	4.3	5.2	3.6	9.5	9.5	9.5	9.5	7.94	9.5	7.94	8
23	3.4	4.1	2.9	7.5	6	6	6	4.77	6.7	4.77	5.7
24	2.8	3.5	2.5	6.3	4.8	4.8	4.8	4.8	5.5	4.8	2.8
Peak load	–	–	–	12.8	10.5	11.9	10.5	9	10.5	9.7	–

Adjustable loads are dishwasher (0.7 kW), washer (0.24 kW), dryer (1 kW), water heater (1.4 kW), and PHEV (1.5 kW). Each appliance's duty cycle is 2 h except PHEV, which has a 4-h duty cycle. These appliances can be used at any time throughout the day, but for optimal scheduling the start/end hour timing of each device is considered as 20/24, 11/17, 13/19, 1/16 and 1/8 respectively.

5. Differences between past, present, and future grids

In the conventional grid, power flow is one way—i.e., from utility to load. Power plants are too far from load centers, and power is transferred through transmission lines and then distributed to load, which results in power losses. In the smart grid, power flow is two-way, on either side between customers and utilities. There are many differences between normal and smart grids. Further comparisons are given in Table 2.3. In this discussion, past grid means old power grid under the control of central or state governments at the beginning stage of the power sector. Present grid is after reforms or deregulation of the power sector, and smart grid is the future grid.

There are a number of advantages of the smart grid such as improvements in power quality, self-healing capabilities for power system faults, easy adaptation of new technologies and configuration, choice of supply to customer, integration of renewable energy resources, improvements in security against risks, energy-storage facilities, energy-management techniques,

TABLE 2.3 Comparison between past, present, and future grids [15,16].

Technology/Parameter	Past grid	Present grid	Future grid
Information sharing/ consumer participation	No	No	Yes
Communication	No	One-way	Two-way
Adoption of new techniques or services	No	Limited	Yes, compatible
Quality of power	Highly affected	Limited to outage and blackout	As per standards
Performance optimization	No	Limited	Maximum
Self-healing capability	No	No	Yes
Cybersecurity requirement	No	Limited to top level	Yes
RES/DG support	No	Limited	Plug-and-play model
Storage medium	No	For ancillary services or emergency	Yes
Command response	Poor	Slow	Fast
Reliability	Poor	Moderate, manual	Full, automatic
Human interventions	More	Moderate	Minimum

improved overall efficiency by reducing losses, and protection of the environment. These advantages make it suitable for future use.

On the other hand, the concept of the smart grid is still in a developing stage. It will go through many technological and policy change programs. At present, there are a number of challenges associated with the smart grid—e.g., transmission capacity, capability of integrating renewable energy resources to a high penetration level, huge communication infrastructure, data security, cyberattacks, data management, active customer participation, advanced monitoring and control techniques, and competitive pricing. These challenges must be resolved to make the grid smarter.

6. Conclusion

This chapter is a continuation of the introduction to smart grid-I. The concept of smart grid technologies and types of energy storage are presented in Chapter 1. In this chapter, an overview of the energy management system, demand/load-side management, and the differences in past, present, and future grids are presented. To understand the concept of the energy management system, a case study is also included. The smart grid is the future of the power sector. The smart grid allows renewable energy generation or distributed generation, which helps in reliable, efficient, and flexible operation while reducing greenhouse gas emissions. Besides these benefits, switching to new technology always faces some issues, and to resolve these issues, transformation and adaptation of new technology are required.

References

[1] M. Ganesh, R.K. Patnaik, Innovation in electrical power engineering, communication, and computing technology, in: Lecture Notes in Electrical Engineering 630, 2020, pp. 29–44.

[2] G. Dileep, A survey on smart grid technologies and applications, Renew. Energy 146 (2020) 2589–2625.

[3] R. Bayindir, I. Colak, G. Fulli, K. Demirtas, Smart grid technologies and applications, Renew. Sustain. Energy Rev. 66 (2016) 499–516.

[4] M. Yılmaz, Energy management and smart grids for energy productivity, Int. J. Econ. Finance Manag. Sci. 3 (5–1) (2015) 7–12. Special Issue: Energy and Manufacturing Process Management.

[5] B. Zhou, W. Li, K.W. Chan, Y. Cao, Y. Kuang, X. Liu, X. Wang, Smart home energy management systems: concept, configurations, and scheduling strategies, Renew. Sustain. Energy Rev. 61 (2016) 30–40.

[6] C.W. Gellings, The Smart Grid: Enabling Energy Efficiency and Demand, CRC Press, 2009.

[7] R. Govindarajan, S. Meikandasivam, D. Vijayakumar, Energy management techniques in smart grid, Int. J. Appl. Eng. Res. 10 (15) (2015) 35720–35724.

[8] M.A.S. Masoum, et al., Mitigation of harmonics in smart grids with high penetration of plug-in electric vehicles, in: Conference: Power and Energy Society General Meeting, IEEE, 2010.

[9] S. Maharjan, Q. Zhu, Y. Zhang, S. Gjessing, T. Başar, Demand response management in the smart grid in a large population regime, IEEE Trans. Smart Grid 7 (1) (2016) 189–199.

[10] W. Li, et al., Demand response management for residential smart grid: from theory to practice, IEEE Access 3 (2015) 2431–2440.

[11] D. Al Abri, A.S. Malik, M. Albadi, Y. Charabi, N. Hosseinzadeh, Smart grid, in: W.Y. Chen, T. Suzuki, M. Lackner (Eds.), Handbook of Climate Change Mitigation and Adaptation, Springer, 2017.

[12] M.H. Albadi, E.F. El-Saadany, A summary of demand response in electricity markets, Elec. Power Syst. Res. 78 (2008) 1989–1996.

[13] A.R. Khan, A. Mahmood, A. Safdar, Z.A. Khan, N.A. Khan, Load forecasting, dynamic pricing and DSM in smart grid: a review, Renew. Sustain. Energy Rev. 54 (2016) 1311–1322.

[14] M. Ganji, M. Shahidehpour, Development of a residential microgrid using home energy management systems, Appl. Smart Grid Technol. (2018) 173–192.

[15] J.A. Momoh, Smart Grid: Fundamentals of Design and Analysis, IEEE Press, Piscataway, NJ, 2012.

[16] S.K. Salman, Introduction to the Smart Grid: Concepts, Technologies and Evolution, Institution of Engineering and Technology, London, United Kingdom, 2017.

Chapter 3

Smart grid power system

Neeraj Gupta[1], Navdeep Kaur[2], Sanjay K. Jain[2], Karan Singh Joshal[1]
[1]*Department of Electrical Engineering, National Institute of Technology, Srinagar, Jammu and Kashmir, India;* [2]*Department of Electrical and Instrumentation Engineering, Thapar Institute of Engineering and Technology, Patiala, Punjab, India*

1. Introduction

Large generators are mostly used in conventional power systems, where generated electricity is transmitted through high-voltage transmission networks to low-voltage distribution networks (DNs) to end consumers. In such conventional arrangements, large generators usually are based on nonrenewable sources located mostly away from load centers. In the past few years, renewable energy sources (RESs) have contributed substantially to total power production and are usually near load centers. Such power is commonly known as dispersed generation. The concept of integrating dispersed generation to low-voltage DNs is a paradigm that differs from the central generation-based paradigm used in conventional power systems.

Power produced by RESs is increasing across the world for several reasons. The major factors contributing to the growth of RES are the reduced cost of power production from RESs and regulations and policies for reducing pollution from nonrenewable-based generation. Although not all RESs have intermittent natures, the most dominant sources of power are solar and wind, both of which have variable output. Power output from solar and wind depends on weather conditions and is variable at all times. Due to variable power, it is a challenge to integrate RESs in national grids from the perspective of grid security, especially when the frequency band is very narrow.

Integration of dispersed generation complements efforts to improve the energy efficiency and reliability of power systems. Further, it is poised to address the concerns about gaseous emissions, technoeconomic difficulties in implementing large power plants, and diversification of energy resources to meet energy requirements. With the usage of dispersed generators (DGs), increasing load demand can be met without much expansion of existing transmission systems. Dispersed generation refers to the use nondispatchable resources for electric power generation. These resources may be renewable,

Advances in Smart Grid Power System. https://doi.org/10.1016/B978-0-12-824337-4.00003-5

nonrenewable, or combined reheat plants. The technical benefits of integrating DGs include diminution of losses, enhancement in the voltage profile, and improvements in power quality and reliability [1,2]. The Indian perspective of dispersed generation has been achieved through captive cogeneration with unbundling of the power sector [3].

2. Dispersed generation

Electric power systems are mostly based on central generation, where large generators feed power to a high-voltage transmission network through transformers to transport it over considerable distances. This power is then extracted through distribution transformers for delivery to end users. This conventional arrangement of an interconnected power system is advantageous, as it allows for minimal reserve requirements, the most efficient generating plant, which can be dispatched at any time, and bulk power that can be transported over large distances with limited electrical losses. However, factors such as concerns about gaseous emissions, technoeconomic difficulties in implementing large power plants, rational use and diversification of energy resources to meet national requirements, deregulation, and competition policies have resulted in increased interest in connecting generation into DNs, which is referred to as ***dispersed*** generation. Dispersed generation often offers a valuable alternative to traditional sources and arrangements for electric power generation. Dispersed generation has technical as well as environmental benefits, as most dispersed generation is based on renewable sources. The comparison of central generation with dispersed generation is shown in Fig. 3.1.

In central generation, the DN behaves as a passive network, as power is supplied from a bus and power flow is always from higher to lower voltage levels to supply loads. In such networks, the bus voltage drops when moved away from the source. With significant dispersed generation, the system behaves like an active network where power flows and voltages are determined by the main generator as well as the DG.

Generally, the term dispersed or distributed generation refers to electric power production technology integrated within the DN close to the point of use. DGs are small generators with typical generation ranging from a few kW to 100 MW that are connected directly to medium- or low-voltage grids/DNs. The problem of constantly growing load demand in DNs can be conquered with dispersed generation by meeting the increased load demand. Dispersed generation is contemplated as an attractive option to defer transmission and distribution upgrading and to support reliability and system performance. Increased interest in dispersed generation can be further attributed to the availability of modular generating plants, ease of finding sites for smaller generators, short construction times and lower capital costs of smaller plants, generation that may be sited closer to loads, and government support for

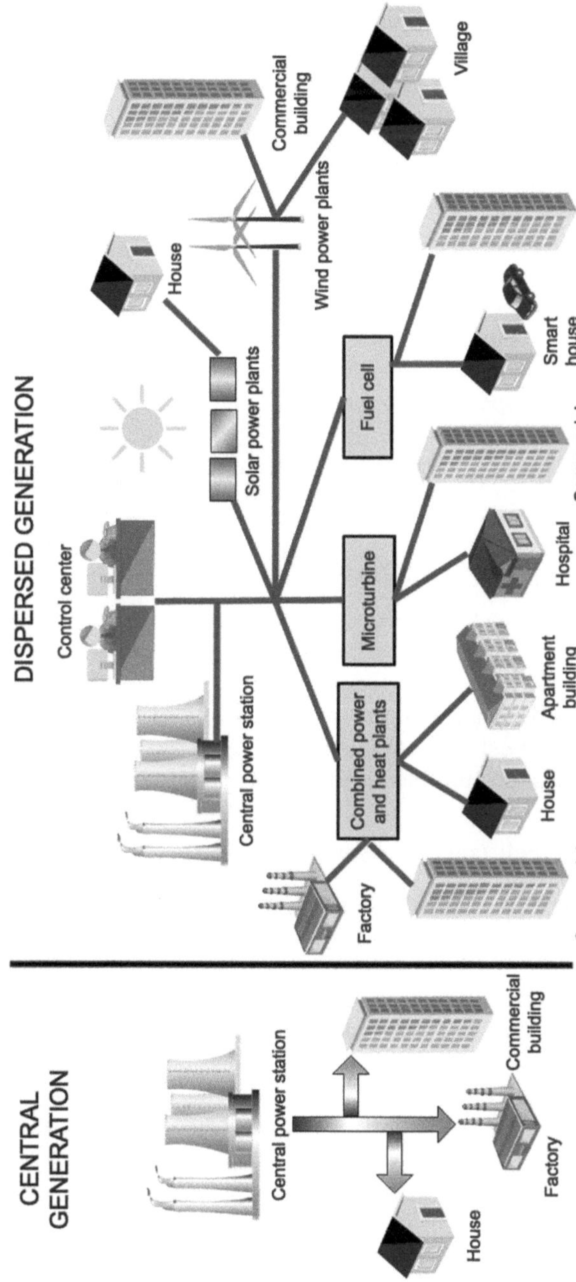

FIGURE 3.1 Central generation versus dispersed generation [4].

harnessing renewable resources, as most DGs are based on renewable sources. Mostly, renewable generation plants are not planned by the utility but developed by entrepreneurs. These are not centrally dispatched but generate whenever the energy source is available. The cost of renewable energy technologies is on a falling trend, as demand and production are increasing. The deregulated environment and open access to the DN are likely to provide greater opportunities for dispersed generation.

The exact definition of dispersed generation is inconsistent and varies somewhat between sources and capacities. However, it is generally and summarily defined as any source of electric power of limited capacity directly connected to the power system DN where it is consumed by end users. Various definitions of dispersed generation are based on related issues such as location, rating, purpose, environmental impact, technology, power delivery area, mode of operation, ownership, and penetration. "The *Electric Power Research Institute* defines the dispersed generation as generation from a few kW up to 50 MW" [5].

The following sections summarize DG technologies, DG types, and benefits of DGs.

3. Dispersed-generator technologies

Environmental concerns, like gaseous emissions from fossil fuels, have led to the exploitation of renewable energy resources for dispersed generation. These resources have lower energy densities than those of fossil fuels, so the generation plants are smaller and more widely spread geographically. DGs may be based on fuel cells, photovoltaic (PV) systems, wind turbines, mini/micro-hydroturbines, gas turbines, and microturbines.

3.1 Photovoltaic cell (solar power generation)

The energy contained in sunlight can be converted into electricity by employing PV cells. PV cells consist of discrete multiple cells connected in series/parallel combinations depending on voltage/current ratings, respectively. As the output of a PV cell is DC, PV cell−based DGs are connected to the grid via a power electronics interface. Solar power plants can be categorized as off-grid, on-grid, and hybrid. Off-grid solar plants are battery-based plants in which the priority is to feed the connected load, while additional power is stored in batteries. The priority of on-grid solar plants is the same as that of the off-grid plant, but excess power in off-grid solar plants is fed to the grids. The combination of off-grid and on-grid is termed a hybrid solar plant, where surplus power can be stored locally in batteries as well as fed to grid. A typical hybrid solar power generation plant connected to a grid as well as supplying power locally is shown in Fig. 3.2. The major benefits of

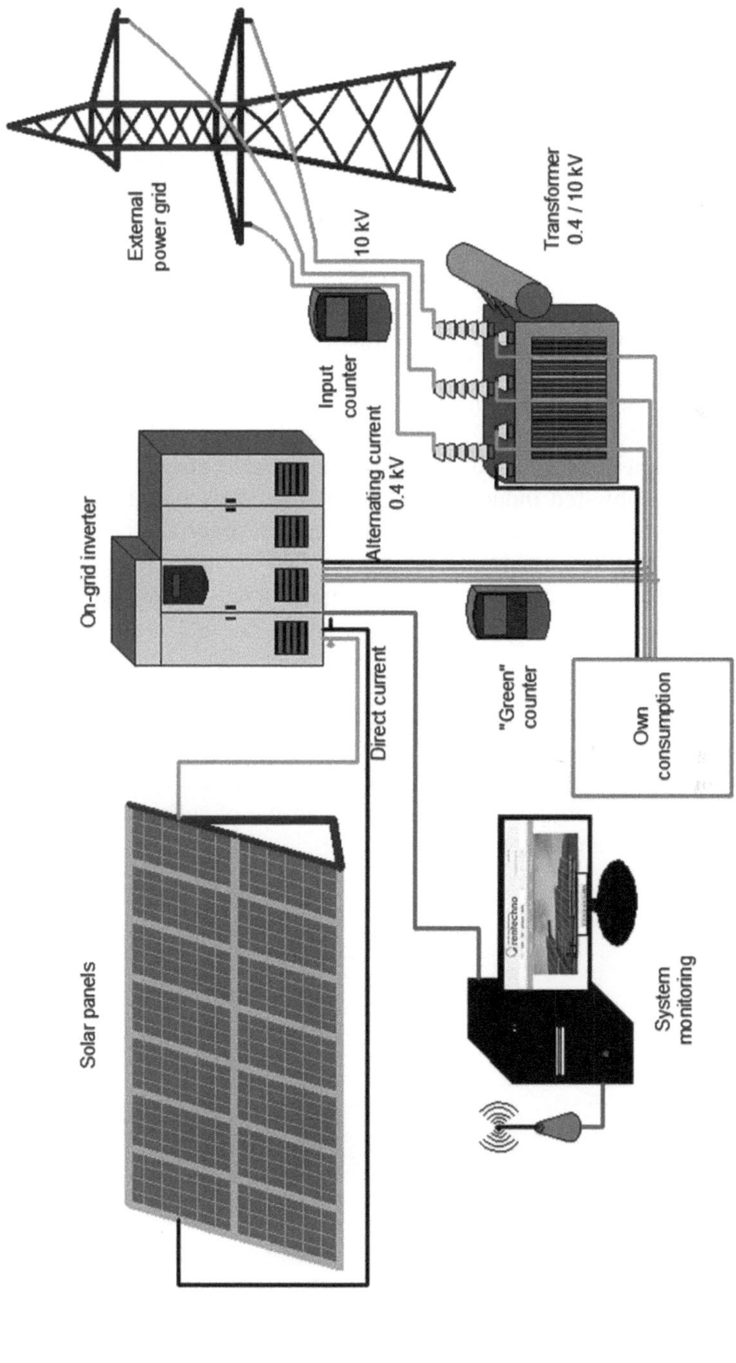

FIGURE 3.2 Hybrid solar power generation.

PV cells include lower maintenance and no emission, while the major issues are intermittent nature and the introduction of harmonics due to the use of inverters for grid connection.

3.2 Wind power generation

The energy of wind can be converted into electrical energy by utilizing wind turbines. Usually wind turbines operate in a group to form a wind power plant to produce electrical power, which is termed a wind farm. It is a free, clean, and renewable source of energy. Although wind energy is free of cost, the initial investment is very high for wind power generation compared with that of fossil-fueled generators. Another major issue in wind generation is weather dependency, as wind speed always varies and hence so does the electrical power output.

Grid-connected wind turbines are either fixed or variable speed. In fixed speed, a propeller through a gearbox drives the cage induction generator directly connected to the grid, whereas for variable speed, a synchronous generator or double-fed induction generator is used. Although the output of wind turbine-based DGs is AC, a power electronic interface is still required because output voltage and frequency are variable due to variable wind speed. Fig. 3.3 shows a grid-connected wind turbine generation system.

3.3 Fuel cell

Fuel cells convert the chemical energy of a fuel directly to electrical energy by an electrochemical process. Unlike solar and wind energy sources, fuel cells are capable of supplying power consistently and hence are more reliable. Fuel cells have a variety of applications and are capable of supplying large power stations as well as small computers.

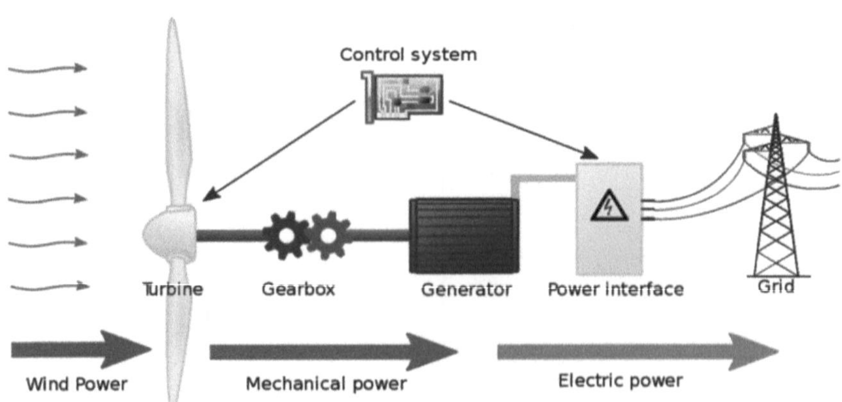

FIGURE 3.3 Grid-connected wind power generation.

They are generally characterized by electrolyte materials and operate like batteries; however, they differ from batteries, as a continuous supply of fuel and oxygen is necessary in fuel cells to sustain the chemical reaction. The most common fuel utilized in fuel cells is hydrogen, and they are categorized into different categories based on the electrolyte used. A typical fuel cell using hydrogen as fuel is shown in Fig. 3.4.

Similar to PV systems, the output of the fuel cell is also DC, and therefore fuel cell–based DGs are connected to the grid via a power electronics interface. Fuel cells have quieter operation, and due to the absence of a mechanical process, their efficiency is higher compared with that of other thermo-mechanical processes. Although the power generated by fuel cells is reliable and free from emissions, the major concern is fuel cells' higher cost.

Variations in solar light intensity and wind speeds lead to fluctuations in the voltage and frequency of electrical power generated through solar and wind power plants. Moreover, harmonics and nonlinearities are introduced to the power electronics interface. The energy of DGs may be injected into the grid via AC generators connected directly to the grid or via a power electronic interface depending on the technology. DGs based on RESs and their interface methodologies for connecting to the grid are summarized in Table 3.1.

4. Dispersed-generator types

DGs are based on the different DG technologies described in Section 3.3. Depending on characteristics of the various technologies and the output and methodology of interconnection to the grid, active power is either injected by the DG or may be zero, but reactive power can be injected, absorbed, or zero.

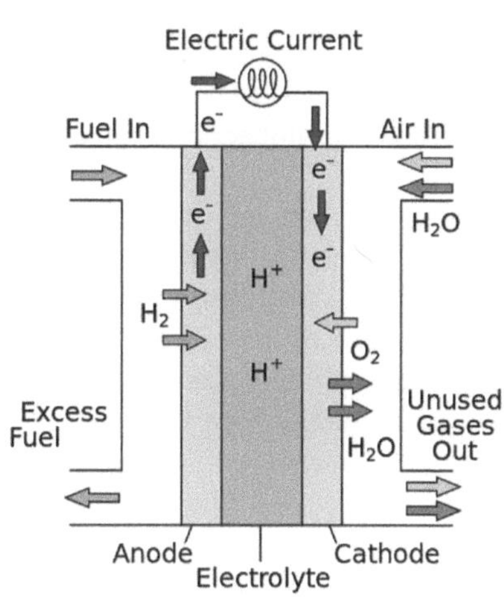

FIGURE 3.4 Fuel cell operation.

TABLE 3.1 Renewable energy source–based dispersed-generator technologies summary.

Technology	Output type	Methodology for grid connection
Fuel cell	DC	Power electronic interface
Photovoltaic cell		
Wind turbine	AC	Directly or through power electronic interface

DGs are classified into four types based on the nature of active and reactive power, and hence the power factor [6]:

- **Type-I** DGs are capable of injecting only active power—e.g., generation based on fuel cells and PV systems.
- **Type-II** DGs are capable of injecting reactive power only—e.g., synchronous compensator, and capacitor.
- **Type-III** DGs are capable of injecting both active and reactive power and based on synchronous machines. These generators have a lagging power factor.
- **Type-IV** DGs are capable of injecting real power but absorb reactive power—e.g., wind turbine-driven induction generators. These generators have a leading power factor.

5. Benefits of renewable energy source–based dispersed generation

Dispersed generation is contemplated as an attractive option to defer transmission and distribution upgrading and support reliability and system performance [1]. Economic and environmental aspects are the main drivers for the development of DGs. DGs have an impact on grid voltage quality, safety, and protection [7].

The integration of DGs has technical, economic, and environmental benefits as shown in Fig. 3.5. The technical benefits include diminution of losses, enhancement in the voltage level of all buses, and improvement in the power quality of the DN. The diminution of losses also leads to savings in the cost of energy losses, making DG integration economically favorable. Integration of DGs might change the characteristics of the DN, and therefore there are associated challenges.

5.1 Technical benefits

- ***Loss diminution:*** RESs are not centrally planned and frequently connected in DNs. DNs are characterized by high *R/X* ratio, lower voltage levels, and

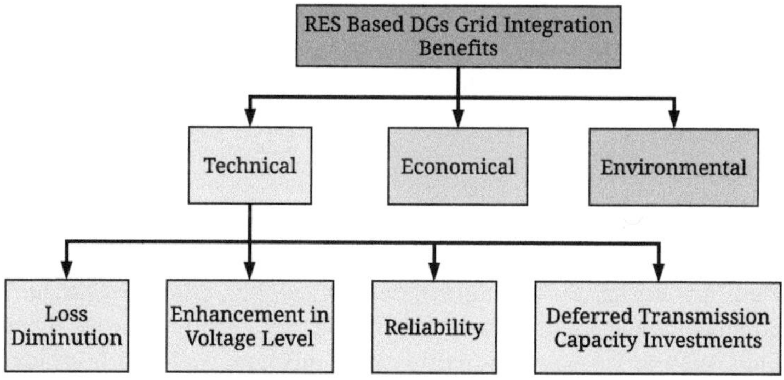

FIGURE 3.5 Benefits of renewable energy source—based dispersed-generation integration.

higher current levels; therefore, losses are highest in DNs compared with those of generation or transmission systems. Moreover, DNs are always under stress due to a constantly growing load demand. Due to heavy loads, the voltage levels of all buses fall, currents increase, and hence cause incremental real power losses. As DGs are integrated into DNs, the voltage levels of different buses in DNs are enhanced, which further leads to reduced current levels and losses. Therefore, the major benefit of DG integration is diminution of real power losses, which reduces the cost of lost energy.

- ***Enhancement in voltage level:*** The decreasing voltage profile of all buses is a concern in grids, especially in DNs, which may lead to voltage instability. This is due to the shortage of reactive support at the distribution level. In the past, capacitors were employed for the support of reactive power and hence enhancing the voltage profile. The integration of DGs in DNs is expected to enhance the voltage level of all buses by locally meeting demand and reducing the burden on the substation and hence on the grid.
- ***Reliability***: Reliability is associated with interruptions in power supply due to decreases in voltage or voltage collapse or outages. As the assimilation of DGs enhances the voltage levels of all buses, reliability of the power supply is improved with the integration of DGs in grids. Moreover, the load can be met locally.
- ***Deferred transmission capacity investments***: As the load continuously increases due to increasing dependency on the electricity, there is a need to expand and reform transmission systems. The investment to enhance transmission capacity can be deferred by assimilation of DGs, as they can meet the increased load demand.

5.2 Economic benefits

The economic benefits of DG assimilation in DNs are linked to technical benefits. The lower power loss reduces the cost of power lost. The cost of transmission system expansion can be reduced without expanding it by meeting load demand locally. Consequently, grid economics can be improved by reducing the costs of power loss and expansion.

5.3 Environmental benefits

Fossil fuel-based conventional power plants have three major emissions—i.e., carbon dioxide, sulfur dioxide, and nitrogen oxide. These emissions should be minimized according to the agreements of global climate change. These environmental concerns are major forces for the deployment of renewable-based DGs such as photovoltaic cells, wind turbines, and hydroturbines.

6. Challenges in renewable energy source–based dispersed generator integration

Most RESs are variable sources of power and nondispatchable. The power of RESs such as solar and wind is controlled by weather conditions. The electrical power produced by such sources has an intermittent nature. This intermittent nature is a major challenge for RES integration in grids, which operate at constant voltage and frequency. Fig. 3.6 shows the main challenges involved in integration of RES-based DGs.

6.1 Power quality issues

The efficiency and stability of a grid is ensured by power quality, which is an imperative factor in power systems. A grid with ideal power quality has high reliability of power and lower costs, whereas a grid with poor power quality has deleterious effects on the grid. The most common power quality issues are:

- ***Voltage fluctuation:*** Voltage output from solar or wind plants is always uncertain, as it is decided by solar irradiance and wind speed, respectively.

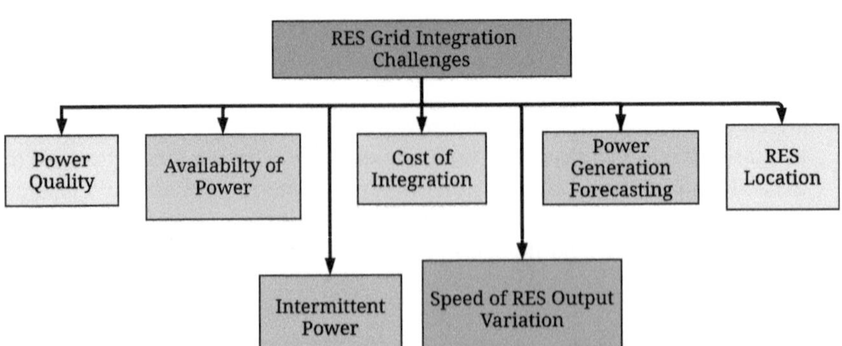

FIGURE 3.6 Challenges of renewable energy source–based dispersed generator integration.

Solar irradiance and wind speed are always variable as well as time and weather dependent, leading to fluctuations in output voltage and poor power quality in the grid.

- ***Frequency fluctuation:*** The band for frequency variations is very narrow for successful operation of grids. Similar to voltage fluctuations, uncertain solar irradiance and wind speed also lead to frequency fluctuations, resulting in bad power quality and grid failures.
- ***Harmonics:*** As stated in 1, the integration of RESs to the grid is through power electronic devices, which affects power quality by introduction of harmonics in the grid. In addition to harmonics, DC components may also be introduced in an AC grid, as the output of both solar plants and fuel cells is DC.

These power quality issues are major challenges for the integration of RESs in the national grid.

6.2 Availability of power

Generation from RESs is dependent on natural resources for which the major concern is that it is not possible stabilize the output from these sources manually. In PV plants, generation is achievable only during daytime when sunshine is available, and power output is zero at night. Similarly, the power output from wind power plants is zero when wind speed is very low and unable to rotate the wind turbine. The uncertain power production of RESs makes the assimilation of RES-based DGs more complicated and can have deleterious effects on grid operations.

6.3 Variable power generation

The generation power levels for both active and reactive power are different for RESs. The fluctuation in power can be short term or long term due to daily input variations and seasonal differences, respectively. The level of power generation in solar plants is much higher in summer than in winter due to a greater number of sunny hours during the summer. Wind turbine power also fluctuates significantly due to significant differences in peak and low-level wind speeds. Moreover, there is a deficiency of reactive power support in RES-based DGs.

6.4 Speed of variation

The speed of power fluctuations is also a big challenge for integration of RESs in the grid. Fig. 3.7 shows fluctuations in wind power output over a short period. It is evident from Fig. 3.7 that within 24 h, power increases to

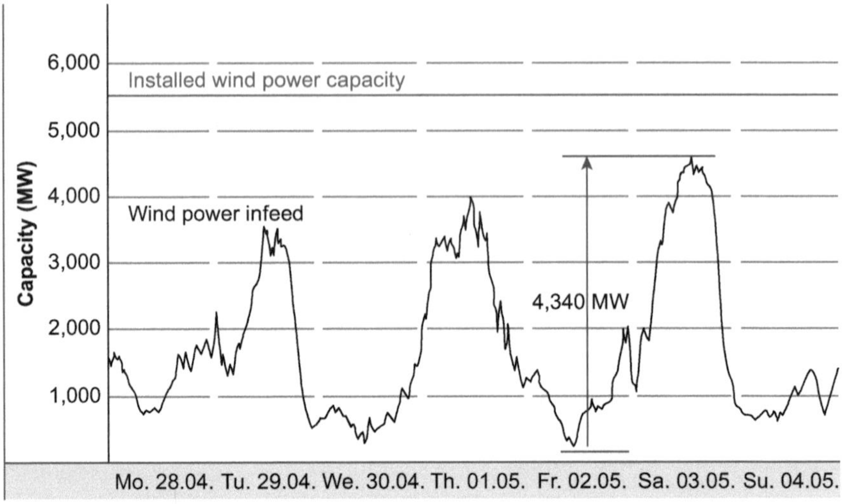

FIGURE 3.7 Wind power variation in E.ON Netz (Germany) [8].

4240 MW from a minimum level of 450 MW. The variation in wind power can appear every day, hour, minute, or even every second. Hence the speed of power variation is another challenge for wind power-based DG integration in the grid, especially when a power plant has a large capacity.

6.5 Power generation forecasting

Forecasting plays a vital role in grid power management by predicting future loads and power production. It is necessary for reliable and stable grid operation because power should be generated as per-load demand. When power generation is from RESs, it becomes mandatory to forecast generation, as RESs are controlled by weather conditions. Accurate prediction of power generation is a difficult task, which makes RES assimilation a big challenge for grids. Moreover, the forecasting methodologies for different RESs are distinct due their different characteristics. In addition, the inaccuracy of long-term forecasting is also an imperative issue for grid integration of RESs.

6.6 Renewable energy source location

The selection of sites for power plants based on RESs is also a challenging task. The distance between the RES plant and grid is a major concern, as it affects cost as well as efficiency. Moreover, not all RESs are present in all regions. For example, wind energy is more effective in coastal areas in comparison with the plains, so it is more effective to integrate wind power plants there.

6.7 Cost of integrating renewable energy sources

The economical operation of a grid is always imperative in grid integration and grid operation, because the cost of power should be as low as possible. Difficulties in the economic operation of RESs arise when the distance between large RES power plants and the grid is significantly long. This is due to higher transmission costs and losses over long distances. In addition, higher costs are involved in RESs, as energy storage is necessary in an RES plant due to variable power, which must be stored in large batteries before integration to the grid. Therefore, the higher cost of RES integration in the grid is also a challenge.

6.8 Protection challenges

The integration of RES-based DGs in DNs changes their nature from passive to active, which changes short-circuit current levels and may lead to reverse power flows and even islanding. The changed nature of DNs leads to upgrading of the existing protection system so it can handle bidirectional power flows. The design of such a protection system is also a challenging factor for the grid integration of RESs.

7. Storing energy for renewable energy sources

Energy management systems require proper energy-storage techniques for their efficient utilization. On the other hand, the environmental and economic problems associated with fossil fuels for producing electricity have led to requirements for renewable energy sources. These renewable energy sources are now becoming an essential part of electricity generation in our power systems. At the same time, renewable sources have the disadvantage of being intermittent energy sources and therefore require significantly more energy-storage technology for proper utilization. Also, renewable energy sources not only are producing electricity at main generating sites but also are being integrated with the distribution as well as the load side, further increasing the need for storage techniques. This section is further divided into different subsections that provide an overview of the different types of energy storage technologies used for RESs [9].

7.1 Batteries

Electrochemical cells are stacked together to form the electrical batteries required for storing electricity based on electrochemical energy storage. The significant benefits of rechargeable batteries include economical recharging and repeated use. They have a nonpolluting nature with minimal impact on the environment. Their energy flow depends on charging and discharging.

The various important factors related to the efficiency and life expectancy of the rechargeable battery include temperature effect, depth of discharge, and recharge cycle. Being sensitive to the environment, the battery can be adversely affected by temperature changes caused by charging and discharging. Also, a battery's life cycle is inversely proportional to the depth of discharge. Lead-acid, lithium-ion, redox-flow, and sodium-sulfur are some of the types of batteries that will be discussed and compared in subsequent sections [9].

7.1.1 Lead-acid battery

The lead-acid battery is one of the most broadly used rechargeable batteries. It is made up of lead and lead dioxide as cathodes in an electrolytic solution of sulfuric acid. The flow of electric current is due to conduction between anode and cathode during the discharging process. Benefits include economically low cost, good efficiency, easy recyclability, and appropriateness for uninterruptible power supply applications [9]. It has the disadvantage of lead being a poisonous metal that can be harmful. It has a low life span with about a 1000-cycle (charging and discharging) life, which becomes serious when it comes to large-scale energy storage applications. Lithium–ion has a low discharge depth (about 50%)—i.e., it should not be discharged more than 50%, as that can affect its life span—and due to its low cost, it is are best suited for off-grid home energy systems [10].

7.1.2 Lithium-ion battery

Lithium-ion batteries are more commercialized batteries with major application areas covering electronic devices like smartphones and laptops. With nearly twice the voltage (3.7 V), the lithium-ion battery is a better option than a lead-acid battery. It has a three-layer design with the first layer of lithium compound (anode), the second layer of graphite (cathode), and the third layer of an insulator placed between the first and second layers. Its benefits include light weight, fast response, a low self-discharge rate, and less maintenance. However, lithium-ion batteries face cost- and safety-related problems that hinder their growth in power systems. Compared with lead-acid batteries, lithium batteries have higher efficiency, a longer life, and higher discharge depth (about 80%). But lithium batteries are not cost-effective compared with lead-acid batteries, which limits their usage for large-scale wind power support. Lithium-ion batteries are most suitable for home energy storage systems using solar panels, although that is not economically reasonable compared with other batteries [11].

7.1.3 Redox-flow battery

Similar to electrochemical cells, redox-flow batteries contain an ionic solution containing electroactive elements. But the electrolyte is not kept around the electrodes but instead is stored in a tank outside the cell. The electrolyte (containing electroactive elements) then flows into the cell from outside the tank to generate electricity [11], and the amount of generated electricity depends on the size of the tank. Its life span is very long, greater than 10,000 life cycles, and it is capable of handling large-scale storage applications. However, it suffers from relatively low energy density together with lower efficiency.

7.1.4 Sodium-sulfur battery

In sodium-sulfur batteries, the electrolyte is in solid state but both electrodes are in molten states—i.e., molten sodium and molten sulfur as electrodes. From a technological point of view, the sodium-sulfur battery is very promising as it has very high efficiency (about 90%), high power density, a longer lifetime (4500 cycles), and 80% discharge depth. Operation of sodium-sulfur batteries requires a high temperature to liquefy the sodium, which is very difficult to operate and increases the cost of operation. Sodium batteries pose a risk to the atmosphere, as it is dangerous when liquid sodium comes in contact with water that may be present [11].

7.2 Pumped hydro storage

In a pumped hydro storage system, electrical energy is stored in the form of potential energy from water stored in a high-level reservoir that is pumped from a low-level reservoir. The water pumps used to raise the water level (from the lower-level to higher-level reservoir) use electrical power during off-peak hours, which can be considered the charging process. During the discharging process or high electricity demand, water from the higher-level reservoir is released and passes through hydroturbines to the lower-level reservoir. These hydroturbines are attached to generators to produce electricity (as shown in Fig. 3.8). Since it uses off-peak hours to store energy, it is cost-effective for storing large-scale electrical energy, but its capital cost is still a critical factor to consider. It is the most capable technology for promoting renewable energy in our power systems. A hybrid power station includes wind energy farms together with a pumped hydro storage system, creating more opportunity for renewable energy integration [12]. The benefits of this storage system include a higher life span (greater than 15,000 cycles), good efficiency (70%−80%), very low self-discharge, and a low energy installation cost. Wind power with a pumped hydro storage system shows the abundance of wind potential to a great extent. There are some disadvantages too, which include geographical restrictions for selecting a suitable site, a long construction time, and a high initial investment cost.

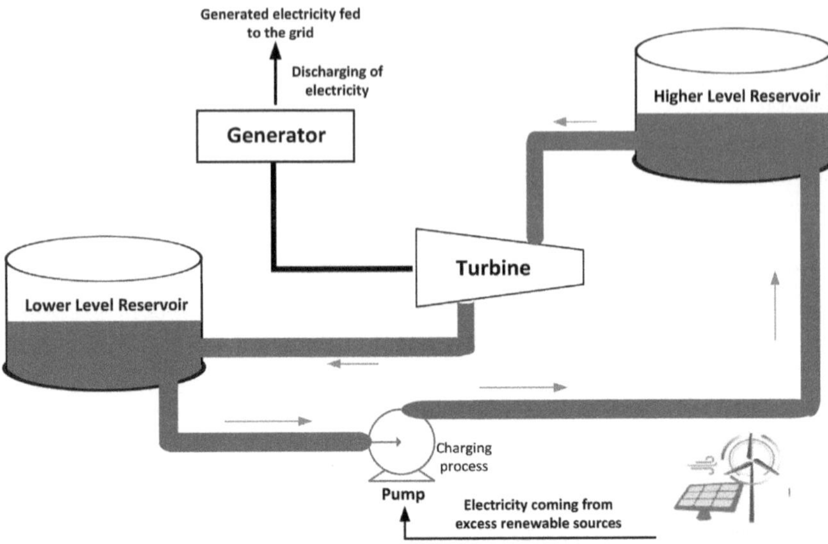

FIGURE 3.8 Pumped hydro storage.

7.3 Compressed-air energy storage

The compressed-air energy storage (CAES) system uses off-peak electrical energy to compress air underground or in a surface vessel. This compressed air inside underground structures (abandoned mines, aquifers, rock structures) or surface vessels is stored with high pressure. Whenever electricity demand is high, this compressed air is burned with a fuel and is expanded in order to pass through turbines to generate electricity. This technology to store electricity is not feasible for the environment, as it produces pollution from fuel combustion with stored compressed gas. Also, underground structures require digging and mining operations that lead to further erosion and noise pollution in the surrounding environment. Its main advantages include a high life span (greater than 10,000 cycles), small self-charge losses, and good efficiency [9]. After the pumped hydro storage system, the CAES system is another option for wind energy systems because of its high capacity and high power ratings.

7.4 Flywheel energy storage

Flywheel energy storage (FES) is one of the most efficient technologies for storing electric energy in the form of kinetic energy by constantly spinning a disk or the rotor of a flywheel. The key components of FES systems are a rotating cylinder, bearings, a generator or motor, and a container to accommodate the flywheel [9]. The charging process requires high acceleration for spinning of the rotor by acquiring the electrical energy given to the motor. This electrical energy is stored in the flywheel by keeping the body rotating at a constant speed. During the discharge process—i.e., when electrical energy is required, the disk rotates the shaft connected to the generator to produce electricity. An FES system can be used as another option for batteries. An FES system comes with a long life, low maintenance, and high power density. The short duration of operation and high losses due to self-discharge are the major drawbacks of this system. Compared with batteries, this storage system can operate under a wider range of temperatures and has no toxic impact on the environment.

7.5 Superconducting magnetic-energy storage

To store electrical energy, the superconducting magnetic energy-storage (SMES) system uses magnetic fields created by DC power passing through superconducting coils. The coils are cryogenically cooled to below the critical temperature for superconducting, so there are no losses due to the passing of currents through them. Other than superconducting coils, the SMES system comprises a power conditioning system, a cryogenic system (refrigerators, vacuum pumps, and helium tank), and a control unit [13]. The DC fed to superconducting coils comes from AC power (during off-peak hours) converted to DC using a power conditioning system. This stored DC electric energy can be supplied back to the grid system by transforming it to AC power using the same power conditioning system and control unit. This control unit sets up a link between the grid and SMES system.

The main advantage of this system includes its very fast response—i.e., fast charging and discharging rates. Therefore, with this storage system, the electric power required is transferred almost immediately on request, which improves the stability and quality of the electric power supplied. It is easier to deploy an SMES system where batteries or CAES is not feasible. One disadvantage of SMES is that it requires high power to keep the coils at low temperature. Also, the overall cost for an SMES unit is high, and for higher power applications, the SMES system requires higher units.

7.6 Supercapacitors

Appropriate for lower-power applications, the supercapacitor is a short-duration energy-storage device. Supercapacitors can be considered a hybrid between capacitors and batteries. Supercapacitors comprise an electrolyte between two electrodes plates. During the charging process, electric energy is stored in the form of potential energy in the electric field developed between ions and electrons when the potential is applied across the electrodes. The charging and discharging process is shown in Fig. 3.9.

Unlike batteries, no chemical reaction takes place in supercapacitors. Supercapacitors have about 1000 times more charge storage capacity than

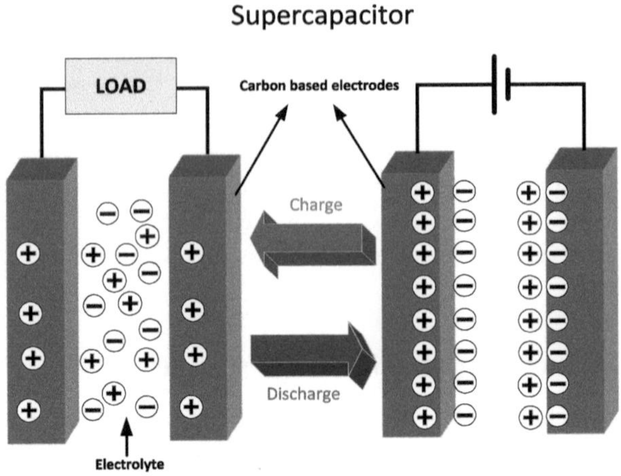

FIGURE 3.9 Supercapacitor.

normal capacitors. Its advantages include fast charging and discharging, a long lifetime, a wide operating temperature range, environmental friendliness, and easy recyclability. On the other hand, its disadvantages are low energy density, high self-discharge, and the highest dielectric absorption of any capacitor. All these energy-storing technologies are important depending on the type of application. The different parameters of all the above-mentioned energy-storage technologies are compared in Fig. 3.10.

8. Small-scale renewable energy sources

Growing populations and higher living standards are the main reasons for the high rate of increasing demand for electrical power. Even if we fulfill this power demand, we still cannot risk polluting our environment. The world is focused on clean energy not only because fossils fuels are being depleted but also to achieve minimal to zero negative impact on our environment. Our main concern is to produce electricity using renewable energy sources that can become substantial enough to completely meet electricity demands on their own. Fig. 3.11 shows the global scenario of solar PV capacity (by country and region) from 2008 to 2018 [14].

There is a considerable amount of growth in solar PV capacity for producing electricity, but it was roughly just 10% (in 2018) of the installed capacity of fossil fuels. It is therefore very clear that we need more efforts to increase power generation from renewable sources. One of the efforts is to accept small-scale schemes that include renewable and low-carbon energy equipment at domestic and nondomestic premises next to large-scale renewable energy development. Low-carbon energy installations include biomass systems, ground- or air-source heat pumps, and combined heat-and-power systems.

Energy Storage System	Lifespan (in cycles)	Depth of discharge	Efficiency (%)	Energy Density (Wh Kg^{-1})	Response time	Advantages	Disadvantages
Lead- Acid Battery	250-1000	50%	75-85	30-50	< 10 s	Low cost	Low life span
Lithium – Ion Battery	400-1200	80%	80-90	100-200	< 10s	High efficiency and high energy density	High production cost
Redox – Flow Battery	>10000	75%	60-70	10-50	< 5s	High lifespan and high capacity	Low energy density
Sodium-sulfur Battery	2500-4500	80%	70-85	100-250	< 10 s	High energy density and high efficiency	High cost and needs high temperature for its operation
Pumped Hydro Storage	> 15000	-	70-80	0.5 - 3	1-10 min	High lifespan and high capacity	Special site requirement and Low energy density
CAES	>10000	-	70-80	30-60	1-10 min	High lifespan and low self charge losses	Special site requirement and adverse effect on Environment
FES	>20000	-	85-95	5-50	< 1 s	Long life and low maintenance	High self discharge losses
SMES	100,000	-	80-90	1-10	< 5ms	Ability to respond very fast	Low energy density and high production cost
Super capacitor	10000	-	85-98	5-10	< 1s	High power density and quick response	Low energy density

FIGURE 3.10 Comparison of different parameters for various energy-storage systems.

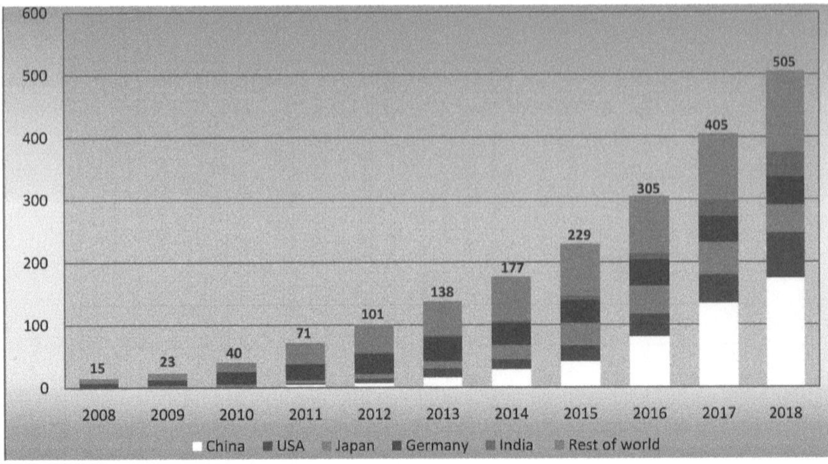

FIGURE 3.11 The global scenario of solar photovoltaic capacity (by country and region) from 2008 to 2018 [14].

Small-scale renewable energy generation has the key characteristic of including local-scale generation of heat or electricity (using renewable sources), and a small amount of electricity is generated compared with that of centralized generation. Fig. 3.12 shows various energy sources and related energy technologies for small-scale energy generation [15]. Renewable energy generation projects can be large in scale, such a size is not well received in densely populated areas, whereas small-scale renewable projects are nearer to consumers and well accepted. As shown in Fig. 3.13A "Mandironwala bhuddi" is a village in India having no electricity poles but still fulfilling its electricity needs using small-scale solar power alone [16].

Wind	Micro-wind power	Electricity
Hydro Power	Small hydro-turbines	Electricity
Biomass : Wood & Wood pellet	Fireplaces, wood burning boiler & wood pellet boilers	Heat (air & water)
Various sources (e.g. biogas, wood, wood pellet)	Micro-CHP/Stirling Engine	Heat & Electricity
Hydrogen (from renewable sources)	Micro-CHP/Fuel Cell	Heat & Electricity

FIGURE 3.12 Various energy sources and related energy technologies [14].

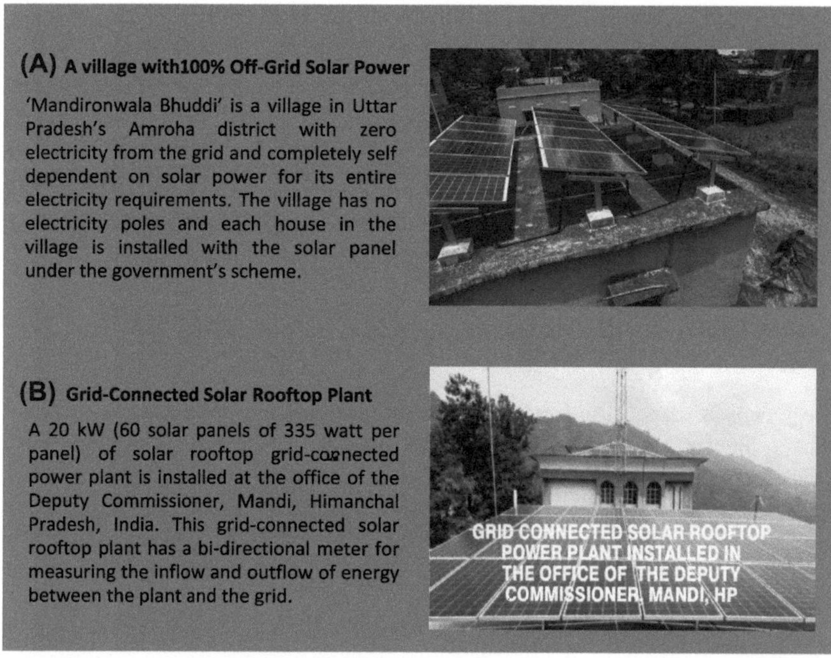

FIGURE 3.13 Off-grid and grid-connected small-scale solar power generation in India [16].

This small-scale solar PV usage is an example of off-grid solar energy generation. In another example, shown in Fig. 3.13B "Himurja," a government of India scheme under the Ministry of New and Renewable Energy, is giving 60% financial assistance for installing grid-connected solar rooftop power plants for government sector usage [17].

Electricity generation from small-scale renewable sources has great possibilities not only for fighting climate changes but also in providing affordability and sustainable economic revival to local energy needs. The possible benefits from using small-scale renewable energy systems are as follows:

1. Productive small-scale renewable operations at areas incompatible for large-scale plants
2. Reduced dependency on central energy generation
3. Reduced losses at transmission and distribution levels
4. Reduced negative impact on the climate from electricity production
5. Increased employment in the fields of research, design, installation, and management of these small-scale renewable projects
6. Local community benefit from services provided by nearby small-scale projects

7. Reduced electricity demands directly at the local level
8. Reduced capital risks and costs of electricity at the local level
9. Can be used as grid backup to important local loads in cases of grid failure
10. Electricity generation at the local community level makes consumers more aware of their energy consumption patterns

In developing countries, more than a billion people worldwide still lack access to electricity, and small-scale renewable energy generation is becoming a more feasible solution than large-scale centralized generation.

9. Reliability of distribution network in presence of dispersed generators

The DN is a direct connection to the main consumers, which has a strong impact on the reliability of the power system. Reliability is considered a major factor in planning DNs. Most power system failures are initiated from the distribution side. DGs, which are usually integrated in DNs, can improve the reliability of DNs by meeting load demand locally, enhancing the voltage profile and diminishing losses. These advantages can be achieved if DGs of optimal size are located at the best sites in DNs. Many researchers have planned the placement of DGs in DNs for improved DN reliability [18–21].

The assimilation of DGs in DNs changes their structure and operation. Therefore, a DN's reliability evaluation needs to be performed in the presence of DGs. The reliability of DNs is measured using various reliability indices. The major indices for reliability measurment are as follows:

1. **System average interruption duration index (SAIDI)** is the measure of the average duration of interruptions faced by each customer measured in time. It can be mathematically expressed as

$$\text{SAIDI} = \frac{\sum \mu_i N_i}{\sum N_i} \tag{3.1}$$

where N_i is the number of customers at i^{th} load point, and μ_i is outage time.

2. **System average interruption frequency index (SAIFI)** is the measurement of average sustained interruptions by each customer. It is expressed as

$$\text{SAIFI} = \frac{\sum r_i N_i}{\sum N_i} \tag{3.2}$$

where N_i is the number of customers at i^{th} load point, and r_i is failure.

3. **Customer average interruption duration index** is the ratio of SAIDI and SAIFI and can be viewed as average restoration time. It indicates average duration of interruptions experienced by any customer. It is given as

$$\mathrm{CAIDI} = \frac{\sum \mu_i N_i}{\sum r_i N_i} \tag{3.3}$$

4. **Customer average interruption frequency index** is another index used in power systems for reliability measurement. It indicates the trend in customer interruptions by showing the number of customers affected from a customer base. It is expressed as

$$\mathrm{CAIDI} = \frac{\text{Total Number of Customer Interruptions}}{\text{Number of Distinct Customers Interrupted}} \tag{3.4}$$

5. **Average service availability index** is the ratio of the total number of customer hours for which power was available to total customer hours demanded. It can be measured on a monthly or yearly basis and is mathematically expressed as

$$\mathrm{ASAI} = \frac{\sum N_i \times 8760 - \sum \mu_i N_i}{\sum N_i \times 8760} \tag{3.5}$$

where μ_i is the annual outage time in hours.

6. **Average service unavailability index** is the ratio of the total number of customer hours for which power was not available to total customer hours demanded and is expressed as

$$\mathrm{ASUI} = \frac{\sum \mu_i N_i}{\sum N_i \times 8760} \tag{3.6}$$

It is expected that the integration of DGs in DNs improves the reliability of power systems.

10. Concluding remarks

This chapter has presented a review of different DG technologies based on RESs that can be integrated in the national grids. The benefits and challenges associated with grid integration of RESs have also been discussed. It is concluded that although RESs are environmentally friendly, uncertainty in power production from RESs still poses a big challenge for grid integration.

References

[1] G. Pepermans, J. Driesen, D. Haeseldonckx, R. Belmans, W. D'haeseleer, Distributed generation: definition, benefits and issues, Energy Pol. 33 (6) (2005) 787–798, https://doi.org/10.1016/j.enpol.2003.10.004.

[2] N. Gupta, Gauss-quadrature-based probabilistic load flow method with voltage-dependent loads including wtgs, pv, and ev charging uncertainties, IEEE Trans. Ind. Appl. 54 (6) (2018) 6485–6497, https://doi.org/10.1109/TIA.2018.2855164.

[3] S. Mukhopadhyay, B. Singh, Distributed generation - basic policy, perspective planning, and achievement so far in India, in: IEEE Power Energy Society General Meeting, 26-30 July 2009, Calgary, AB, Canada, 2009, pp. 1–7, https://doi.org/10.1109/PES.2009.5275741.

[4] R. Viral, D. Khatod, Optimal planning of distributed generation systems in distribution system: a review, Renew. Sustain. Energy Rev. 16 (7) (2012) 5146–5165, https://doi.org/10.1016/j.rser.2012.05.020.

[5] T. Ackermann, G. Andersson, L. Soder, Distributed generation: a definition, Electr. Power Syst. Res. 57 (2001) 195–204.

[6] D.Q. Hung, N. Mithulananthan, R.C. Bansal, Analytical expressions for DG allocation in primary distribution networks, IEEE Trans. Energy Convers. 25 (3) (2010) 814–820, https://doi.org/10.1109/TEC.2010.2044414.

[7] J. Driesen, R. Belmans, Distributed generation: challenges and possible solutions, in: IEEE Power Engineering Society General Meeting, 18-22 June 2006, Montreal, Que., Canada, 2006, pp. 1–8, https://doi.org/10.1109/PES.2006.1709099.

[8] G. Boyle (Ed.), Renewable Electricity and the Grid: The Challenge of Variability, Earthscan Publications Ltd, London, UK, 2007. URL, http://oro.open.ac.uk/8394/.

[9] H. Chen, T.N. Cong, W. Yang, C. Tan, Y. Li, Y. Ding, Progress in electrical energy storage system: a critical review, Prog. Nat. Sci. 19 (3) (2009) 291–312.

[10] Y. Chang, X. Mao, Y. Zhao, S. Feng, H. Chen, D. Finlow, Lead-acid battery use in the development of renewable energy systems in China, J. Power Sources 191 (1) (2009) 176–183.

[11] IEC, Electrical energy storage white paper, Tech. Rep (2011).

[12] S. Rehman, L.M. Al-Hadhrami, M.M. Alam, Pumped hydro energy storage system: a technological review, Renew. Sustain. Energy Rev. 44 (2015) 586–598.

[13] C. Das, An investigation into the energy storage technologies available for the integration of alternative generation techniques, Tech. Rep. (2007).

[14] REN21, Renewables 2019: Global status report, REN21 Secretariat (2019).

[15] M.F. Akorede, H. Hizam, E. Pouresmaeil, Distributed energy resources and benefits to the environment, Renew. Sustain. Energy Rev. 14 (2) (2010) 724–734.

[16] Akshay Urja, Volume 12, Issue 5, March-April 2019, Tech. Rep.

[17] Akshay Urja, Volume 12-13, Issue 6, June-August 2019, Tech. Rep.

[18] L. Wang, C. Singh, Reliability-constrained optimum placement of reclosers and distributed generators in distribution networks using an ant colony system algorithm, IEEE Trans. Syst. Man Cybern. C (Appl. Rev.) 38 (6) (2008) 757–764, https://doi.org/10.1109/TSMCC.2008.2001573.

[19] A.A. Chowdhury, S.K. Agarwal, D.O. Koval, Reliability modeling of distributed generation in conventional distribution systems planning and analysis, IEEE Trans. Ind. Appl. 39 (5) (2003) 1493–1498, https://doi.org/10.1109/TIA.2003.816554.

[20] N.R. Battu, A.R. Abhyankar, N. Senroy, DG planning with amalgamation of operational and reliability considerations, Int. J. Emerg. Elec. Power Syst. 17 (2) (2016) 131–141, https://doi.org/10.1515/ijeeps-2015-0142.

[21] G.H. Reddy, P. Chakrapani, A.K. Goswami, N.B. Choudhury, Optimal distributed generation placement in distribution system to improve reliability and critical loads pick up after natural disasters, Eng. Sci. Technol. Int. J. 20 (3) (2017) 825–832, https://doi.org/10.1016/j.jestch.2017.05.001.

Chapter 4

An introduction to smart grid and demand-side management with its integration with renewable energy

Bharti Koul[1], Kanwardeep Singh[2], Y.S. Brar[3]
[1]*Department of Electrical Engineering, NIT Hamirpur, Hamirpur, Himachal Pradesh, India;* [2]*Department of Electrical Engineering, Guru Nanak Dev Engineering College, Ludhiana, Punjab, India;* [3]*Department of Electrical Engineering, IKGPTU, Kapurthala, Punjab, India*

1. Introduction

Electricity is the most vital entity in today's world, although approximately a billion people are deprived of its usage [1], and demand for electricity is projected to grow considerably in coming years. Demand-side management (DSM) integrated with SG helps to adjust electricity demand to the power system's creation, transmission, and distribution of electricity.

The evolution of DSM for electric conveniences and development of this concept for demand management, strategic conservation, and marketing were coined in 1973 and 1979 [2]. In the 1980s, the Electric Power Research Institute introduced DSM publicly in the literature [4]. Demand response (DR) refers to "electricity consumption changes by end-use customers from their normal consumption patterns in response to changes in the price of electricity over time, DR can be also defined as the incentive payments that are designed to induce lower electricity usage at times when wholesale market prices are high or when system reliability is jeopardized" [2,3]. The concern it emphasizes is that customers and power producers require more passive and efficient participation in power markets.

Various DSM procedures and algorithms of DR programs are stated in the review papers, with devices managed by less price-dependent DR programs and with incentive-dependent DR programs. In former algorithms, user devices within a home energy management system worked with various pricing situations such as time-of-use [7] and real-time pricing [8]. Such schemes have the objective of being user-centric, which means optimizing the

Advances in Smart Grid Power System. https://doi.org/10.1016/B978-0-12-824337-4.00004-7

consumer energy bill [7,8], optimizing discomfort initiated due to DR involvement, decreasing the monetary risk involved in device scheduling under real-time pricing schemes [9] or a combination of these [9,10]. The schemes having utility-based goals provide incentive-based DR programs. Such programs are introduced by utilities to minimize costs for the utility [11,12], reduce peak demand [13], improve valley filling [14], enhance reliability [15,16], and improve system security [17].

1.1 The need for smart grids and demand-side management

In an SG, DSM focuses on variation in demand profiles to balance usage and generation. This is done by modeling the DSM strategy to prioritize either consumer demand patterns with respect to generating variations, price variations, or both. In DSM, users have incentives to benefit from the lower-cost schemes prevalent for various time-varying tariffs. Currently in the SG, variation in demand profiles according to the availability of power generation is of utmost importance, as the amount of existing, immediate energy is restricted.

By using DSM, system efficiency can be improved widely by making it consumer-friendly. Such improvements can provide more reliable distribution in the power industry. The requirement for DSM lies in the desire to reduce energy use, improve utilization to its optimal level, and hence enhance system efficiency.

Various DSM techniques (see Fig. 4.1) are supportive for driving required changes in the demand curve known as peak clipping, valley filling, load shifting, strategic conservation, strategic load growth, and flexible load-shape. The techniques for DSM policies—energy efficiency, time of use, DR, and spinning reserve—are available for studying various load-demand patterns.

Furthermore, the integration of DSM and load-shifting techniques also benefits utilities and consumers:

- Reductions in total power consumption and reduced losses in transmission and distribution systems so that mutual profit for power utilities and consumers is achieved.
- The aforesaid benefit also results in a reduction in power generation.
- Successful implementation of a DR scheme reduces the need to initiate expensive-to-run power plants to meet peak demands while it empowers energy providers to meet their pollution commitments.

SGs, being intelligent power grids, are capable of managing uncertainties by incorporating renewable energy (RE) sources as they practice bilateral drift of power and data to construct an extensively distributed automated energy network. In the SG system, the consumer is treated as a prosumer, as the user participates at the generating level in addition to consuming energy. This requires the creation of proper communication among the utility and users to ensure more efficient consumption and transmission of energy. The bilateral

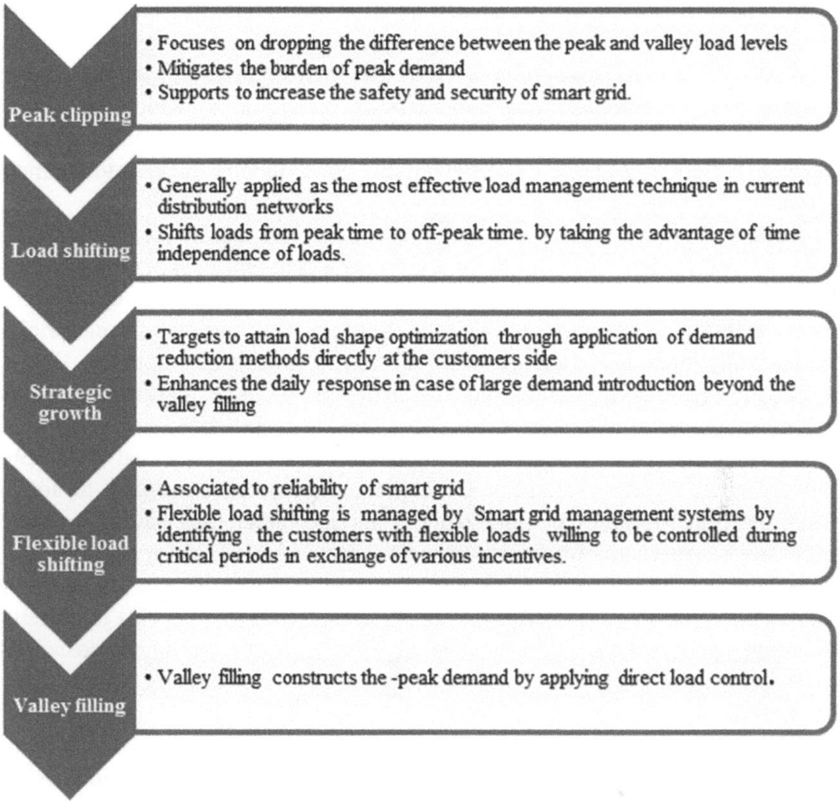

FIGURE 4.1 Demand-side management techniques.

flow of data among power producers and users lets users manage their respective loads comfortably and retains power system limits within prescribed ranges.

The prime purpose of the SG is to provide quality, uninterruptable power. As increasing demand has resulted in total power-network breakdown, so the best way to manage grid issues is by limiting overuse of power and helping with load-shifting from the user side, which is the DSM.

1.2 Demand-side management

DSM actions benefit from prospects for enhancing the competence of energy services. DSM measures can be used to activate cost-effective savings in electricity and peak demand that involve the use of distinct DR programs. Such programs support the overcoming of several obstacles that inhibit cost-dependent DSM measures from being approved. Such obstacles persist even in countries with totally advanced marketplaces. In this respect, the open image of a utility company can thus be enriched by promoting DSM.

DSM is beneficial to power producers, utilities, customers, and society and can be appreciated for its many advantages (see Figs. 4.2 and 4.3).

Another way to revolutionize the distribution structure is to charge consumers for the actual price of electricity, a method by which the power being sold is priced to current market values. If a commodity is limited, rates increase, sellers try to trade extra power, and consequently buyers reduce their usage.

Because power is a temporary good and cannot be commercially stored (it must be used when generated), marketplaces frequently experience short-duration deviations as capacity varies from spare to scarce because of hourly and day-to-day variations in load. Static pricing schemes for electricity are basically ancient and create cross-subsidies between customers.

It can be seen (Fig. 4.2) that incentives depend on price, and improvements in RE volumes will alter cost elasticities. The vertical demand curve for consumers indicates that they are constrained from reacting to actual electricity delivery costs in power markets. By implementing pricing matched to actual costs, the price becomes flexible at the load side rather than having constant price levels. Alternatively, the improved quantity of RE could restructure this curve-defining source. At the instant of scarcity of green energy, the rate for identical energy improves, thereby moving the curve upwards. This is because RE—e.g., photovoltaic (PV) and wind—has a lower operating cost. DR programs permit users to contribute and benefit economically while being eco-friendly.

Reimbursement for resource investment is of great concern while scheduling the network extension. In this case, DSM plays a prime part in improving investment returns, with its implementation of resource applications making the power industry much more impressive for investors.

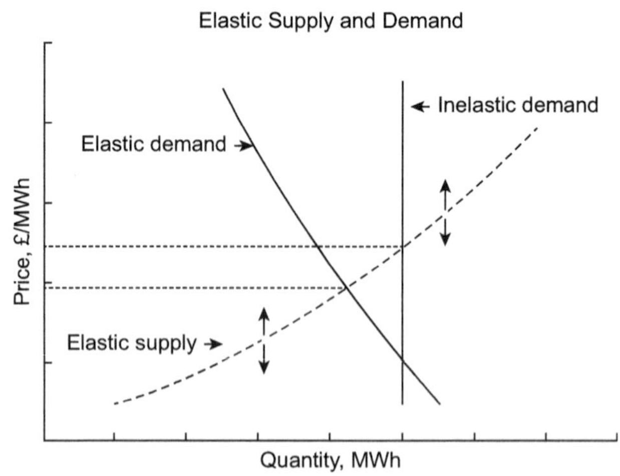

FIGURE 4.2 Demand and supply curves of electricity.

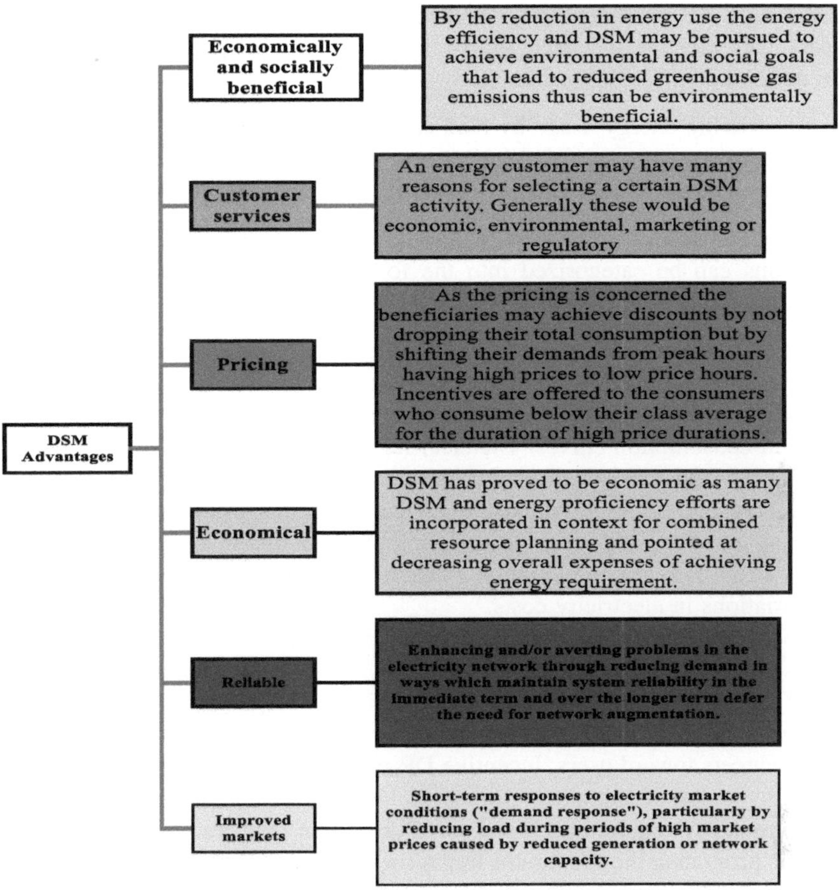

FIGURE 4.3 Demand-side management advantages.

1.3 Demand response

Reactive adjustments to electricity consumption are managed with the help of DR techniques that manage electricity usage with the help of power system operations. A variety of DR resource types are responsible for attaining DR by the application of DG, shiftable-load, loading, and other assets that can help amend the electrical energy provided by the main power grid. DR programs follow mechanisms that encourage customers to reduce loads to curtail maximum load, although they can likewise support increases in load during times of peak production and fewer loads. It is truly worth mentioning that while DR might not consider customer ease, it could restrict the amount of time that can lead to such discomfort by avoiding more drastic adjustments (e.g., blackouts). Mechanization, monitoring, and regulator expertise are hence the essential components for achieving this energy-usage process that thus makes DR less hampering for the user. Three conceivable ways in which users can alter their usage of electrical energy by participating in DR programs [5,6]:

- Decreasing energy usage by load-curbing schemes
- Shifting energy usage to other time slots
- Using DG and thus restraining dependency on the main power grid.

Though various categories of DR are noted in the literature [7], DR programs can be categorized into the following groups depending on who performs the load-reduction task [6,8]:

- Price-dependent DR programs: DR is executed in this category with the help of sanctioned utility charges in a deregulated environment in which energy charges differ for particular durations to inspire users to modify load-consumption curves. The electricity price changes dynamically daily, weekly, and yearly and also might vary based on the already prevailing reserve margin. Accordingly, users must pay the maximum cost during peak durations and lower costs during off-peak durations. The cost should be fixed daily for the duration of each hour, allowing users to respond to variations in electricity costs.
- Event-dependent DR programs: these DR programs provide incentives for users to minimize demand by requesting or allowing administrator control of consumers' consumption of electricity. Load-minimization graphs, in terms of intentional load minimization requirements, are conducted by producers toward users. Incentive DR programs can be used in reaction to many types of initial conditions including grid congestion, functioning consistency requirements, and local or system temperature [6,8]. In such DR programs, the maximum limits for the time spans of such events and cumulative yearly event-time are constant. Such bounds arise because of restrictions on user accessibility with the prevailing tools of DSM programs.
- Demand-minimization bids: for this type, consumers take the initiative to participate in DR programs and send load-minimization bids to the aggregator [6]. These bids comprise existing load-reduction capability and bid cost. Such a program motivates big consumers to perform load minimization at costs where they are ready to be curtailed and also at positions where they can curtail the already mentioned price [9].

2. Integration of demand-response programs with smart grids

DR is the utmost economical and consistent remedy for flattening of the load curve when there is any congestion in the system. The integration of DR in

SGs creates an alternative that is applied to influence deviations in user timing in reply to benefits from variable electricity pricing.

2.1 Smart grid architecture

New necessities and requirements drive power industries, investigatory establishments, and government agencies to reconsider and thus develop real opportunities for the SG. To comprehend this new grid model, the National Institute of Standards and Technology has delivered a concrete model that references several portions of the power system in which SG regularization occurs. The concrete architecture splits the SG into various zones. Each zone encompasses a couple new SG performers comprising devices, structures, or programs that perform judgments and record interchange statistics essential to executing the applications. The explanations of various zones and performers are described (see Fig. 4.4) [10].

2.2 Smart grid and demand response: benefits and features

An outline of DR programs and devices centered on a report by the US Department of Energy (2012) is discussed in this section (see Fig. 4.5). The current segment describes the various prime components of SG and DSM:

- Smart meters: The SG comprises various segments, and the smart meter is considered one of the prime elements for the SG, allowing for uninterrupted metering and recording of energy usage remotely. To provide different incentives to users that show real consumption data and have flexible energy loads, the reading frequency should be hourly at a minimum.
- Aggregation: The aggregator model means compiles power users and can manage demand remotely. The end user has incentive to defer the usage of electricity during peak-load durations, and distribution agencies and the aggregator are paid by the grid operator for reducing demand. In considering the dynamic pricing model, users are helped by the aggregator to balance their energy usage based on price, leading to reduction of electricity bills for users, thus creating an economic incentive for reducing demand.
- Electrical vehicles: As far as the SG and DSM are considered, improvements in the electrical vehicle segment are a significant concern. Electrical vehicle charging is economically preferably during night hours, as it is important to not disrupt the existing electricity system with electrical vehicle loads and to have lower prices during off-peak hours.
- Standardization: Flowing directly from generation to residential customers (RCs) with various performers in the chain, standardization is known as a significant component for SG. According to news presented in Ref. [11], interoperability is a crucial issue for SG standardization.

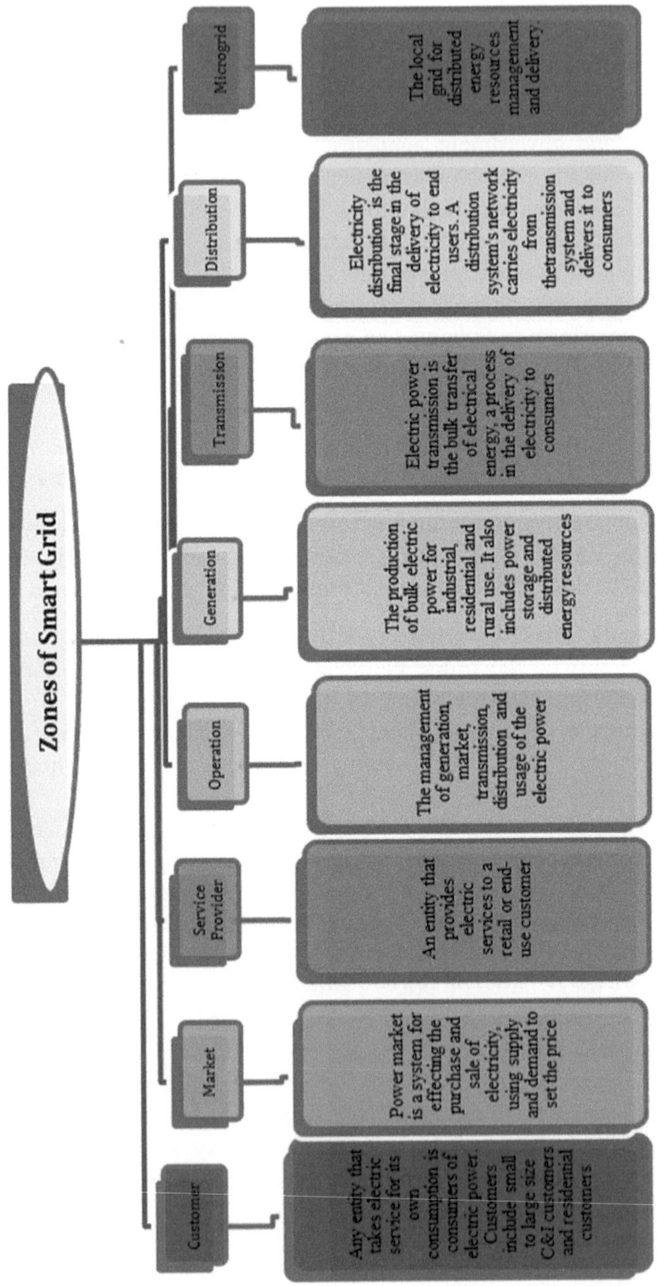

FIGURE 4.4 Various zones of smart grids.

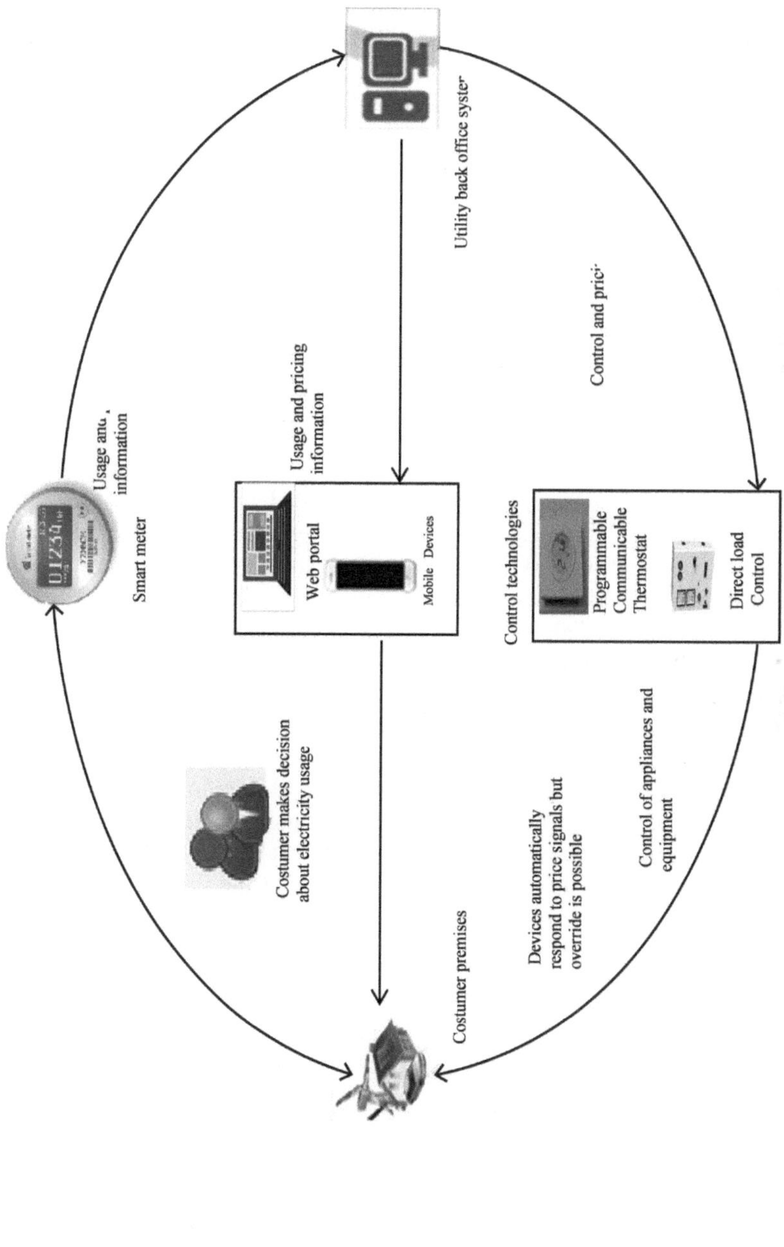

FIGURE 4.5 Elements of the smart grid and demand response.

Standardization concerns safeguarding the problem that various products planned to justify a particular threshold must possess indicators and boundaries that match so they may work in synchronization.

- Indirect and direct control: There are two methods—i.e., direct and indirect control—by which consumer usage can be managed. Direct control comes into the picture if the consumer provides direct control to usage of energy output, whereas indirect control refers to various categories of agreements that provide gains to inspire the consumer to adjust electricity usage. As such, consumers do not have control, but the utility can judge this based on knowledge and previous data. The remedy to such a problem is to provide a home area network to the user so devices can be managed accordingly with a smart network.
- Tariffs and pricing models: There is mostly inconsistency in tariffs, as for certain circumstances only network charges are considered, and in other cases all charges for power usage are comprised in network charges. In this case, only network charges are considered. The peak-load durations use time-of-use tariffs, real-time pricing, and critical peak pricing.

The SG systems that are emerging help to achieve uninterrupted balancing of the network comprising many characteristics: enhanced forecasting, smart inverters, integrated storage, DR, and real-time systems.

Firstly, enhanced forecasting includes extensive instrumentation and innovative computational systems that help participants better predict and manage RE inconsistency and intermittency.

Secondly, with the help of inverters and electronic devices, proper control of the system along with automation also provides support to the grid.

Another aspect is DR is advanced metering interfacing coupled with some smart devices, which permits the balancing of supply and demand from the load side as well.

Next, with combined storage, smoothing of short-duration deviations in RE and managing of supply—demand mismatches can be achieved.

Lastly, instrumentation and control equipment across transmission and distribution networks helps real-time management of the system including the potential for analyzing grid behavior.

3. Influence of demand response on smart grids

The DSM has pivotal importance in today's power industry for sustaining stability of the supply—demand ratio. The DSM motivates consumers to minimize usage of electricity during heavily loaded durations or to transfer consumption of power usage to less burdened durations to minimize peaks during heavy loads. To maintain the increased attention to the study of DR modeling, it must ease dependency in residential demand systems. The aim of the mentioned case study is to show the growth in such demand strategies from shifting device time zones. This is how consumers are also able to contribute to the bidirectional flow of information in the upcoming electrical grid known as the SG. In this paper, DSM uses a load shifting algorithm incorporated into residential loading to minimize consumer utility bills through optimal load shifting. The paper also includes a comparison of different categories of demand shifting: (1) desirable demand shifting and (2) undesirable demand shifting based on the comfort and urgency of RCs.

3.1 Introduction to load shifting

The prevailing power system has become complex and is facing numerous challenges for its optimal, secure and effective operation. These problems arise due to the increased interconnections and loading of the current power grid. In this reference the SG initiates various remedies for overcoming these challenges. SG performs new expertise, like quick and automatic regulators, innovative software for managing the records, and bidirectional link among utility and users.

The concept of DSM embraces all events that target the modification of user load profiles according to their choice and comfort.

Furthermore, incorporation of DSM and load shifting techniques has other benefits for utilities and consumers:

Decreased total power consumption and decreased losses with respect to transmission and distribution systems so that common revenue for producers and users is attained [12].

The successful enactment of the DR scheme reduces the requirement to initiate costly-to-run projects to attain high demands, and this empowers the utility to meet their environment prevention commitments [13,14].

The advanced aim of DSM is therefore to smoothen demand for time durations by minimizing peak values and maximizing low-value durations or by allocating flexible demand away from peak times as much as possible into periods of minor activity. A consistent and efficient methodology would make streamlining of current generation achievable without affecting total power consumption. This enables the implementation of climate-related production of electricity by changing the demand to other durations.

3.2 Demand-response programs and load shifting (a case study)

The proposed formulation for the SG presents an inclusive day-ahead DSM approach using a load shifting technique that results in reduced prices for energy usage along with conforming physical characteristics of end users and end-user preferences. In the given work, the DSM approach is studied for a proposed formulation for a variety of devices for RCs, commercial consumers (CCs), and industrial consumers (ICs). The load shifting technique applied in this study is of two types, desirable load shifting and undesirable load shifting.

Desirable demand shifting: In desirable load shifting, load shifting is performed according to the ease and operational roster of the consumer.

Undesirable demand shifting: In undesirable demand shifting, loads are shifted to the lowest priced time slots without considering ease for the consumer. The purpose of the proposed strategy is to attain greater efficiency and savings in the context of economics and energy use. Precisely, it accomplishes savings by shifting domestic shiftable loads to flatten the electricity load shapes of a particular household. So the implementation of such a shifting by several households can progress the efficiency of handling energy demand, resulting in substantial cost savings for consumers.

The proposed DSM technique can be mathematically formulated as

$$\text{Min} \sum_{t=1}^{n} \text{EBill}$$

where

EBill is the energy consumed by RCs

n = total duration (hours)

t = 1, 2 ....24.

The energy consumed is calculated by

$$\text{EBill} = [P_t]^* [N_{kt}]^* [D_{kt}]$$

where $[P_t]$ = Price of device at time t,

$[N_{kt}]$ = Number of device type k,

$[D_{kt}]$ = Load demand of device type k

k = device type (k = 1, 2.......m).

The total utility bill is calculated for 24 h as

$$\sum_{k=1}^{m} \text{EBill}$$

where m is the total number of device types.

Validation of the proposed approach to demand-side scheduling strategy is done by testing on the SG with various devices for RCs, CCs, and ICs. The simulations are carried out for a total duration of 24 h.

In the case of uncomfortable loads, load shifting is done irrespective of user preferences considering only the objective of minimizing the bill.

Conversely, in the case of comfortable load shifting, the simulations are carried out for certain durations depending on user comfort and preferences. For example, in the case of residential loads, devices types like coffeemakers and kettles have minimum prices at the 20th h (i.e., 3 a.m.), but these loads need to be operated during the 1st to 3rd h (i.e., 8–10 a.m.). Day-ahead wholesale costs and forecasts for all zones of the SG are given (see Table 4.1), while Tables 4.2–4.4 provide data for shiftable devices of RCs, CCs, and ICs.

TABLE 4.1 Forecast load and price of energy (Ct/kWh).

		Forecast load demand		
Hours	Wholesale price ct/KwH	Residential	Commercial	Industrial
1	12	729.4	923.5	2045.5
2	9.19	713.5	1154.4	2435.1
3	13.497	713.5	1443	2629.9
4	22.759	808.7	1558.4	2727.3
5	29.502	824.5	1673.9	2435.1
6	30.085	761.1	1673.9	2678.6
7	15.191	745.2	1673.9	2678.6
8	19.041	681.8	1587.3	2629.9
9	16.42	666	1558.4	2532.5
10	10.813	951.4	1673.9	2094.2
11	8.63	1220.9	1818.2	1704.5
12	9.757	1331.9	1500.7	1509.7
13	9.185	1363.6	1298.7	1363.6
14	18.084	1252.6	1096.7	1314.9
15	16.19	1046.5	923.5	1120.1
16	8.87	761.1	577.2	1022.7
17	8.65	475.7	404	974
18	8.11	412.3	375.2	876.6
19	8.25	364.7	375.2	827.9
20	8.1	348.8	404	730.5
21	8.14	269.6	432.9	730.5
22	8.943	269.6	432.9	779.2
23	8.34	412.3	432.9	1120.1
24	9.35	539.1	663.8	1509.7

TABLE 4.2 Data of shiftable devices in the residential zone.

	Hourly consumption of device (KW)			
Device type (K)	1st h	2nd h	3rd h	Number of devices
Dryer	1.2	–	–	189
Dishwasher	0.7	–	–	288
Washing machine	0.5	0.4	–	268
Oven	1.3	–	–	279
Iron	1.0	–	–	340
Vacuum cleaner	0.4			158
Fan	0.2	0.20	0.20	288
Kettle	2.0	–	–	406
Toaster	0.9	–	–	48
Rice cooker	0.85	–	–	59
Hair dryer	1.5	–	–	58
Blender	0.3	–	–	66
Frying pan	1.1	–	–	101
Coffeemaker	0.8	–	–	56

TABLE 4.3 Data of shiftable devices in the commercial zone.

	Hourly consumption of device (KW)			
Device type (K)	1st h	2nd h	3rd h	Number of devices
Water dispenser	2.5	–	–	156
Dryer	3.5	–	–	117
Kettle	3.0	2.5	–	123
Oven	5.0		–	77
Coffeemaker	2.0	2.0	–	99
Fan	3.5	3.0	–	93
Air conditioner	4.0	3.5	3.0	56
Lights	2.0	1.75	1.5	87

TABLE 4.4 Data of shiftable devices in the industrial zone.

	Hourly consumption of device (KW)			
Device type (K)	1st h	2nd h	3rd h	Number of devices
Water dispenser	2.5	–	–	156
Dryer	3.5	–	–	117
Kettle	3.0	2.5	–	123
Oven	5.0		–	77
Coffeemaker	2.0	2.0	–	99
Fan	3.5	3.0	–	93
Air conditioner	4.0	3.5	3.0	56
Lights	2.0	1.75	1.5	87

Different zones of the SG have various types of shiftable loads that are categorized as follows:

Residential zone: Devices in the residential zone have relatively low energy usage and short operating durations. The device types and their demand schedules are tabulated.

Commercial zone: The loads of the CC zone have more usage compared with that of the RC zone; the demand schedules under these loads are given in Table 4.3. A total of 800 shiftable loads are used in the CC zone from a variety of devices.

Industrial zone: This zone has devices with maximum usage of energy with continuous load durations, although loads for the IC zone are fewer but critical between all three zones. Table 4.4 tabulates load schedules for the IC zone. A total of 100 devices are within the six various shiftable loads in the IC zone.

Simulations have been carried out on DSM methodology that accomplished decreased and optimized electricity bills in the three load zones of RCs, CCs, and ICs. The proposed algorithm compares the results of comfortable and uncomfortable load shifting before and after DSM for loads of several types in the SG.

The simulation results obtained for the residential zone are shown (see Fig. 4.6). The reduction in energy cost of the residential zone for a duration of 24 h was from \$230287.9 to \$ 224417.5 after the DSM strategy using load shifting.

The results obtained for the commercial zone are given (see Fig. 4.7). The energy cost of the commercial zone for the 24 h duration was reduced from \$362663.96 to \$354534 with the DSM strategy.

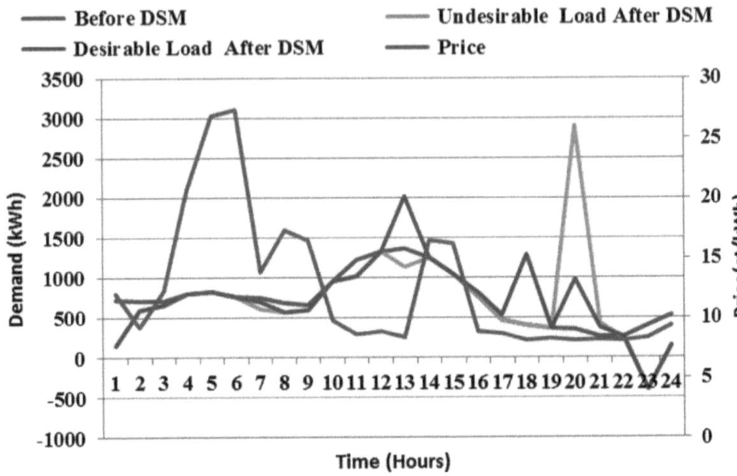

FIGURE 4.6 Demand-side management results for the residential zone.

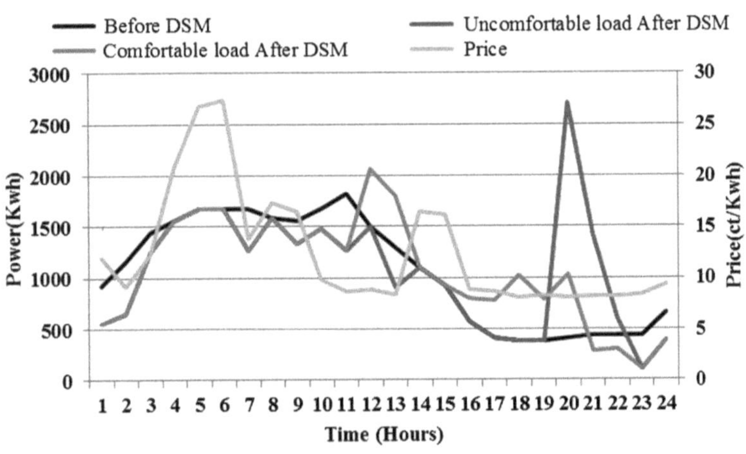

FIGURE 4.7 Demand-side management results for the commercial zone.

The simulation results obtained for the industrial zone have reduced energy costs for a duration of 24 h from $571204.7 to $491700.9 after the DSM strategy using load shifting (see Fig. 4.8).

The simulation carried out with this DSM strategy accomplished minimization of the entire energy price in the residential zones. The proposed algorithm compares the results of desirable and undesirable load shifting before and after DSM for loads of several types in the SG.

Fig. 4.6 compares residential loads before and after DSM, depicting the growing load requirement when demand increases during periods when price is at or near minimum before DSM—e.g., between the 3rd and 7th h.

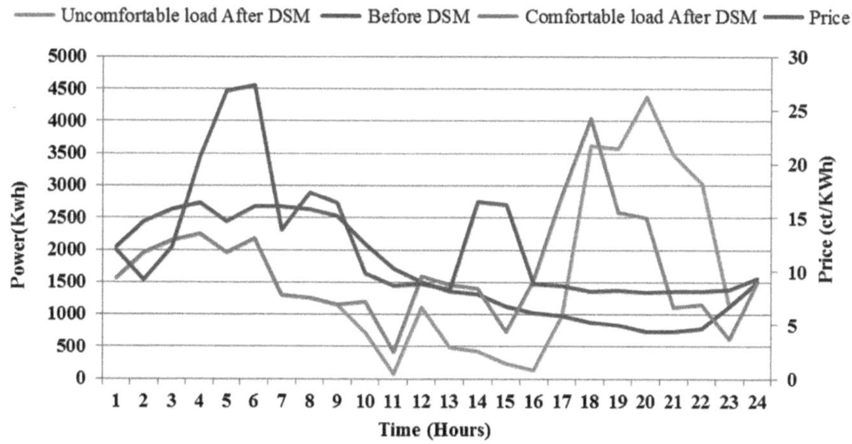

FIGURE 4.8 Demand-side management results for the industrial zone.

After DSM, there is a hike in price in response to increased demand, and as a result some devices are shifted to lower-priced periods (e.g, from the 17th to 21st h), which results in bill minimization as mentioned previously.

4. Smart grid scenario in the Indian power market

When the Electricity Act of 2003 in India was introduced, a comprehensive modification occurred in the Indian power industry, and producers have witnessed a restructuring process with modifications to minimize total procedural and commercial losses, expand consumer comfort, achieve better power quality, justify the electricity tariff, and improve the reliability of the power supply.

4.1 Smart grid scenario in India

As far as the desire for the SG in India is concerned, the power industry is presently confronted with several key challenges. The Indian network is the third-largest main transmission and distribution system in the world, with several challenges: insufficient access to the power system, supply deficits, massive losses, etc.

According to statistics from the Ministry of Power, transmission and distribution losses in India are close to the maximum for countries globally, averaging 26% of gross power production with a maximum of 62% for various states. Theft of power and other losses that are not technical are not included, and when these losses are added, the average value of these losses increases to 50%. India is among the fragile countries with respect to weaker electric grids globally, so it suffers monetarily for each unit of electricity sold. The various technical gaps in the Indian electricity grid are because of no preplanned and

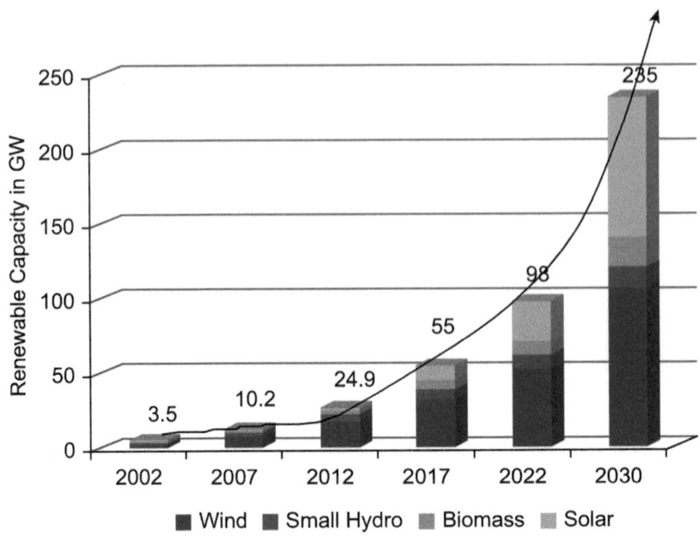

FIGURE 4.9 Renewable generation growth trend.

automated distribution network and also congestion of network components, deficiency of reactive power maintenance and instruction facilities, and less meter reading capability for electricity bill collection.

Progress to a SG would address these concerns and renovate the prevailing power grid in an effective, less constrained, bidirectional way. The grid, with improved value and security, would support electricity access for all.

As far as implementation of SG for India's present power situation, the main obstacles have been highlighted and remedies proposed [70].

The RE growth inclination in India is shown (see Fig. 4.9). The foremost impact in RE can be seen through hydro, wind, and PV energy. In India, this is most intense in states like J& K, Andhra Pradesh, Himachal Pradesh, Gujarat, Rajasthan, Karnataka, and Tamil Nadu.

4.2 Strategy to implement smart grids in India (a case study)

The government of India has initiated implementation of the SG [11]. Monetary support is provided by the government to support SG. Symposiums, discussions, and workshops are planned, and a policy has been prepared that views the current power situation as well as social and financial conditions. Considering all these parameters, implementation of SGs in India is a tedious job. In order to avoid this problem, a phase-wise implementation with suitable preparation is compulsory. Such a plan is shown in Fig. 4.10. For implementation of this plan,

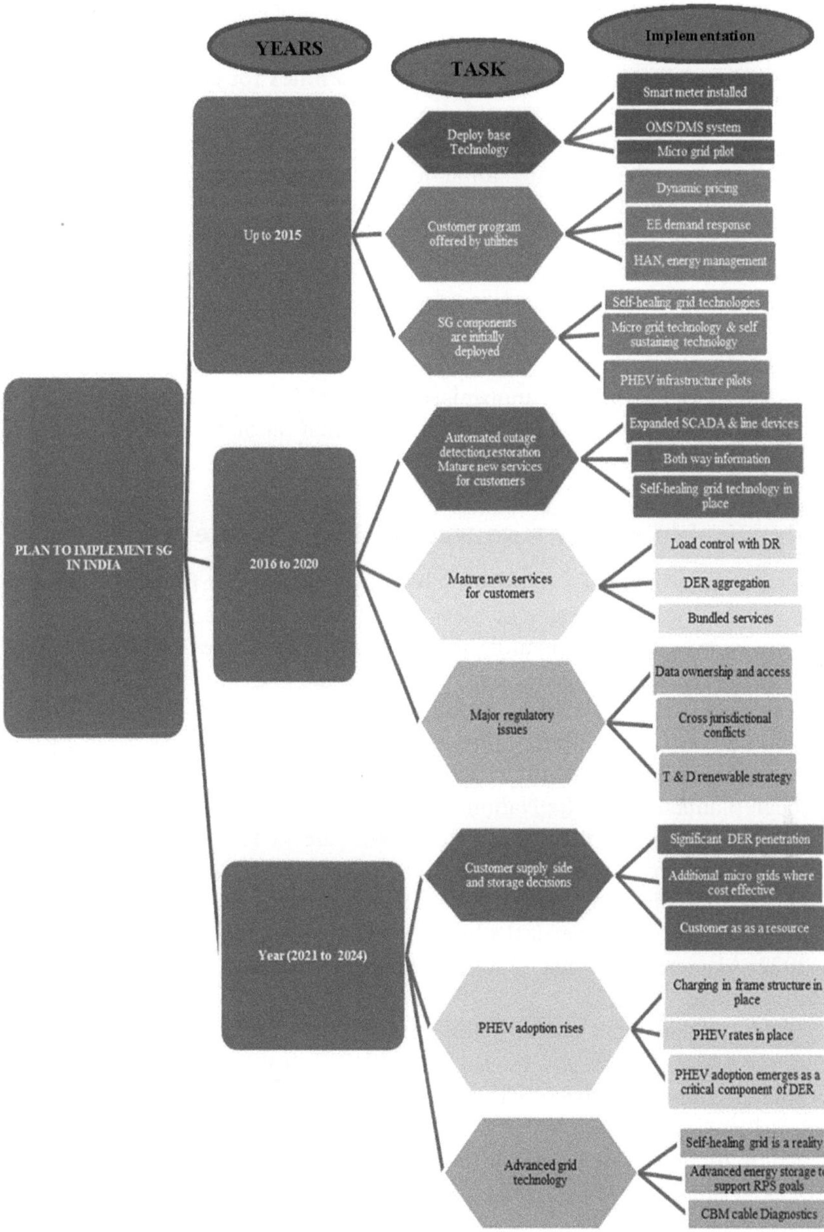

FIGURE 4.10 Strategy to implement smart grids in India.

firstly SG constituents like self-sustaining community concept, self-healing grid technologies, microgrid technology, PHEV infrastructure pilots, and smart meters are installed. After this, some recent facilities for consumers and main regulating policies can be arranged. For the accomplishment and implementation of SG in India, the duration is until 2024. The action plan shown in Fig. 4.10 explains the implementation of the action plan for SG under the Indian scenario.

Presently in India, execution of a SG is in its preliminary phase. Accounting for all parameters such as grid, management, environmental, and market levels as well as features of the Indian electric network, SG is to be accomplished in phases. Considering the existing condition of the electric network, its future aim can still be improved greatly. In view of current power transmission, operation, control, and recent strategies, the upcoming objective in implementation of SG is illustrated (see Fig. 4.11).

5. Renewable energy resources in India (a case study)

This case study has been performed based on data taken from the year-end 2018 review by the Ministry of New and Renewable Energy (MNRE). This case study provides information regarding the present scenario of SG and implementation of RE resources in India. Some RE implementations have been studied in this case study as follows:

PV: The revision of grid-connected PV plants has been targeted by the government to increase from 20,000 to 100,000 MW by year 2021–22. This has been planned under the National Solar Mission of the Indian government. Some of the measures being taken by MNRE are as follows:

In India, there is substantial installed solar capacity, and it is presently ranks fifth globally in terms of solar production with overall installed capacity of 24.33 GW as of October 2018 in comparison with the objective of 100 GW by 2022. Also, an additional 22.8 GW of capacity is under execution and in the planning stage.

Many policies were made by the ministry for bidding PV capacity during 2018–19 and 2019–20 and envisioned 100 GW of total capacity.

An inverse e-auction is used with the help of a competitive bidding procedure by which grid-linked PV project tariffs are defined. The mentioned e-auction method plays a significant role in considerably minimizing this tariff. The minimum value of the PV tariff noted in July 2018 was ₹2.44/kWh in PV projects in India for interstate transmission systems–based bidding. There is a decline in the PV tariff from ₹18/kWh (approximately) in 2010 to

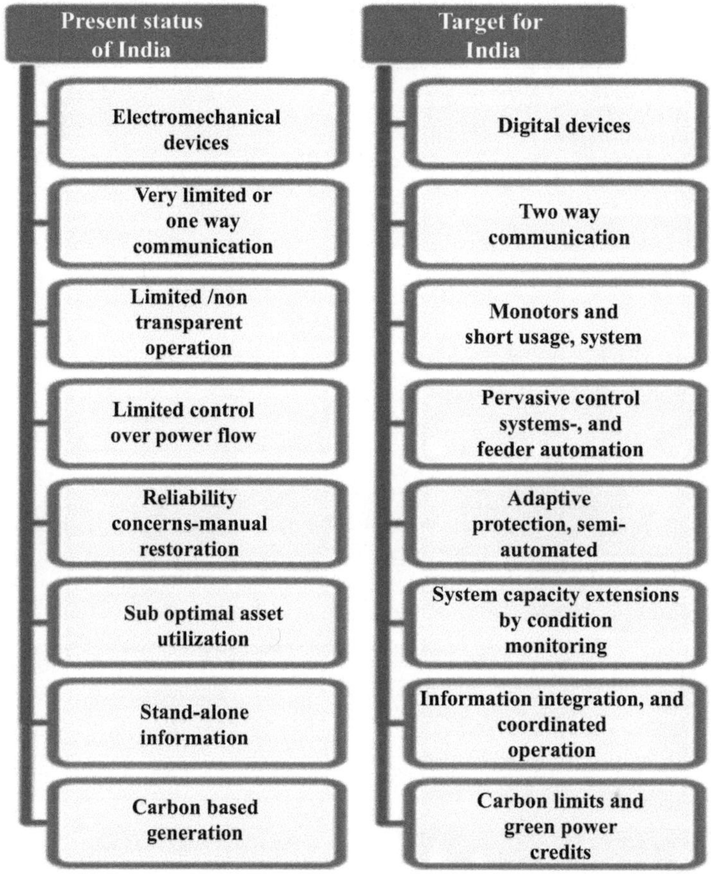

FIGURE 4.11 Current status and upcoming goals for implementation of smart grids.

₹2.44/kWh in 2018. The decline happened for several reasons such as economies of scale, guaranteed accessibility of terrestrial and energy withdrawal structures, and so on.

PV parks are being established throughout the country. Through November 2018, a total 47 PV parks with a cumulative capacity of 26,694 MW had been sanctioned in 21 states. An area of 100,000 lakh acres has been set aside with 75,000 acres to be developed for this purpose. PV projects with a total capacity 4195 MW have been installed within several PV parks.

THE PHOTO VOLTAIC (PV) SOURCE

In India there is the maximum solar installed capacity in the world and the nation presently has the fifth number in this generation having overall installed capacity of 24.33 GW as on October, 2018 in comparison to an objective having 100 GW by 2022.

- There is the establishment of PV parks being done all over the country. Till November, 2018, a total 47 PV parks having cumulative capability 26,694 MW has been sanctioned in 21 States. The area of 100000 lakh acres is recognized for several PV parks among 75,000 acres are developed for this purpose.

THE WIND ENERGY SOURCE:

In present scenario India possess wind capacity of 34.98 GW till October 2018 which is at the fourth highest position globally. It has also set a target of attaining 60 GW by 2022 for which a capacity of about 9.4 GW is under execution.

- Until 2017 capacity additions the Feed in Tariff procedure was used and consequently this tariff mechanism has been replaced by the bidding system.
- In May 2018 the National Wind-Solar Hybrid Policy has been issued with the aim of providing a background to promote huge grid-connected wind-solar PV hybrid structure in optimal and effective operation of wind energy and PV resources, transmission set-up and land.
- In this context a notification in October 2015 was reported according to National Offshore wind energy policy. The aim of this policy was to develop the offshore wind energy in the Limited Economic Zone of India and the coastal zone also.
- LiDAR was commissioned on the monopile platform in November 2017 at Gulf of Khambhat, off Gujarat coast for wind resource assessment.
- NIWE floated Expression of Interest (EoI) for establishment of 1 GW offshore wind farm in Gulf of Khambhat region off Gujarat coast. 35 parties (both national and international) showed interest.

BIO POWER: Waste to Energy projects are also being set up for generation of Energy from Urban, Industrial and Agricultural Waste / Residues such as municipal solid wastes, vegetable and other market wastes, slaughterhouse waste, agricultural residues and industrial wastes & effluents.

- A total capacity of 9.54 GW of grid cotinnected bio-power has been installed in the country as on October 2018 against a target of 10 GW bio-power by 2022.
- This includes 8.73 GW from bagasse cogeneration, 0.68 GW from non-bagasse cogeneration and 0.13 GW from waste to energy.

SMALL HYDRO POWER

A total capacity of 4.5 GW of grid connected small hydro power has been installed in the country as on October 2018 against a target of 5 GW small hydro power by 2022. Further, 126 no. of projects of capacity 0.73 GW are under various stages of implementation.

ENERGY STORAGE

- Energy Storage is one of the crucial & critical components of India's energy infrastructure strategy and also for supporting India's sustained thrust to renewables and electric mobility.
- National Energy Storage Mission (NESM) has been developed. The Mission focuses on demand creation, indigenous manufacturing, innovation and necessary policy support for proliferation of Energy Storage in the country.
- Keeping in view our commitment to a healthy planet and our Nationally Determined Contributions as per the Paris Accord on Climate Change, India made a pledge that by 2030, 40% of installed power generation capacity shall be based on clean sources, it was determined that 175 GW of renewable energy capacity will be installed by 2022

The wind energy source: India possessed wind capacity of 34.98 GW in October 2018, the fourth-highest position globally. It has set a target of attaining 60 GW by 2022, for which a capacity of about 9.4 GW was in execution. In this regard, MNRE arranges bidding to meet the 60 GW capacity by bidding for 10 GW wind power capacity for the years 2018–19 and 2019–20 with a margin of 2 years.

According to a recent calculation by National Institute of Wind Energy, total wind power potential is 302 GW in India over ground level at a value of 100 m.

For implementation of a bidding system, the government has fixed some already decided measures cited as “Guidelines for Tariff Based Competitive Bidding Process for earning of power from grid-linked Wind Power Projects,” based on a notice in December 2017. The aim was to offer a structure for attaining wind energy through an impartial bidding process that further has given rise to detection of the lowest tariff for wind energy. The countrywide goal set by the ministry for tidal energy is 5 GW in 2022 with a capacity of 30 GW by 2030.

The growth in wind-source manufacturing has ensured the establishment of a robust ecology, plant-functioning abilities, and a developed base. Very reliable and efficient expertise is offered for the production of wind turbines nationwide. Various main participants globally in the respective areas are involved across the country. About 24 various prototypes of these wind turbines have been produced with the help of various manufactures across the nation. Several technologies and wind turbines are traded to various parts globally such as the United States, Asian countries, various parts of Australia, some European states, and Brazil. Almost 80% indigenization is attained along with a robust local trade in the area of wind generation.

6. Biopower

The MNRE has taken the initiative to promote platforms to endorse the cogeneration of biomass power and bagasse in India with the aim of using the country’s existing biomass capital such as bagasse, straw, cotton stalk rice husks, coconut shells, and others to generate power.

7. Small-scale hydropower

By October 2018, an entire capacity of 4.5 GW in grid-linked minor hydropower plants had been fitted in India toward the goal of 5 GW of minor hydropower by 2022. Additionally, a total of 126 plans for hydropower plants with a capacity of 0.73 GW were in several phases of execution.

8. Energy storage

Taking aim to endeavor governance in the field of energy storage by making a permitting strategy and regulatory outline, an inclusive National Energy Storage Mission has been established. The task emphasis is on demand creation, local trade, novelty, and essential strategy funding for the creation of energy storage within the country.

The significantly higher capacity goal confirms better energy safety, enhanced energy admission, and improved occupation openings. With the achievement of this aspiring goal, India can be among the major green energy manufacturers globally, exceeding the outputs of various other developed countries. The segment of RE as a a share of total mounted capacity in India as of October 31, 2018 is given in Tables 4.4 and 4.5.

Worldwide, India is fifth in total installed RE capacity, fourth for wind power, and fifth for solar PV.

According to inverse auctions conducted by the PV Energy Corporation of India during May 2017, India listed the lowest ever PV tariffs at ₹2.44 per unit for 200 MW and another time as 600 MW in July 2018. According to statistics for December 2017, India had registered its lowest wind tariff at ₹2.43 per entity in a tender of 500 MW (Fig. 4.12).

Collective RE mounted capacity had improved from 35.51 to 73.35 GW from March 31, 2014 to October 1, 2018 with an update of around 106% in the past few years. The increase in capacity of 37.84 GW of grid-linked RE has been accomplished throughout the past few years (2014–15 to 2018–19) (as in Fig. 4.13). Data indicate 21.7 GW from PV, 13.98 GW from wind, 0.7 GW from minihdryopower plants, and 1.5 GW from bioenergy sources.

Generation of an aggregate of 101.83 billion units in India through 2017–18 from RE compares with 61.78 BU produced during 2014–15, a hike

TABLE 4.5 Renewable energy in overall installed capacity [18].

Source of energy	Mounted capacity (GW)	Percentage
Thermal	221.76 GW	63.84%
Nuclear	6.78 GW	1.95%
Hydro	45.48 GW	13.09%
Renewable	73.35 GW	21.12%
Total	347.37 GW	−100%

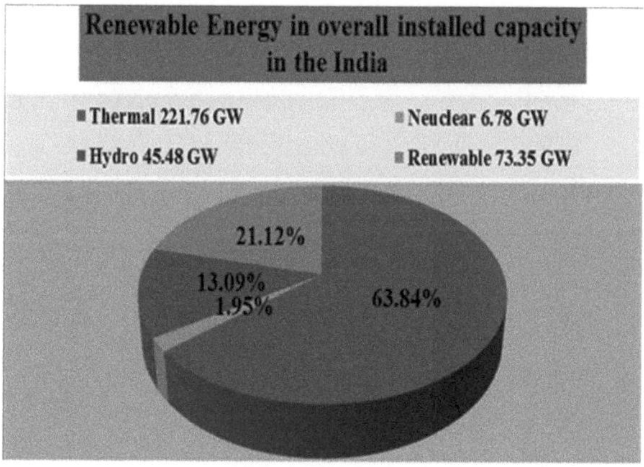

FIGURE 4.12 Segment of renewable energy as a percentage of total installed capacity.

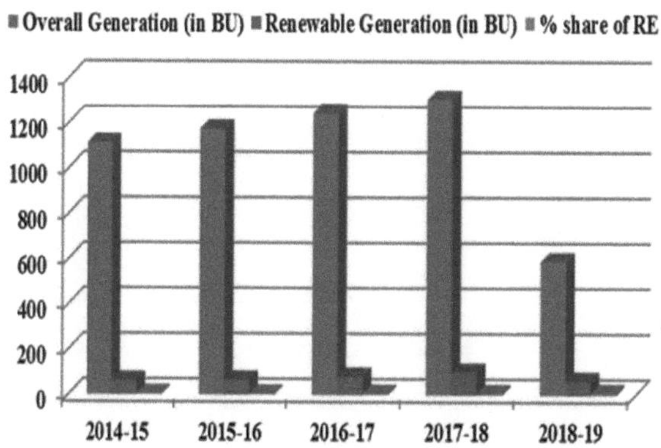

FIGURE 4.13 Yearly data on renewable energy generation.

of about 65%. The share of RE as a percentage of total power production was extended during 2014–15 to about 8% from 5.5%.

Power totaling 62.66 BU was produced in 2018–19 through August 2018. The data related to this statistic of RE are presented in Table 4.6.

An aggregate of about 73.35 GW of RE capacity had been mounted in India by October 2018 from RE sources comprising about 34.98 GW from wind, 24.33 GW from PV, 4.5 GW from minihydropower plants, and 9.54 GW from bioenergy sources. Also, plans for about 46.75 GW of capacity were bid out or were under implementation. A path was set by the government to bid out 60 GW of capacity from PV sources and 20 GW from wind by March 31, 2020. A PV plant of 30 GW and a wind-source plant of 10 GW were to be bid out for 2018–19 and 2019–20 as shown (see Fig. 4.14).

TABLE 4.6 Yearly regeneration data.

Years	Total production (BU)	Renewable (RE) production (BU)	RE portion (%)
2014–15	1110.18	61.78	5.56
2015–16	1172.98	65.78	5.6
2016–17	1241.38	81.54	6.56
2017–18	1303.37	101.83	7.81
2018–19	590.04	62.66	10.62

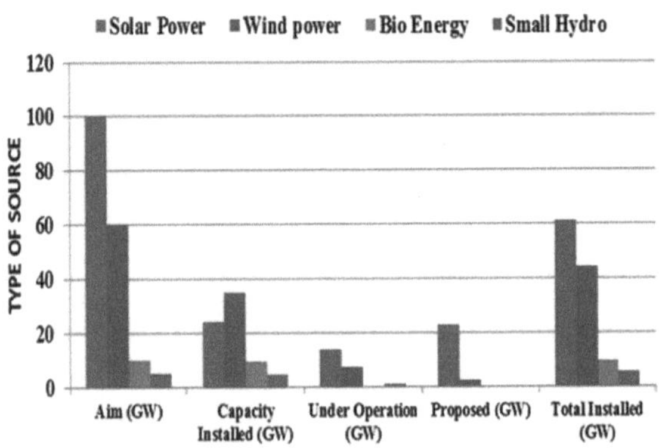

FIGURE 4.14 Status of renewable projects in India.

TABLE 4.7 Statistics for status of projects.

Type of source	Aim (GW)	Capacity installed (GW)	Under operation (GW)	Proposed (GW)	Total installed (GW)
Solar	100	24.33	13.8	22.8	60.93
Wind	60	34.98	7.02	2.4	44.4
Bioenergy	10	9.54	0	0	9.54
Small-scale hydro	5	4.5	0.73	0	5.23
Total	175	73.35	21.55	25.2	120.1

The strategy can provide information to RE developers and stakeholders about the extensive assurances and preparations of government entities within the RE zone to reassure the provision of threat-free funds. The information and statistics regarding plan statuses as of October 2018 are provided in Table 4.7.

9. Conclusion

DSM is excellent for allocating energy sources for power distribution to reduce customer energy charges. The DSM approach using a load shifting technique can be employed in the SG to minimize electricity bills and load and thus be beneficial to both users and utilities. This strategy is used for comfortable and uncomfortable load shifting and solved for bill minimization in this case study. The simulations were performed on an SG system containing various customer areas. This study resulted in reduction in the peak load and minimization of the system's electricity bill for a 24 h duration.

The emergence of SG is essential for converting/transforming the congenital grid into an intelligent and smarter one. This chapter has identified the benefits of SG and various SG solutions for balancing demand and generation of the power system. Also, from an environmental perspective, it is discussed that reduced load due to enhanced efficiency also minimizes impacts on the ecosystem and power usage linked to a specific level of generation.

As the forthcoming study is concerned, it is beneficial to emphasize the novel communication structure for isolated grids and some research on power quality studies and to work to enrich the efficiency of the current system.

RE resources are intermittent, and incorporating them into the grid solves some uncertainties by concentrating every time duration and more practical methods for large-scale RE system integration that will help SG research to be more effective in attaining energy sustainability and an intelligently integrated future grid.

References

[1] www.iea.org/publications/freepublications/publication/2013_AnnualReport.pdf.

[2] C.W. Gellings, Then and now: the perspective of the man who coined the term 'DSM', Energy Pol. 24 (4) (1996) 285–288.

[3] DOE Report, Benefits of Demand Response in Electricity Markets and Recommendations for Achieving Them, 2006.

[4] W. Li, T. Logenthiran, W.L. Woo, et al., Implementation of demand side management of a smart home using multi-agent system, in: 2016 IEEE Congress on Evolutionary Computation (CEC), Vancouver, BC, Canada, 2016, pp. 2028–2035.

[5] N.G. Paterakis, O. Erdinç, A.G. Bakirtzis, et al., Optimal household appliances scheduling under day-ahead pricing and load-shaping demand response strategies, IEEE Trans. Ind. Inf. 11 (6) (2015) 1509–1519.

[6] T. Cortés-Arcos, J.L. Bernal-Agustín, R. Dufo-López, et al., Multiobjective demand response to real-time prices (RTP) using a task scheduling methodology, Energy 138 (2017) 19–31.

[7] M. Ali, J. Jokisalo, K. Siren, et al., A user-centric demand response framework for residential heating, ventilation, and air-conditioning load management, Elec. Power Compon. Syst. 44 (1) (2016) 99–109.
[8] N. Kinhekar, N.P. Padhy, F. Li, et al., Utility oriented demand side management using smart AC and micro DC grid cooperative, IEEE Trans. Power Syst. 31 (2) (2016) 1151–1160.
[9] M.T. Bina, D. Ahmadi, Stochastic modeling for the next day domestic demand response applications, IEEE Trans. Power Syst. 30 (2015) 2880–2893.
[10] T. Coletta, R. Delabays, L. Pagnier, et al., Large electric load fluctuations in energy-efficient buildings and how to suppress them with demand side management, in: IEEE PES Innovative Smart Grid Technologies Conf. Europe, Slovenia, 2017, pp. 1–6.
[11] M.Z. Degefa, A. Safdarian, M. Fotuhi-Firuzabad, et al., Distribution network reliability improvements in presence of demand response, IET Gener. Transm. Distrib. 8 (12) (2014) 2027–2035.
[12] C. Vivekananthan, Y. Mishra, G. Ledwich, et al., Demand response for residential appliances via customer reward scheme, IEEE Trans. Smart Grid 5 (2) (2014) 809–820.
[13] A.A. Hourani, M. AlMuhaini, Impact of demand side management on the reliability performance of power distribution systems, in: 2016 Saudi Arabia Smart Grid, Saudi Arabia, 2016, pp. 1–5.
[14] Z. Wang, C. Gu, F. Li, et al., Active demand response using shared energy storage for household energy management, IEEE Trans. Smart Grid 4 (4) (2013) 1888–1897.
[15] US Department of Energy, Benefits of demand response in electricity markets and recommendations for achieving them, report to the United States congress, Available online, http://eetd.lbl.gov, February 2006.
[16] N. Motegi, M.A. Piette, D.S. Watson, S. Kiliccote, P. Xu, Introduction of Commercial Building Control Strategies and Techniques for Demand Response", Ernest Orlando Lawrence Berkeley National Laboratory, 2007. Report no. LBNL-59975.
[17] S. Mohagheghi, J. Stoupis, Z. Wang, Z. Li, H. Kazemzadeh, Demand response architecture: integration into the distribution management system, in: Proceedings of the First IEEE International Confe Communications (SmartGridComm), 2010, pp. 501–506.
[18] A.J. Conejo, J.M. Morales, L. Baringo, Real-time demand response model, IEEE Trans. Smart Grid 1 (2010) 236–242.

Further reading

[1] P. Palensky, D. Dietrich, Demand side management :demand response, intel- ligentenergy systems, and smart loads, IEEE Trans. Ind. Inf. 7 (3) (2011) 381–388.
[2] M. Reid, R. Levy, A. Silverstein, Coordination of Energy Efficiency and Demand Response, National Laboratory, Charles Goldman, 2010.
[3] J. Han, M.A. Piette, Solutions for summer electric power shortages :demand response and its applications in air conditioning and refrigeration systems, J. Refrig. AirCond. Electr. Power Mach. 29 (1) (2008) 1–4.
[4] National Institute of Standards and Technology, NIST Framework and Roadmap for Smart Grid Interoperability Standards, Release 1.0, January 2010.
[5] http://www.elstandard.se/nyheter/nyhet.asp?NewsID=1366625904.
[6] A.-H. Mohsenian-Rad, V.W.S. Wong, J. Jatskevich, R. Schober, A. Leon-Garcia, Autonomous demand-side management based on game-theoretic energy consumption scheduling for the future smart grid, IEEE Trans. Smart Grid 1 (3) (December 2010) 320–331.

[7] T. Logenthiran, D. Srinivasan, T.Z. Shun, Demand side management in smart grid using heuristic optimization, IEEE Trans. Smart Grid 3 (3) (2012) 1244–1252.

[8] https://pib.gov.in/PressReleaseIframePage.aspx?PRID=1555373.

[9] P. Agrawal, Overview of DOE microgrid activities, in: Proc. Symp. Microgrid, Montreal, QC, Canada, 2006.

[10] Available from [Online]: http://der.lbl.gov/2006microgrids_files/USA/Presentation_7_Part1_Poonumgrawal . Pdf.

[11] R. Kumar, Smart Grid Maturity Model (SGMM), International workshop on challenges in Smart Grid and Renewable Resources, May, 2010, Mysore, India.

Chapter 5

Approaches to smart grid network communication and security

Dharmbir Prasad[1], Rudra Pratap Singh[1], Sushri Mukherjee[2], Sumana Chattaraj[3], Koushik Sarkar[4], Md. Irfan Khan[4]
[1]*Asansol Engineering College, Asansol, West Bengal, India;* [2]*Indian Institute of Technology Delhi, Hauz Khas, New Delhi, India;* [3]*Indian Institute of Technology Jodhpur, Rajasthan, India;* [4]*Supreme & Co. Pvt. Ltd., Kolkata, West Bengal, India*

Glossary of abbreviations

2,-3,-4,-5G Generations of wireless technology for cellular network
3-D Three-dimensional
3P Three-phase
ACS Aluminum-clad steel
ADSL Asymmetric DSL
ADSS All-dielectric self-supporting
AGS Armor rod grip suspension
AMI Advanced metering infrastructure
AMR Automatic meter reading
ANN Artificial neural network
AON Active optical network
BB Broadband
BBO Biogeography-based optimization
BCH Bose Chaudhuri Hocquenghem
BDSFTR Bayesian Dempster-Shafer fuzzy-based trusted routing
BDST Bayesian Dempster Shafer theory
BGAN Broadband global area network
BIA Burned-in address
BPON Broadband PON
BS Base station
CAB Collective animal behavior
CAE Computer-aided engineering
CE Consumer electronics
CL-AS Certificate-less aggregate signature scheme
CPE Customer premises equipment
CRN Cognitive radio network
CRSN Cognitive radio sensor network
CT Clevis thimble
CUSUM Cumulative sum
DDoS Distributed DoS
DE Differential evolution
DER Distributed energy resource
DisCos Distribution companies
DoF Degree of freedom
DoS Denial of service
DR Demand response
DSL Digital subscriber line
EADDE Evolving ant direction DE
EGA Enhanced GA
EGA-DQLF EGA with decoupled quadratic load flow
EL Extended link
ELM Extreme learning machine
EPON Ethernet PON
ESFCM ELM-based semisupervised Fuzzy C-Means

Advances in Smart Grid Power System. https://doi.org/10.1016/B978-0-12-824337-4.00005-9

EWMA Estimated weighted moving average
FAN Field area network
FDIA False data injection attack
FEA Finite element analysis
FOT Future-oriented technique
FSK Frequency shift keying
FTTH Fiber to the home
G2V Grid to vehicle
GA Genetic algorithm
GenCos Generation companies
GEO Geostationary earth orbit
GIS Geographic information system
GOOSE Generic object oriented substation event
GPON Gigabit PON
GSA Gravitational search algorithm
HAN Home area network
HAR Highway advisory radio
HD-PLC High definition PLC
HDSL High bit digital subscriber
HSPA High-speed packet access
IaaS Infrastructure as a service
IBFD In-band full duplex
ICO Information Commissioner Office
ICS Industrial control system
ICT Information and communications technology
IDR Incentive-based demand response
IEC International Electrotechnical Commission
IGA Improved GA
IIP Intelligent information processing
IoT Internet of Things
IP Internet protocol
ISO International Standardization Organization
ITU International Telecommunication Union
KNX Konnex
LEO Low-earth orbit
LL Line-to-line
LTE Long-term evaluation
MAC Medium access control
MDE Modified DE
MEO Medium-earth orbit
MIMO Multiple input multiple output
MLRM Multi-linear regression model
MPDS Mobile packet data system
NAEMI Nonlinear autoregressive exogenous multivariable input
NB Narrow band
NILM Non-intrusive load monitor
NS Network simulator
OPF Optimal power flow
OPGW Optical power ground wire
OTDR Optical time domain reflectometer
P2MP Point-to-multipoint
PGCIL Power Grid Corporation of India Limited
PHY Physical
PLC Power line communication
PMU Phasor measurement unit
PoF Power over optical fiber
PON Passive optical network
PQ Power quality
PR Power reliability
PRIME PoweRline Intelligent Metering Evaluation
PSO Particle swarm optimization
PUF Physical unclonable function
QoS Quality of service
RE Renewable energy
RISI Repository of industrial security incident
RoW Right of way
SCADA Supervisory control and data acquisition
SDN Software defined networking
SDO Standards developing organization
SDR Software defined radio
SDSS Saving decision support system
SEAL SEcure and AgiLe
SG Smart grid
SLD Single line diagram
SNR Signal to noise ratio
SQL Structured query language
SSH Secure shell
T&D Transmission and distribution
TEL Twisted eye link
TIA Transient ischemic attack
TransCos Transmission companies
TSTM Time series trust model
UHF Ultra-high frequency
UIO Unknown input observer
UMTS Universal mobile telecommunication system
UTS Ultimate tensile strength
V2G Vehicle to grid
VAR Volt ampere reactive

VDSL Very high bit rate digital subscriber line
VHF Very high frequency
WAN Wide area network
Wi-Fi Wireless fidelity
WiMAX Worldwide interoperability for microwave access
WSN Wireless sensor network

List of symbols

$\boldsymbol{u}$ Control variable vector
$\boldsymbol{v}$ Random vector of small length
$\boldsymbol{v}$ Poisons ratio
$\boldsymbol{x, y, z}$ Direction of stress resolutions
$\boldsymbol{a_i}$ Animal position in D-dimensional vector
A Initial sorted animal positions based on the fitness function (dominance)
$\boldsymbol{b_p}$ (mm) Width of the TEL plate
$\boldsymbol{b_t}$ (mm) Thread section width at the root
$\boldsymbol{B}$ (S) Best positions may be chosen and stored in memory $\mathbf{M}_g$ and $\mathbf{M}_h$.
$\boldsymbol{B_{ik}}$ (S) Transfer susceptance
$\boldsymbol{d_{CT_1}}$ (mm) Inner diameter of the CT ring section
$\boldsymbol{d_m}$ (mm) Mean diameter of the turnbuckle
$\boldsymbol{d_r}$ (mm) Diameter of the TS riveting
$\boldsymbol{d_{TEL_1}, d_{TEL_2}}$ (mm) Inner and outer diameters of the TEL plate
$\boldsymbol{d_1, d_2, d_c}$ (mm) Minor, major and core diameters of the TEL bolt
$\boldsymbol{f_{EL}}$ (N) Force acting to the EL
$\boldsymbol{FC}$ (\$/h) Fuel cost
$\boldsymbol{g_k}$ (mho) Conductance of branch k
$\boldsymbol{G_{ik}}$ (siemens) Transfer conductance
$\boldsymbol{H_{CAB}}$ Threshold value to determine individual best position
$\boldsymbol{i}$ Individual indices
$\boldsymbol{j}$ Parameter indices
$\boldsymbol{J}$ Objective function to be minimized
$\boldsymbol{k}$ Number of buses
$\boldsymbol{m}$ (kg) Mass of the pole clamp
$\mathbf{m}_g^{nearest}$ Nearest elements of $\mathbf{M}_g$ to $\mathbf{X}_i$
$\mathbf{m}_h^{nearest}$ Nearest elements of $\mathbf{M}_h$ to $\mathbf{X}_i$
$\mathbf{M_g}$ Best found positions in each generation
$\mathbf{M_h}$ Best history locations while evolutionary process completion
$\mathbf{M_U}$ Memory elements after merging elements of $\mathbf{M}_g$ and $\mathbf{M}_h$
$\boldsymbol{n}$ No of bolt thread
$\boldsymbol{nl}$ Number of transmission line
$\boldsymbol{NB}$ Total number of bus
$\boldsymbol{NC}$ Number of shunt VAR compensators
$\boldsymbol{NFFE_{max}}$ Maximum value of net fitness function evaluation
$\boldsymbol{NG}$ Number of generators
$\boldsymbol{NL}$ Number of load buses
$\boldsymbol{N_p}$ Number of animals
$\boldsymbol{NT}$ Number of tap changing transformers
$\boldsymbol{p}$ (mm) Pitch of the screw bolt
$\boldsymbol{p_m}$ Probability to govern chance of mutation
$\boldsymbol{P_b}$ (N) Load appear to the TEL bolt
$\boldsymbol{P_{CAB}}$ Position probability of animal
$\boldsymbol{P_{CT}}$ (N) Applied load on the CT, while force acting on both sides
$\boldsymbol{p_d}$ Dilation (scale) parameter
$\boldsymbol{P_{D_i}}$ (MW) Active power demands at bus i
$\boldsymbol{P_G}$ (MW) Active power outputs of the generator
$\boldsymbol{P_{G_i}}$ (MW) Injected active power
$\boldsymbol{P_{G_i}^{\max}}$ (MW) Upper limit of injected active power
$\boldsymbol{P_{G_i}^{\min}}$ (MW) Lower limit of injected active power
$\boldsymbol{P_i}$ (N) Initial tension to the TEL bolt
$\boldsymbol{P_{Loss}}$ (MW) Real power loss
$\boldsymbol{P_{TEL}}$ (N) Applied load at the TEL
$\boldsymbol{P_{TS}}$ (N) Tensile load appearing on the TS
$\boldsymbol{Q_C}$ (MVAr) Outputs of shunt VAR compensator
$\boldsymbol{Q_{C_i}^{\min}}$ (MVAr) Minimum outputs of shunt VAR compensators

$Q_{C_i}^{max}$ (MVAr) Maximum outputs of shunt VAR compensators
Q_{D_i} (MVAr) Reactive power demand
Q_G (MVAr) Reactive power outputs of the generator
Q_{G_i} (MVAr) Injected reactive power
$Q_{G_i}^{max}$ (MVAr) Upper limit of injected reactive power
$Q_{G_i}^{min}$ (MVAr) Lower limit of injected reactive power
r Uniformly generated random number
r_m Uniform random generated number within the range [0,1]
S_{l_i} (MVA) Transmission line power flow capacity in branch l of i-th bus
$S_{L_i}^{max}$ (MVA) Transmission line upper limit of power flow capacity in branch l of i-th bus
T Tap settings of the tap changing transformer
T_{TEL} (Nm) Torque applied at the TEL bolt
t_{CT} (mm) Thickness of the CT ring section
T_i^{max} Upper limit of tap settings of the tap changing transformer in the ith line
T_i^{min} Lower limit of tap settings of the tap changing transformer in the ith line
t_p (mm) Thickness of the TEL plate
T_{EL} (N) Tension appearing along extension link
V_i and V_j (V) Voltage magnitudes at the buses i and j, respectively
V_G (V) Terminal voltages at the generator bus
$V_{G_i}^{max}$ (V) Upper limit of terminal voltages at the generator bus
$V_{G_i}^{min}$ (V) Lower limit of terminal voltages at the generator bus
V_L (V) PQ bus voltage
V_{L_i} (V) Line voltage for ith bus
$V_{L_i}^{max}$ (V) Upper limit of line voltage for ith bus
$V_{L_i}^{min}$ (V) Lower limit of line voltage for ith bus
W_{EL} (N) Load on the eye bolt/pole clamp
θ (°) Angle of load appearing to the shackle
θ_{ij} (°) Phase angle difference between the voltages at buses i and j
θ_{ik} (°) Phase angle difference between the voltages at buses i and k
α (°) Helix angle of the tension pole assembly
$\alpha_i, \beta_i, \gamma_i$ Cost coefficients of the i-th generator
σ_H (Nm^{-2}) CT circumferential stress
σ_L (Nm^{-2}) CT longitudinal stress
$\sigma_x, \sigma_y, \sigma_z$ ($kgm^{-1}s^{-2}$) Normal stresses in the direction of x, y and z, respectively
$\in_v$ Volumetric strain
τ (Nm^{-2}) Torsional shear stress at the TEL bolt
τ_r, τ' (Nm^{-2}) Shear stress at the riveting point and on the TS

1. Introduction

Rapid population growth in the contemporary world in concert with relentless increases in the number of power-feeding appliances causes a steep rise in power demand. The term smart grid (SG) signifies a modernization of the power system with various associated technologies-e.g., dispersed generation, loads dispatch, storage devices, and communication systems that operate in islanded and grid-connected modes. In practice, a multitude of complex power systems are cumulatively interconnected, and any disruption of certain assets may lead to serious power outages. Thus, the operational resilience of the grid infrastructure is affected in the absence of automated monitoring and diagnosis of the system and ubiquitous communications infrastructure. In this regard, many electric utilities and information and communications technology (ICT) companies are striving to provide last-mile communication facilities to improve livelihoods of the people. Simultaneously, active and reactive optimal power flow (OPF) limits are important in reducing emissions (carbon, NO_X, SO_X, etc.), and fuel cost minimization for energy savings to address the OPF problem will further decrease the global warming effects on SGs.

The remaining portion of this chapter is elaborated as follows: in Section 2, we review SG communication and OPF. General aspects of communications and operational integrity for SGs are presented in Section 3. Section 4 is a briefing on a feasibility study of all-dielectric self-supporting (ADSS) cable networking for SGs. In Section 5, finite element analysis (FEA) results for the proposed hardware are presented, and Section 6 discusses OPF test systems vis-à-vis simulation results. Finally, a summary of the current study and corresponding future scope of ADSS cable hardware accessories and OPF is given in Section 7.

2. Review of communication and optimal power flow in smart grids

At present, the numerous applications of SGs are based on an information architecture that confirms the transfer of information between multiple stakeholders and incorporates its input from heterogeneous information collection along with its distribution and analysis. In general, the SG communication system relies on two major subsystems, viz. the middleware platform and communication infrastructure [1]. Here the communication infrastructure is equipped with multiple leapfrogging technologies, a resilient networking infrastructure, and applicable communication protocols among appliances or subsystems of the grid to facilitate information sharing and execution of commands. Middleware platforms consist of software layers and underlying infrastructure. The incorporation of a pervasive-monitoring and real-time-responsive communication network is essential for any SG to get system status updates, gather sensor information, control signal propagation, and allow power flows for distributed smart applications as presented in Fig. 5.1. However, the availability of fast communication depends on the robustness of the physical infrastructure and its available technologies as briefed in Table 5.1.

2.1 The smart grid optical communication network

Previously, the optical mode of communication has been extensively adopted by utilities to lay interconnecting communication networks in substations with control centers (as portrayed in Fig. 5.2). For instance, quite a number of protection and restoration methods have been applied for power grids. In addition, a few special types of cable-e.g., ADSS and optical ground wire

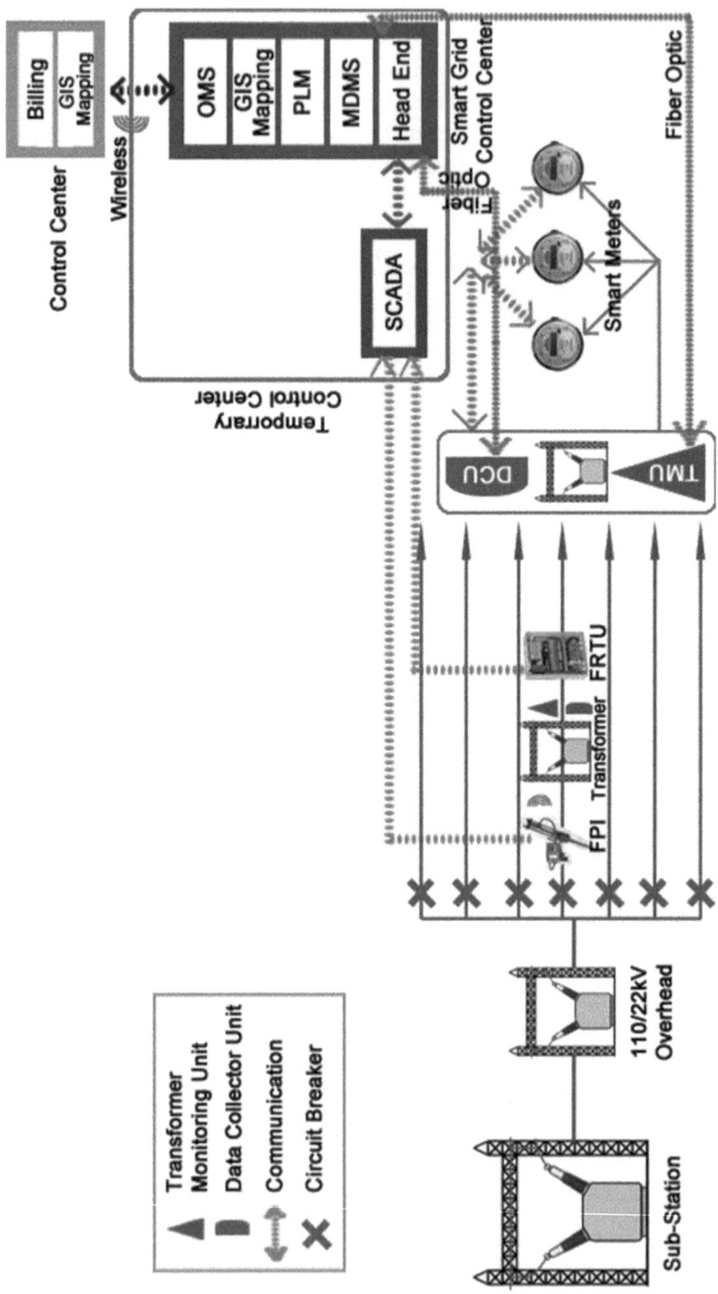

FIGURE 5.1 Communication architecture for a smart grid.

TABLE 5.1 Smart grid wired-communication technologies.

Family	Standards	Coverage	Data rate	Scope
PLC	• NB-PLC: ISO/IEC 14908-3 (Lon works), ISO/IEC 14543-3-5 (KNX), CEA-600.31 (CE bus), IEC 61334-5 (FSK and spread FSK), IEC 61334-3-1 • BB-PLC: TIA-1113 (Home Plug 1.0), IEEE 1901, ITU-T G.hn (G.9960/G.9961) • non-SDO NB-PLC: Insteon X10, G3-PLC, PRIME • non-SDO BB-PLC: Home Plug AV/Extended HomePlug Green PHY, HD-PLC	• NB-PLC: 150 km or more • BB-PLC: 1.5 km	• NB-PLC: 1–10 kbps for low data rate • PHYs, 10–500 kbps for high data-rate PHYs • BB-PLC: 1–10 Mbps (up to 200 Mbps over very short distance)	• NB-PLC: Large-scale AMI, FAN, WAN • BB-PLC: HAN/small-scale AMI
Optical fibers	• AON: IEEE 802.3ah • PON: ITU-T G.983 (BPON), ITU-T G.984 (GPON) IEEE	• IEEE 802.3ah (AON): Up to 10 km • BPON, GPON: Up to 20–60 km • EPON: Up to 10–20 km	• IEEE 802.3ah (AON): 100 Mbps up/down • GPON: 155–2448 Mbps up, 1.244–2.448 Gbps down • BPON: 155–622 Mbps up/down	• WAN • AMI (with FTTH systems)

Continued

TABLE 5.1 Smart grid wired-communication technologies.—cont'd

Family	Standards	Coverage	Data rate	Scope
Optical fibers	• 1901, IEEE 802.3ah (EPON)		• EPON: 1 Gbps up/down	
DSL	• ITU G.991.1 (HDSL) • ITU G.992.1 (ADSL), ITU G.992.3 (ADSL2), ITU G.992.5 (ADSL2+) • ITU G.993.1 (VDSL), ITU G.993.1 (VDSL2)	• ADSL: Up to 4 km • ADSL2: Up to 7 km • ADSL2+: Up to 7 km • VDSL: Up to 1.2 km • VDSL2: 300 m (maximum rate)—1 km (50 Mbps)	• ADSL: 8 Mbps down and 1.3 Mbps up • ADSL2: 12 Mbps down and up to 3.5 Mbps up • ADSL2+: 24 Mbps down and up to 3.3 Mbps up • VDSL: 52–85 Mbps down and 16–85 Mbps up • VDSL2: Up to 200 Mbps down/up	• AMI • FAN

(OPGW)-avail long distance communications at a higher data rate in transmission and distribution (T&D) lines [1,2].

Key advantages:

- Communication over long distances
- UH bandwidth range (feasible for residential multimedia support services)
- Less prone to radio interference and electromagnetic effects

Common limitations:

- Network deployment is costlier
- Equipment associated with terminal is costlier
- Upgrading difficulty

It is notable that the use of FTTH could be possible by the introduction of passive optical network (PON) tools [1]. Moreover, PONs facilitate a single optical fiber according to different network topologies to provide various sites in P2MP connectivity. One of the PON variants, EPON, facilitates interoperability with available IP-based networks.

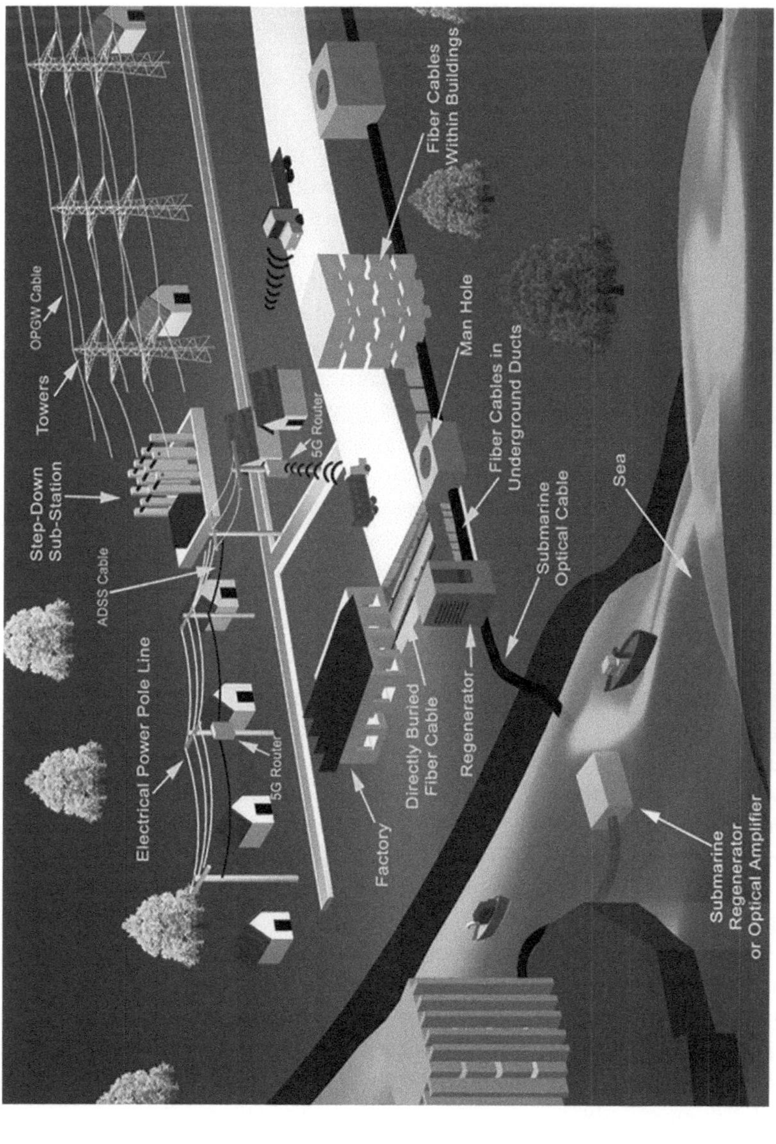

FIGURE 5.2 Typical layout of fiber optic network.

3. Smart grid communications integrity

Deployments of SG projects incorporating distributed computing schemes in grid interoperations have brought more efficiency to power systems. However, these techniques have triggered various security concerns. Thus, considering the operational sensitivity and resilient communication needs of any SG, the following are the major points to be considered.

3.1 Privacy concerns with smart grids

The SG integrates the communication system in any grid with a centralized control center along with smart appliances for intelligent management of electricity by the facilitators. The SG, smart appliances, and each node will lead the creation of massive amounts of indigenous data [3]. These data reveal energy consumption patterns for households in the grid and can be exploited. Privacy maintenance of these data is an utmost requirement for reliable and efficient operation of the SG. Disclosure risk or data misuse creates a new set of worries for clients and privacy professionals. Much research work has been carried out recently on privacy issues of SGs. In 2013, Döbelt et al. studied access control, decentralized data storage, privacy enhancement, and information transparency aspects of the Austrian service-oriented architecture [4]. In that case, encrypted data of smart meters are communicated to the operation center via an aggregator as studied in Ref. [5]. Baloglu and Demir [6] have proposed perturbation with a cryptosystem technique in combination with a lightweight data aggregation method. Its privacy has been evaluated by Holt–Winters and seasonal trend decomposition using the Loess prediction method. Work by the authors of Ref. [7] presented a user privacy protection scheme for SG incorporating a solar-powered system and rechargeable battery. Dimitriou et al. [8] have produced an incentive on generation and its accumulation principle, namely REWARDS, which obstruct linking of various communication tokens along with restraining user identity disclosure. Another privacy idea is presented by Ref. [9] following three measurement divisions for (1) operational metering (attributable), (2) electricity incentives (fine-grained measurements), and (3) electricity billing (attributable measurements). Hmielowski et al. [10] have examined various demographic aspects of inclusion for smart meters in power systems such as social norms, a technology readiness index, and privacy. A real/random oracle model with relevant hardware has been presented for maintaining intractability and anonymity of smart meters by swift mutual authentication between the service provider and meter [11]. Kong et al. [12] have highlighted recent privacy aspects, viz. mathematical modeling and algorithms for data aggregation and the blind signature technique for authentication of user anonymity.

3.2 Hacking of smart grids

The energy-efficient smart system automatically adjusts usage within homes and business with bidirectional communication. Energy management, reliability, and resilience enhancement with integration of intermittent renewable energy generation and storage are crucial for smooth operation of SGs. The major problem is unknown vulnerabilities that can be exploited by intruders through hardware and software manipulation of equipment to take control over the energy supply by gaining command over power grid operations. In recent years, many studies have been made to avoid hacking in the SG paradigm. However, it has been seen that the most common ways of network penetration by hackers are SQL injection and in certain instances, file-sharing sites [3]. As such, under a DDoS attack, the target web site becomes inaccessible. However, functional spin-offs like MalSec and LulzSec may be used as the common antidote for such issues [13]. An RISI was conducted by Ref. [14] to reveal impacts of malware infections in the segmented network and resist its detection by customary defenses.

3.3 Emerging wireless communication technologies for smart grids

Wireless communication is works as the spine of the complete T&D infrastructure of the SG. The various modes of faster wireless communication with active response are given in Table 5.2. Transition from GSM-based communication to 5G through mobile operators and Wi-Fi and WiMAX play vital functions in the implementation of SGs, which has leaped from AMR to AMI. Low latency in the communication infrastructure enhances centralized control and data aggregation keeping the load-end device simpler

TABLE 5.2 Wireless communication technologies summary for smart grids.

Family	Standards	Coverage	Data rate	Scope
WPAN	• IEEE 802.15.4 • Non-SDO: ZigBee, wireless HAR, ISA 100.11a (all based on IEEE 802.15.4)	• IEEE 802.15.4: Between 10 and 75 m	• IEEE 802.15.4: 256 kbps	• V2G • HAN • AMI
Wi-Fi	• IEEE 802.11e (QoS enhancement) • IEEE 802.11n (ultra-high network throughput) • IEEE 802.11s (mesh networking) • IEEE 802.11p (WAVE- wireless access in vehicular environments)	• IEEE 802.11e/s/n: Up to 300 m (outdoors) • IEEE 802.11p: Up to 1 km	• IEEE 802.11e/s: Up to 54 Mbps • IEEE 802.11n: Up to 600 Mbps	• V2G • HAN • AMI

Continued

TABLE 5.2 Wireless communication technologies summary for smart grids.—cont'd

Family	Standards	Coverage	Data rate	Scope
WiMAX	• IEEE 802.16 (fixed and mobile broadband wireless access) • IEEE 802.16j (multihop relay) • IEEE 802.16 m (advanced air interface)	• IEEE 802.16: 0–10 km • IEEE 802.16 m: 0–5 (optimum), • 5–30 (acceptable), 30–100 (reduced performance) km	• IEEE 802.16: 128 Mbps down and 28 Mbps up • IEEE 802.16 m: 100 Mbps for mobile users, 1 Gbps for fixed users	• AMI • FAN • WAN
3G/4G	• 3G: UMTS (HSPA, HSPA+) • 4G: LTE, LTE-Advanced	• HSPA+: 0–5 km • LTE-Advanced: 0–5 (optimum), 5–30 (acceptable), 30–100 (reduced performance) km	• HSPA: 14.4 Mbps down and 5.75 Mbps up • HSPA+: 84 Mbps down and 22 Mbps up • LTE: 326 Mbps down and 86 Mbps up • LTE-Advanced: 1 Gbps down and 500 Mbps up	• V2G • HAN • AMI
Satellite	• LEO: Iridium, Globalstar, • MEO: New ICO • GEO: Inmarsat, BGAN, swift, MPDS	• depend on number of satellites and their beams	• Iridium: 2.4–28 kbps • Inmarsat-B: 9.6 up to 128 kbps • BGAN: 384 up to 450 kbps	• AMI • WAN

and more robust. In this context, [15] have given low-power wireless link reliability analysis for underground network transformer vaults, indoor power control rooms, and 500 kV outdoor substations to evaluate the impact of different sensor radio parameters.

Likewise, in 2014, Sahin et al. proved SG reliability and timeliness application for QoS-aware multi- and single-path routing techniques [16]. The smart home energy management is studied using an effective load-shedding algorithm [17]. Faheem and Gungor [18] have introduced Esti-Net9.0 for QoS (e.g., memory utilization, packet delivery ratio, performance metrics, network lifetime, residual energy, reduced packet, error-rate, throughput, reduced end-to-end delay, and overhead message control) simulation. In the industry 4.0 paradigm, [19] employed ICT, IIP, and FOT to monitor and control power at various levels is found to be economical and effective in energy savings. The ongoing technological innovations in IoT with two-way communication features enable PQ and PR improvement among utilities and consumers during power transactions [20]. In 2019, Aydeger et al. studied autonomous frameworks to detect and restore either Wi-Fi or LTE-based link failures for internal communications in substations through SDN. They adopted GOOSE to ascertain the recovery performance for tele-protection. They are applied Mininet emulator and its integration with the NS-3 simulator for Wi-Fi and LTE/Wi-Fi, respectively [21]. The resilience of the TSTM is checked for data loss and offset faults in smart energy meters [22]. In this paper, five different algorithms (viz. wavenet, tree partition, sigmoidal, cascade forward net and feed-forward net) have been examined to check prediction accuracy of power generation and consumption. Emerging 5G-based IoTs and corresponding networks may boost DR infrastructure for high reliability, fast transfer speed, robust security, numbers of connections, and low power consumption. The 5G technology utilizes a maximum bandwidth of 30–300 GHz on the radio frequency spectrum, also called millimeter waves, available in the space that makes for faster communication than optical fiber for short distances. These short millimeter waves have limitations in traveling at longer distances and can be restricted by buildings and trees, as they need a clear line of sight for better connectivity. Penetration of OPGW and ADSS cables has made it possible to have small cell networks with low-power base stations for short distances. Research over PoF shall also contribute greatly to powering up these mini base stations for communication. 5G plans to have a massive MIMO station with a beam-forming method of communication to individual users reducing the interference between individual stations and streamlining transmission. These beam formations controlled by the base station shall be smart enough to track the movement and direction of the transmitter and receiver and back to each user with a full duplex method reciprocating the signal direction. Agencies are working rigorously to roll out the 5G network, which provides faster communication for the bulk of data. Hence, better connectivity with low latency shall provide better control and faster response over distances [23].

3.4 Cognitive radio-based smart grid communications

The CR is an apt SDR technology that facilitates dynamic, reliable, and efficient use of the less popular radio spectrum by restructuring functionalities and operating parameters in real-time radio environments [1]. The CR communication is also known as the IEEE 802.22 standard. The bands planned to be used by IEEE 802.22 are UHF/VHF bands between 54 and 862 MHz and their guard bands [3]. Literature surveys of recent development of CRs suggest its wide application for SG communication. [24] successfully integrated and realized CRSN to reduce energy losses with cheap communication infrastructure for deployment of an SG network in Pakistan. The PUF was proposed for privacy-preserving of SG surveillance over CRSN for secure transmission of data over a stipulated time period with minimum channel interference. Premarathne et al. [25] have proposed a reliability consensus plan. Its reliability improvement for SG is examined while comparing with local estimation [26]. Next, an adaptable channel selection method is tested to check the performance of its selection potential based on the type of transmitted data for cognitive radio-based last-mile SG communications [27]. A BCH-based stenographic technique is tested to find and recover any loss from the sensitive secret information without disclosing the privacy of any genuine recipients. In this study, minimum distortion with high randomization is achieved by adopting a 3-D wavelet transforming decomposed normal readings to their frequency domain [28]. In paper [29], the high-quality link between sensor nodes in the SG spectrum is preserved, and packet loss probability is simultaneously reduced by the undertaken algorithms. Khan et al. have applied a fuzzy inference-based algorithm to find available idle channels for SG applications [30]. In comparison with HD–CRN, IBFD–CRN is found to utilize a higher spectrum for wireless communications [31].

3.5 Big data, privacy, and security in smart grids

The grid requires information from multiple sources including sensors, PMUs, energy market pricing data, weather data, population data, GIS mapping data, and bidding data gathered through smart meters. Millions of smart meters transmit data to thousands of nodes and data aggregators and then those are forwarded to the control center, resulting in the generation of billions of scripts every hours; this information needs to be identified and stored securely on the server for a resilient and efficient grid. Big data provides fundamental storage of these data in the cloud and are utilized to forecast energy trends in the particular grid. This information is sensitive, as it may be exploited to create blackouts or have an impact on energy trading. In this regard, [32] have presented a data aggregation method based on a secret sharing idea. Likewise, Marksteiner et al. have presented a Microsoft threat-modeling tool to transfer power from producer to consumer for a smart low-voltage distribution grid in a controlled and efficient manner [33]. In Ref. [34], the authors have conceptualized a CL-AS algorithm for concurrent privacy preservation and batch-wise authentication. Then in paper [35], a data-perturbation scheme named SEAL (based on Laplacian noise and Chebyshev interpolation) is examined to find a balancing solution between utility and privacy aspects.

3.6 Attacking smart grid network communications

The resilience of SG communication networks is becoming a matter of concern for software and network components that operate the grid. Monitoring and altering the messages communicated over the infrastructure are most obvious to hackers. Most devices are redundant to these attacks and check for irregularities, sending an alarm about the intruder. A device master is an interesting target for potential attackers for multiple reasons. Likewise, snooping into the communication network through any open links or nodes is possible through a few open source protocols. Simultaneously, the SG ecosystem may be attacked—various device manufacturers, service providers, network operators, power plants, and utility centers are the main targets as seen in recent literature. In this context, Eder-Neuhauser et al. have studied a life-cycle model of malware-based cyberattacks to investigate convenience in mitigating such issues by a qualitative evaluation [36]. Resiliency from the most vulnerable attacks (viz. price manipulation and DoS), the SG is analyzed for abnormal behavior detection and CUSUM techniques [37]. The BDSFTR is experimentally validated using NS for secure-packet transmission while handling cross-layer metric heterogeneity to avoid communication uncertainty [38]. Wang et al. [39] have proposed a UIO-based FDIA technique to restrain false-data injection in the SG. Lee and Hu [40], evaluating sequential collapsing aspects of the power system communication topology. In this context, a few significant nodes are large-branch tree-shaped root node, long single loop, and star-shaped node, which further help in the optimal location of substations along with network optimization [40]. In Ref. [41], the authors present an attack-prediction technique (viz. EWMA filter) for real-time network traffic. Considering distributed characteristics of the SG to detect and isolate any unforeseen attacks, Luo et al. [42] have proposed BIA and signature judgment logic for both local and global energy management. Then [43] suggested a game theory approach for three security levels of the SG network, viz. generation, transmission, and distribution, characterizing both attacker and defender.

3.7 Data analytics in smart grids

Accumulation of the huge amount of data stored on the server needs to be visualized through data management. Data analytics plays a pivotal role in recent power systems, where it breaks down chunks of data into an information system. The introduction of ICT into T&D networks for storage and analysis of collected data is helped by extensive smart meters and sensor installations to simplify the overall data management process. Data analytics compile the data and simplify it to visualize and extract results. A large volume of data is gathered from two-way communication of SGs at various time resolutions. Nowadays, advanced data analytics are needed to mine valuable information not only for billing information but also for keeping updated about the electricity network's status. In continuation, [44] have proposed the SDSS framework, which includes an SG web-based portal, big data collection layer, and analytics bench for energy consumption optimization and prediction accuracy. PMU is an integral component of the SG, improving its data flow, identification of unknown events, and the relationship between potential anomalies, as performed in Ref. [45]. Munshi et al. [46] studied SG big data management while establishing a secure link between the framework's cluster nodes by adopting the SSH version 2 protocol. Likewise, they hosted a Cloudera Manager Hadoop distribution framework on an IaaS Google cloud platform to attain improved accessibility and scalability infrastructure in a cost-effective manner. Next, smart metering of big data oriented challenges was studied for core broker client system architecture in Ref. [47]. Then, Olivares-Rojas et al. presented an edge fog cloud computing architecture for analytic applications like energy consumption forecasting, prediction of power quality, and energy theft prediction [48].

3.8 Supervised and unsupervised learning algorithms for smart grids

Nowadays, the application of learning algorithms in power systems is a trending area of research where different methodologies and algorithms are reviewed. These algorithms may advance performance adaptability as the number of available samples goes up, with increases in the amount of big data, and where machine learning has become a crucial technique to solve problems. Most commonly, two methods of machine learning are involved—supervised learning and unsupervised learning. The supervised technique guides a model with previously known data to foresee outputs effectively. However, the unsupervised technique comes across hidden patterns and intrinsic arrangements within input values. [48] have suggested machine learning data analytics to explore both opportunities and challenges based on prior knowledge of various processes in any plant [49]. An estimation concept of NILM is presented while adapting transition probability according to the power budget. Its activity chain was studied using a Markov model. The semisupervised techniques, viz. ESFCM and ELM, have been successfully tested to handle labeled and faster detection of any issue in the SG, respectively [51]. Likewise, [50] have studied load prediction effectiveness of the machine learning technique while examining MLRM, Ada Boost model, and ANN-NAEMI [52]. Then an incremental semisupervised learning model was tested for intrinsic data structure and network supervision information [53]. In Ref. [54], the NILM is proposed for load disaggregation and operational state detection without deployment of large-scale fine-grained energy meters.

Recently, many optimization methodologies have been applied to solve complex problems. All such methods are required for solving power system–related problems like economic load dispatch, OPF, and optimal reactive power dispatch. OPF was first initiated by Carpentier in Ref. [55]. In continuation, a few algorithms applied for OPF study are briefed as follows. [56] presented IGA to solve a quadratic FC function in both normal and contingency states of the power system. An EGA was tested by Ref. [57] to find OPF problem solutions. Assimilation of PSO as a derivative-free method significantly eased the assumption imposed on the undertaken optimization objectives as shown in Ref. [58]. Furthermore, [59] examined the MDE method to find a solution for OPF issues with nonsmooth cost functions. Another multiobjective evolutionary EGA-DQLF algorithm determined a diverse pareto optimal solution as presented in Ref. [60]. Ela et al. [61] applied a DE approach to find solutions for OPF problems and tested its performance on the standard IEEE 30-bus test system. Then, EADDE was applied by Ref. [62] to solve OPF issues with nonconvex and nonsmooth generator *FC* characteristics. A BBO algorithm was validated on the modified IEEE 30-bus system for solving *FC*, voltage profile, and its stability issues, in Ref. [63]. In the same year, GSA was implemented on the standard IEEE 30- and 57-bus test systems to enhance voltage profile, improve stability, and curtail *FC* (see Ref. [64]). Sumpter [66] has studied animal movements, how they gather to convey information, and collective decision-making approaches. In this chapter, a brief study using one such supervised algorithm (viz. the CAB algorithm) is carried out for an OPF study of the modified IEEE 30-bus test power system. One recently introduced method proposed by Ref. [65] is the CAB algorithm. It applies memory to store the foremost solutions (animal positions) resembling the aforementioned biological processes referred to in Fig. 5.3 (Algorithm 5.1).

A set of operations likely exists that looks like the interaction rules and models in the unified habitude of animals. [66] has investigated how animals move and arrive together, convey information, correlate their activities, make decisions accordingly, and build collective structures. In CAB, [65]'s individual solution confined within the search space constituted an animal position. Then a "fitness value" introduces animal domination with respect to the group. The whole process imitates the unified habitude of animals.

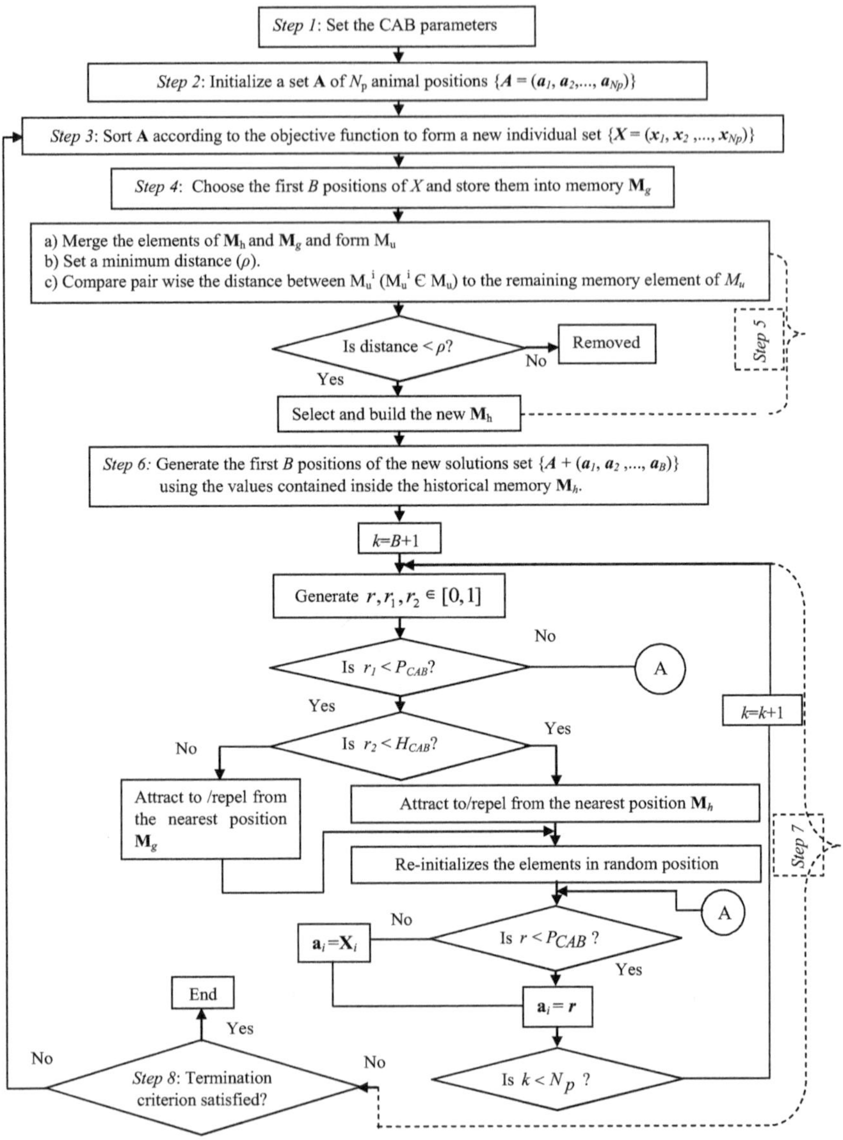

FIGURE 5.3 Computational flowchart of the CAB algorithm (Algorithm 5.1).

4. Feasibility study of all-dielectric self-supporting cable networking for smart grid infrastructure

The landscape of the ICT industry has been transforming abruptly as its paradigm moves toward a human-centric system [67,68]. The ADSS cable can be said to be the next generation of optical fiber cable, as it has plastic on the top layer protecting the optical fiber and core constituting the strength member.

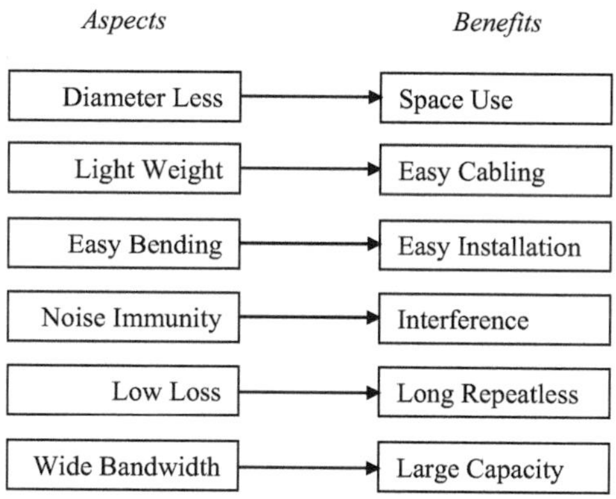

FIGURE 5.4 SG feasibility aspects of ADSS cable.

It makes it versatile for use in T&D line communication [69–71]. Since it lies below the live conductor in the transmission tower, live line installation is possible compared with OPGW, which needs a high outage time [72]. The polyethylene outer sheath also makes it suitable for installation on the distribution line with bare or covered conductor. The number of wireless devices keeps growing to provide flexible application (as shown in Fig. 5.4), and thus the availability of this networking infrastructure could satisfy demand [73].

Nowadays, optical fiber has been widely integrated into electric power cables and is nearly present in the cable for each voltage level [74,75]. The acceptance of ADSS network infrastructure can be developed for the following five key strategic goals. Clean energy-based connectivity is economically feasible with the help of retrofit hardware to bring last-mile connectivity to consumer reach and environmentally safe for dispersing a wide band of data connectivity over the optical network, minimizing radiation while overcoming the technical challenges of installation and penetration into deep congested areas and maintaining ecological balance with minimum RoW clearance and a minimum of damage to flora and fauna. It is durable to be sustained over decades and proven with millions of tests to have over a year's reliability as conveyed in Fig. 5.5. These five strategic approaches assist in getting the last-mile high speed Internet connectivity for integration of smart- and clean-generation technologies. Next, deployment of ADSS cable connectivity caters to the need for present generation, which is the key to SG as well as smart technology, governance, resources, wastes, and transport (as shown in Fig. 5.6). Thus, the value addition with the help of hardware to overcome resource management problems leads to better deployment. The ADSS cable begins to play a role for installations in transmission and distribution environments, even while performing live–line installation work (as presented in Table 5.3).

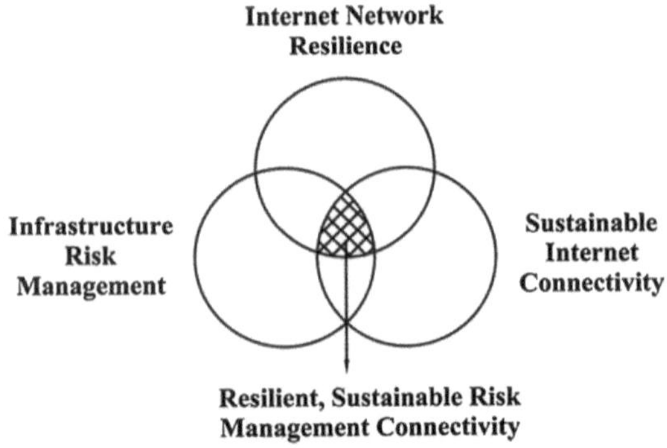

FIGURE 5.5 Internet service risk management.

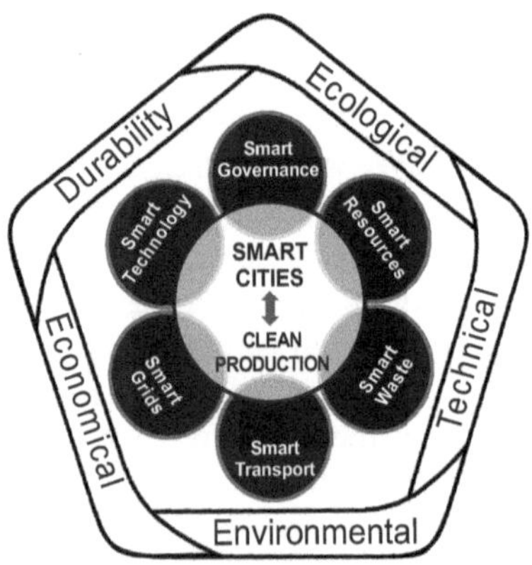

FIGURE 5.6 Five-dimensional sustainability assessment of a smart city.

TABLE 5.3 Accessories required for different cable type.

Application	Cable type	Hardware accessories
Transmission line	OPGW	Helical dead-ends and suspension sets, Stockbridge vibration damper, OPGW downlead clamps for towers.
	ADSS	Helical dead-ends and suspension sets, spiral vibration damper, downlead clamps for towers.
Distribution line	ADSS	Articulated suspensions, helical dead-end and suspension

As the term "ADSS" of the optical cable's variant signifies, there is no messenger wire or any specific support required; hence, its installation is possible in a single pass, making it a simple and economical means of deploying fiber optical networks [76,77]. The cable installation should be carefully made so to avoid damage at strain towers due to partial forces during the line construction and broken fiber is avoided. Thus, taking control of the constant tension at a uniform speed and paying attention to the proper angle and pulling direction of fibers during installation is significant [78]. These factors contribute to this lack of resilience; to address these gaps, this manuscript proposes a resilient design hardware accessory to upgrade the communication network.

4.1 Evaluation of hardware compatibility for smart grid networks

The ADSS cable (as portrayed in Fig. 5.7) is used for overhead installation, and the type of hardware required is categorized into (A) a tension fitting that generally constitutes 30% of the line and is used at angles or junction boxes (as presented in Fig. 5A of the Appendix); (B) suspension fittings that make up the remaining 70% of the line—they are generally simpler and require fewer components than tension fittings and also bear less load (as portrayed in Fig. 5B of the Appendix). Major components associated with this objective are as follows:

(a) Pole clamp: The pole clamp is used for suspension and tension fittings to mount the hardware on poles. There are a few commonly used types of pole clamps as illustrated below:
   - Circular pole clamp (half/full): These are limited to a certain diameter only.

FIGURE 5.7 Cross sectional view of ADSS cable.

- Square pole clamp: It is also suitable with a fixed diameter and shape.
- Universal pole clamp: These types of pole clamps may be installed over a range of diameters for octagonal and conical poles at different heights. Since the ADSS cables are not as bulky as the conductor weights, they can be easily installed with the universal pole clamp.

(b) Wall support bracket: It serves the same purpose as the pole clamp but should be installed on the wall instead of poles. These are installed with the help of anchor bolts to grip the support bracket to the walls. The wall support bracket can also be used as a tension or suspension type, such as different shapes applied for corners where a bend is needed.

(c) Spiral vibration dampers: They belong to a category of helical fittings made of PVC with a rod diameter of 12 mm and length up to 1.6 m to limit vibration caused by wind or snow. Their helical diameters are suited to the diameter of the ADSS to grip it at the hardware side for suspension or tension fittings. The ratio of the length is generally 2:1 for damping and gripping lengths. The spiral vibration damper restricts vibration at both suspension and tension ends.

(d) Helical fittings: These fittings so far have been considered a solution for gripping ADSS cables and conductors as they spread the gripping force over a wider surface area without damaging its layers for ease of installation. This one is used for tension fittings such as dead-end helical fittings and protective helices. They are color-coded for ease of identification for the different sizes of communication cables. Bonding of the helical wires is done with weatherproof glue to last longer at ambient condition.

(e) Protective helices: They are laid over the ADSS cable to increase surface area of the cable for better grip with the cable applying the principle of friction. It also helps distribute compression load over the complete area of the ADSS uniformly to prevent any puncture on the cable from the rough surface of the dead-end grip. It also helps bind other accessories such as the spiral vibration damper, preventing bends due to loads on the ADSS cable, which is more feasible than the protective helix.

(f) Twisted shackle: It is the primary link between the pole clamp and turnbuckle; its twisted nature keeps the orientation of the turnbuckle and pole clamp aligned to the direction of the cable. It also functions to help in ease of installation by locking either component with the rivet and pin. Its body is made from forged steel to withstand loads up to 70 kN, which makes it sufficient for the higher spans of ADSS cable.

(g) Turnbuckle: It is used to deliver the appropriate tension required for ADSS after stringing. Size can be varied depending on the required adjustment with a range of 150 mm, maximum effective length of 505 mm, and minimum effective length of 355 mm. To meet standardization among the hardware, the UTS is 70 kN for the turnbuckle, too.

(h) Extension links: These are fabricated from flat iron and used to connect the turnbuckle with the clevis thimble for the helical dead-end. Extension links are used to maintain a sufficient distance from the dead-end to the downlead clamp, as ADSS cables are prone to bending beyond a certain radius, which may lead to fiber damage and brings down latency.

(i) Clevis thimble: It is semicircular aluminum-casted thimble applied to link the dead-end grip with a certain degree of rotation and fit around the loop of the helical dead-end. It also provides clearance for easy installation. The load-bearing capacity is sufficient to take the load specified by the hardware. Some transformation, such as ribbing the circular area or a flat chord cast between the two ends of a semicircle, is done to improve the load-bearing capacity of the thimble.

(j) Dead-end grips: These are helical grips to lock the ADSS cable with the thimble. They are prone to slip if subjected to 95% load of the cable to prevent the cable from being damaged under certain circumstances. These are installed directly on the wire, but in the case of ADSS cables, they are installed over the protective helix to provide better grip.

(k) Downlead clamps: They are mounted on the poles and sometimes through a hole by a nut-bolt, sometimes by a steel strap, and sometimes by anchor bolts on the wall. It is made suitable for the diameter of the ADSS and is composed of a metal body as a rigid member with a rubber bushing to grip the ADSS. The cable is drawn through the downlead clamp toward the cable tray or termination box.

(l) Twisted eye link: It is used for suspension to attach the pole clamp or load support with the armor-grip suspension. Likewise, twisted shackles help in the orientation of the suspension hardware parallel to the line. They are attached at the AGS and shaped wider at the bottom to restrict free movement of the AGS nut and bolt to avoid damage due to wear and tear.

(m) AGS clamp: The complete set of AGS clamps are aluminum cast to the required diameter of the ADSS with a rubber grip to support with the armor rod. The suspension fitting does not need to withstand heavy loads, as most parts are made of aluminum. Here, armor rods may be made from ACS or Al alloy. The AGS fitting is assembled on the protective helix that covers the ADSS cable. It allows slip at 40%–50% of the cable UTS.

4.2 Mathematical modeling

The present study addresses two different SG issues:

(a) Validation if ADSS cable networking hardware for the SG
(b) Application of a supervised algorithm for OPF in the SG

Mathematical formulations for the aforementioned two studies are presented in detail in the succeeding sections.

(a) Validation of ADSS cable networking hardware for the SG. In this section, mathematical modeling of the proposed ADSS network hardware is made

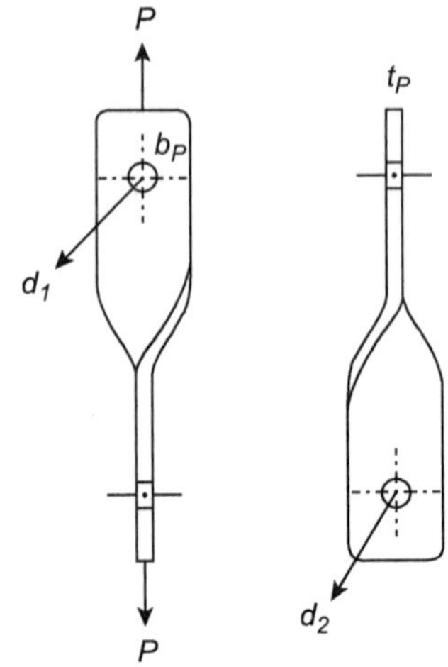

FIGURE 5.8 Schematic of TEL force resolution.

prior to FEA performance [81]. Tensile stress due to stretching of bolts is designed with a high factor of safety for intermediate stresses. The preliminary tension appeared at a bolt can be given by a relation (5.1) [82,83],

$$P_i = 2840\, d_1 \tag{5.1}$$

If the joint is not required as fluid-tight (as presented in Fig. 5.8), then the simplified relation may be revised as in (5.2) [82,83],

$$P_i = 1420\, d_1 \tag{5.2}$$

Torsion shear stress is produced by the frictional resistance of thread while tightening based on torsion theory, and it may be given by (5.3) [82,83]

$$\tau = \left(\frac{16T}{\pi d_c^3}\right) \tag{5.3}$$

Shear stress appearing across the thread of a screw is $\tau_s = \frac{P}{\pi d_c b_t n}$, and for the nut, $\tau_n = \frac{P}{\pi d_2 b_t n}$. These two shear stresses collectively yield (5.4) [82,83]

$$\sigma_c = \frac{P}{\pi (d_2 - d_c) n} \tag{5.4}$$

Thus, the torsion shear stress at d_1 may be given by (5.5) [82,83]

$$\tau_1 = \frac{P}{(b_p - d_1) t_p} \tag{5.5}$$

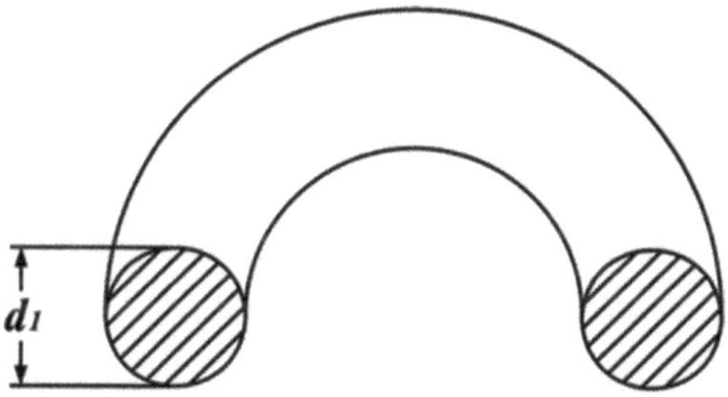

FIGURE 5.9 Sectional view of twisted shackle.

Similarly, the tensile load may be represented in a simplified way using (5.6) [82,83]

$$P = (b_p - d_2)t_p\tau_1 \tag{5.6}$$

When the twisted shackle is under a loaded condition, the simplified formula will be (5.7) [82,83]

$$2P_{TS_1} = \tau_r \pi d_r^2 \tag{5.7}$$

Shear failure at the rivet contact for a twisted shackle (refer to Fig. 5.9) can be mathematically represented by (5.8)–(5.9). Here, at two sections the amount of shearing will be doubled as the area is doubled:

$$2P_{TS_2} = \frac{\pi}{2}\tau' d_1^2 \tag{5.8}$$

$$P = 4d_1^2\tau' \tag{5.9}$$

Circumferential stress of the clevis thimble (refer to Fig. 5.10) may be given as in (5.10) [82,83]:

$$\sigma_H = \frac{P_{CT}}{4\pi h t_{CT}} \tag{5.10}$$

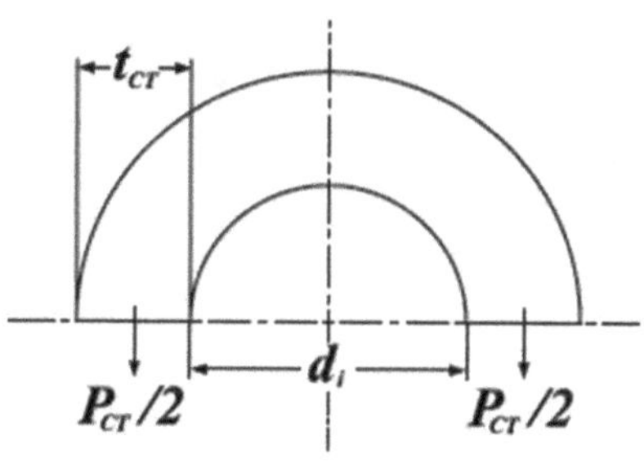

FIGURE 5.10 Schematic of force resolution acting on clevis thimble.

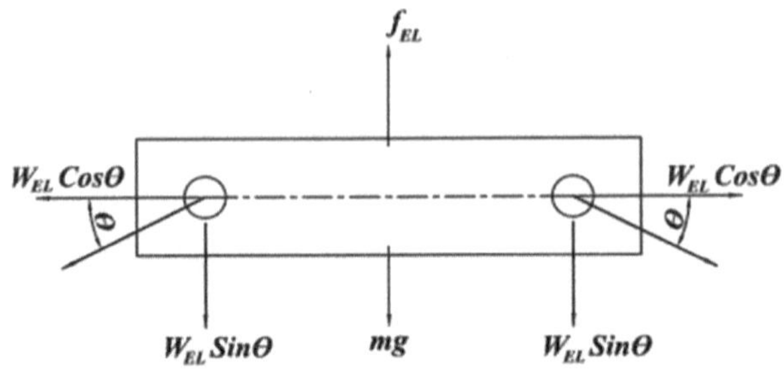

FIGURE 5.11 Schematic of tension pole assembly.

Similarly, longitudinal stress may be formulated as in (5.11) [82,83]:

$$\sigma_L = \frac{P_{CT} d_i}{8\pi t d' h} \tag{5.11}$$

Here, the tension pole assembly (both sides) has a pole clamp to which the twisted shackle is attached, and from that shackle is an assembly of eyebolt and turnbuckle (as shown in Fig. 5.11). When a pole is in a locked condition, tensile force or load will act at the eye bolt and be the same as the tension at the shackle and pole clamp assembly. So here the free body diagram of pole clamp is given as the force required at the circumference of screw/eye bolt and can be mathematically formulated by (5.12)–(5.13) [82,83]:

$$P_{EL} = W_{EL} \tan(\alpha_h + \varphi) \tag{5.12}$$

$$T_{EL} = \frac{W_{EL} d_m (\varphi - \alpha_h)}{2} \tag{5.13}$$

The eye bolt tightens or raises the load based on a mathematical relation (5.14) [82,83]:

$$T_{EL} = \frac{W_{EL} d_m \tan(\varphi + \alpha_h)}{2} \tag{5.14}$$

Whenever the eyebolt loosens or is lowered, the load may be formulated by (5.15)–(5.17) [82,83]

$$f_{EL} = 2W_{EL} \sin\theta + mg \tag{5.15}$$

$$f_{EL} = \frac{2P_{EL} \sin\theta}{d_m \tan(\varphi \pm \alpha_h) + mg} \tag{5.16}$$

$$f_{EL} = \frac{4T_{EL} \sin\theta}{d_m \tan(\varphi \pm \alpha_h) + mg} \tag{5.17}$$

It is notable that the $+ve$ sign of (5.17) is applicable for extension of the turnbuckle; however, its $-ve$ sign allows tightening of the turnbuckle.

The failure criterion selected for the analysis is Von Mises or maximum distortion energy criterion. Now considering the case of a three-dimensional body, strains in the respective directions can be formulated by (5.18)–(5.20) [82,83]:

$$\in_x = \frac{\sigma_x - v(\sigma_y + \sigma_z)}{E} \tag{5.18}$$

$$\in_y = \frac{\sigma_y - v(\sigma_x + \sigma_z)}{E} \tag{5.19}$$

$$\in_z = \frac{\sigma_z - v(\sigma_x + \sigma_y)}{E} \tag{5.20}$$

Now the volumetric strain is given as in (5.21) [82,83]:

$$\in_v = \in_x + \in_y + \in_z \tag{5.21}$$

On using the above Eqs. (5.18)–(5.21), the net volumetric deformation in terms of stresses may be given by (5.22) [82,83]:

$$\in_v = \frac{(\sigma_x + \sigma_y + \sigma_z) - 2v(\sigma_y + \sigma_z + \sigma_x)}{E} \tag{5.22}$$

The volumetric stress can be taken as the average of the stresses in (5.23) [82,83]:

$$P = \frac{\sigma_x + \sigma_y + \sigma_z}{3} \tag{5.23}$$

The volumetric deformation energy for the material as loaded with the above-mentioned conditions will be given in (5.24) [82,83]:

$$U_{vol} = \frac{1}{2}\left\{\frac{(\sigma_x + \sigma_y + \sigma_z) - 2v(\sigma_y + \sigma_z + \sigma_x)}{E}\right\}\left(\frac{\sigma_x + \sigma_y + \sigma_z}{3}\right) \tag{5.24}$$

On solving the above equations, we shall obtain (5.25) [82,83]:

$$U_{Vol} = \frac{(1-2v)(\sigma_x + \sigma_y + \sigma_z)^2}{6E} = \frac{1-2v}{6E}\left\{\sigma_x^2 + \sigma_y^2 + \sigma_z^2 + 2(\sigma_x\sigma_y + \sigma_x\sigma_z + \sigma_y\sigma_z)\right\} \tag{5.25}$$

Now the strain energy is calculated in all directions and added together to obtain the total strain energy as in (5.26) [82,83]:

$$U_0 = \frac{1}{2E}\left\{\sigma_x^2 + \sigma_y^2 + \sigma_z^2 - 2v\,(\sigma_x\sigma_y + \sigma_x\sigma_z + \sigma_y\sigma_z)\right\} \tag{5.26}$$

The distortion energy is the difference between U_{vol} and U_h and is given by (5.27) [82,83]:

$$U_{vm} = U_0 - U_{vol} \tag{5.27}$$

The Von Misses energy using (5.28)–(5.30) [82,83] is calculated while performing the FEA of the proposed hardware:

$$U_{vm} = \frac{1+v}{2E}\left(\sigma_x^2 - \sigma_y^2 + \sigma_y^2 - \sigma_z^2 + \sigma_z^2 - \sigma_x^2\right) \tag{5.28}$$

$$U_{vm} = \frac{1+v}{3E}\sigma_{vm}^2 \tag{5.29}$$

$$\sigma_{vm} = \sqrt{\frac{\sigma_x^2 - \sigma_y^2 + \sigma_y^2 - \sigma_z^2 + \sigma_z^2 - \sigma_x^2}{2}} \tag{5.30}$$

(b) Application of a supervised algorithm for OPF in the SG—OPF is a static nonlinear programming problem that aims to optimize certain objective functions while satisfying a set of operational and physical constraints imposed by security requirements and equipment limitations. Mathematically, it may be given by (5.31)–(5.33) [79]:

$$\text{Minimize } J(x, u) \tag{5.31}$$

$$\text{subject to } eq(x, u) = 0 \tag{5.32}$$

$$\text{and } ieq(x, u) = 0 \tag{5.33}$$

The dependent variables vector $\boldsymbol{x}$ may be formulated by (5.34) [79]:

$$\boldsymbol{x}^T = \left[P_{G_1}, V_{L_1}\ldots\ldots, V_{L_{NL}}, Q_{G_1}, \ldots\ldots, Q_{G_{NG}}, S_{l_1}\ldots\ldots S_{l_{nl}}\right] \tag{5.34}$$

Similarly, the control variables vector $\boldsymbol{u}$ may be given by (5.35) [79]:

$$\boldsymbol{u}^T = \left[P_{G_2}, \ldots\ldots P_{G_{NG}}, V_{G_1}, \ldots\ldots, V_{G_{NG}}, T_1, \ldots\ldots, T_{NT}, Q_{C_1}\ldots\ldots Q_{C_{NC}}\right] \tag{5.35}$$

4.2.1 Constraints

4.2.1.1 Equality constraints

In those constraints of (5.32), *eq* are signifying the load flow equations as formulated in (5.36) and (5.37) [79]:

$$P_{G_i} - P_{D_i} = V_i \sum_{k=1}^{NB} V_k(G_{ik}\cos\theta_{ik} + B_{ik}\sin\theta_{ik}) \quad \text{where i} = 1, 2, \ldots, \text{NB} \tag{5.36}$$

$$Q_{G_i} - Q_{D_i} = V_i \sum_{k=1}^{NB} V_k(G_{ik}\sin\theta_{ik} - B_{ik}\cos\theta_{ik}) \quad \text{where i} = 1, 2, \ldots, \text{NB} \tag{5.37}$$

4.2.1.2 Inequality constraints

Here (5.33) presents the inequality constraints as *ieq*, which has been elaborated as below.

(i) *Generator constraints*: Real and reactive power outputs must be kept within their lower and upper limits for every generator voltage as presented in (5.38)–(5.40) [79]:

$$V_{G_i}^{\min} \leq V_{G_i} \leq V_{G_i}^{\max}, \quad i = 1, 2, \cdots, NG \quad \text{(with slack bus)} \tag{5.38}$$

$$P_{G_i}^{\min} \leq P_{G_i} \leq P_{G_i}^{\max}, \quad i = 1, 2, \cdots, NG \quad \text{(with slack bus)} \tag{5.39}$$

$$Q_{G_i}^{\min} \leq Q_{G_i} \leq Q_{G_i}^{\max}, \quad i = 1, 2, \cdots, NG \quad \text{(with slack bus)} \tag{5.40}$$

(ii) *Transformer constraints:* The minimum and maximum values of tap settings are given in (5.41) [79]:

$$T_i^{\min} \leq T_i \leq T_i^{\max}, \quad i = 1, 2, \cdots, NT \tag{5.41}$$

(iii) *Shunt VAR compensator constraints:* Its reactive power must be constrained within the defined limit as given in (5.42) [79]:

$$Q_{C_i}^{\min} \leq Q_{C_i} \leq Q_{C_i}^{\max}, \quad i = 1, 2, \cdots, NC \tag{5.42}$$

(iv) *Security constraints:* The line flow in each transmission network must be within its ratings. These are mathematically written as in (5.43) and (5.44) [79]:

$$V_{L_i}^{\min} \leq V_{L_i} \leq V_{L_i}^{\max}, \quad i = 1, 2, \cdots, NL \tag{5.43}$$

$$S_{l_i} \leq S_{L_i}^{\max}, \quad i = 1, 2, \cdots, nl \tag{5.44}$$

4.2.2 Objective function

In this chapter, the viability of the undertaken algorithm is checked using two monotype objectives [80]. These are formulated as below.

(i) *Minimization of FC*: The purpose behind this type of problem is to minimize fuel cost while satisfying all equality and inequality constraints as shown in (5.45) and (5.46):

$$\text{Minimize } J(FC) \tag{5.45}$$

$$J(FC) = \left(\sum_{i=1}^{NG} F_i(P_{G_i})\right) = \sum_{i=1}^{NG}\left(\alpha_i + \beta_i P_{G_i} + \gamma_i P_{G_i}^2\right) \tag{5.46}$$

(ii) *Minimization of transmission loss*: Mathematical formulation of this type of objective function is given by (5.47):

$$\text{Minimize } J(P_{Loss}) = \sum_{k=1}^{NTL} g_k\left(V_i^2 + V_j^2 - 2V_iV_j \cos\theta_{ij}\right) \tag{5.47}$$

5. Finite element analysis results of proposed all-dielectric self-supporting networking hardware

The value of the Von Mises stress is lower than the yield stress in order to prevent failure of the material under loading. The FEA results are presented in Tables 5.4–5.21 and robustness of suggested hardware for SG ADSS cabling has been portrayed in Figs. 5.12–5.15 with graphical outputs.

Apart from FEA validation, in-house testing of the proposed hardware was also done to check its robustness to deploy the communication infrastructure. This will demonstrate the feasibility of hardware for communication networks. The helical terminating fittings of five wire rods were wrapped and unwrapped on a piece of aerial cable three times successively, and the slip strength

TABLE 5.4 Material properties for all three structural loadings.

Properties	Value
Material	Steel
Young's modulus	2e+011N_m2
Poisson's ratio	0.266
Density	7860 kg_m3
Coefficient of thermal expansion	1.17e-005_Kdeg
Yield strength	2.5e+008N_m2

TABLE 5.5 Structural meshing details.

Structural loadings	Entity	Size
Vertical	Nodes	8719
	Elements	4510
Single-sided	Nodes	6123
	Elements	2832
Double-sided	Nodes	5343
	Elements	2283

TABLE 5.6 Structural element types.

Structural loadings	Connectivity	Statistics
Vertical	TE10	4510 (100.00%)
Single-sided	TE10	2832 (100.00%)
Double-sided	TE10	2283 (100.00%)

TABLE 5.7 Structural element quality.

Structural loadings	Criterion	Good	Poor	Bad	Worst	Average
Vertical	Stretch	4508 (99.96%)	2 (0.04%)	0 (0.00%)	0.274	0.567
	Aspect ratio	3424 (75.92%)	1086 (24.08%)	0 (0.00%)	4.532	2.183
Single-sided	Stretch	2747 (97.00%)	85 (3.00%)	0 (0.00%)	0.194	0.538
	Aspect ratio	1966 (69.42%)	836 (29.52%)	30 (1.06%)	8.383	2.359
Double-sided	Stretch	1970 (86.29%)	313 (13.71%)	0 (0.00%)	0.062	0.468
	Aspect ratio	1269 (55.58%)	882 (38.63%)	132 (5.78%)	21.677	2.763

TABLE 5.8 Structure and restraint computation.

Parameters	Structural loadings		
	Vertical	Single-sided	Double-sided
Number of nodes	8719	6123	5343
Number of elements	4510	2832	2283
Number of DoF	26,157	18,369	16,029
Number of contact relations	0	0	0
Number of kinematic relations	0	0	0
Parabolic tetrahedron	4510	2832	2283
Number of SPC	1320	474	72

TABLE 5.9 Load computations (applied load resultant).

Components	Structural loadings		
	Vertical	Single-sided	Double-sided
Fx (N)	1.164e−009	0.000e+000	4.470e−008
Fy (N)	−2.319e−008	2.000e+003	1.717e−005
Fz (N)	−2.356e+003	−1.091e+003	−2.000e+003
Mx (Nxm)	2.231e+002	−3.291e+002	7.590e−006
My (Nxm)	1.855e+001	−1.804e+000	−3.603e+000
Mz (Nxm)	3.005e−009	−3.603e+000	−1.308e−007

TABLE 5.10 Structural mass computation.

Parameters	Structural loadings	
	Vertical	Single-sided
Number of lines	26,157	18,369
Number of coefficients	926,385	611,838
Number of blocks	2	2
Maximum number of coefficients per block	499,987	499,974
Total matrix size	10.70 Mb	7.07 Mb
Structural mass	3.446e+000 kg	9.306e+000 kg

TABLE 5.11 Structural coordinates.

Structural loadings	Inertia center coordinates		Inertia tensor at origin (kg.m2)		
Vertical	Xg (mm)	−7.132e−001	7.224e−002	−2.224e−004	−3.673e−004
	Yg (mm)	−7.336e+001 mm	−2.224e−004	5.604e−002	−2.739e−002
	Zg (mm)	−7.801e+001 mm	−3.673e−004	−2.739e−002	3.490e−002
Single-sided	Xg (mm)	−2.551e−002 mm	2.340e+000	1.705e−004	−5.480e-005
	Yg (mm)	3.876e+002 mm	1.705e-004	2.137e−001	6.536e-001
	Zg (mm)	−1.049e+002 mm	−5.480e-005	6.536e−001	2.135e+000

TABLE 5.12 Stiffness computation.

Parameters	Structural loadings		
	Vertical	Single-sided	Double-sided
Number of lines	26,157	18,369	16,029
Number of coefficients	926,385	611,838	512,064
Number of blocks	2	2	2
Maximum number of coefficients per block	499,987	499,974	499,944
Total matrix size	10.70 Mb	7.07 Mb	5.92 Mb

TABLE 5.13 Structural singularity computation data.

Structural loadings	Computational findings	
Common for all three cases	Number of local singularities	0
	Number of singularities in translation	0
	Number of singularities in rotation	0
	Generated constraint type	0
	Total matrix size	MPC

TABLE 5.14 Constraint computation.

Parameters	Structural loadings		
	Vertical	Single-sided	Double-sided
Number of constraints	1320	474	72
Number of coefficients	0	0	0
Number of factorized constraints	1320	432	72
Number of coefficients	0	0	0
Number of deferred constraints	0	0	0

TABLE 5.15 Factorized computations.

	Structural loadings		
Parameters	**Vertical**	**Single-sided**	**Double-sided**
Method	Sparse	Sparse	Sparse
Number of factorized degrees	24,837	17,937	15,957
Number of supernodes	1525	1248	1149
Number of overhead indices	148,899	91,419	71,433
Number of coefficients	4,934,733	2,095,731	1,385,127
Maximum front width	999	486	267
Maximum front size	499,500	118,341	35,778
Size of the factorized matrix (Mb)	37.649	15.9892	10.5677
Number of blocks	5	3	2
Number of Mflops for factorization	1.914e+003	3.949e+002	1.634e+002
Number of Mflops for solve	1.986e+001	8.473e+000	5.620e+000
Minimum relative pivot	2.560e−003	1.025e−004	2.796e−005

was repeated. Torque of 31.2 kN was applied and no physical damage observed on the helix and optical fiber cable. Spiral vibration damper also plays a vital role to prevent vibration on the suspension fittings as given in Table 5.22. Furthermore, the remaining results are briefed as Case 5.1 and Case 5.2.

Case 5.1: Tension fittings with a set of suspension fittings and spiral vibration damper.

The test was carried on a span of 35 m by terminating the approved aerial fiber optical cable on both sides with tension fittings. The setup was done to provide galloping vibration, and the fibers were spliced together to make a single loop. The attenuation of this loop was measured at 1310 and 1550 nm with the help of OTDR and dummy fiber. At a frequency of 33 Hz for 1 million sec, the cable was vibrated with a mechanical device to amplitude the vibration at the antinodal point not less than 100% the diameter of cable. The attenuation of the loop was again measured after the test, and no physical damage was found on cable and hardware accessories at a frequency of 33 Hz (test findings are given in Table 5.23 and its graph is plotted in Fig. 5.16).

TABLE 5.16 Designed pivotal points (vertical loadings).

Value	DoF	Node	x (mm)	y (mm)	z (mm)
Minimum and maximum pivot					
2.9134e+006	Ty	8719	−1.7578e+001	−9.3201e+001	−1.1959e+002
1.2947e+010	Ty	5139	−2.1221e+001	−5.6002e+001	3.8217e+001
Minimum pivot					
4.8133e+006	Tx	8719	−1.7578e+001	−9.3201e+001	−1.1959e+002
7.7006e+006	Ty	8718	−2.0000e+001	−9.3485e+001	−1.2011e+002
Minimum pivot					
1.1940e+007	Tz	8719	−1.7578e+001	−9.3201e+001	−1.1959e+002
5.6723e+007	Tz	8718	−2.0000e+001	−9.3485e+001	−1.2011e+002
1.5613e+008	Tz	4036	−9.0534e+000	−9.5000e+001	−1.1553e+002
1.6270e+008	Ty	4179	5.2230e+001	−8.4811e+001	−1.4233e+002
1.7023e+008	Ty	7658	2.2174e+000	−8.7303e+001	−1.4296e+002
1.7054e+008	Tz	4179	5.2230e+001	−8.4811e+001	−1.4233e+002
1.7259e+008	Ty	4011	2.7112e+001	−8.7788e+001	−1.6893e+002

TABLE 5.17 Designed pivotal points (single-sided loadings).

Value	DoF	Node	x (mm)	y (mm)	z (mm)
Minimum and maximum pivot					
7.6819e+005	Tx	6122	9.6412e+000	3.4352e+002	−6.2349e+001
2.3850e+010	Tx	5108	6.2290e+000	8.6376e+002	−2.5894e+002
Minimum pivot					
8.5914e+005	Tx	5142	1.1204e+001	8.6427e+002	−3.5445e+002
8.6962e+005	Tz	6122	9.6412e+000	3.4352e+002	−6.2349e+001
2.4876e+006	Tz	5142	1.1204e+001	8.6427e+002	−3.5445e+002
2.5182e+007	Tz	5742	−1.4469e-004	6.8474e+002	−2.0885e+002
2.6905e+007	Tz	1321	−1.4000e+001	2.1961e+002	−1.6330e+001
3.2856e+007	Ty	3535	−3.7877e−005	6.5830e+002	−1.9910e+002
5.2088e+007	Tz	1236	−1.3000e+001	4.0728e+002	−7.9492e+001
6.0152e+007	Tz	3534	−4.5243e+000	1.1383e+002	4.8911e−001
6.1528e+007	Tz	1153	2.9490e+001	5.9604e+001	0.0000e+000

TABLE 5.18 Designed pivotal points (double-sided loadings).

Value	DoF	Node	x (mm)	y (mm)	z (mm)
Minimum and maximum pivot					
6.6883e+005	Tx	3683	−1.4199e+001	−3.4030e+002	−8.4574e+001
1.4382e+011	Ty	5205	−6.6491e+000	−1.9081e+002	−2.8368e+001
Minimum pivot					
9.8974e+005	Tx	2344	1.5316e+001	−6.5806e+002	−2.0136e+002
1.0925e+006	Tx	4742	−1.0000e+001	7.1448e+002	−2.3277e+002
1.1971e+006	Ty	2343	1.0000e+001	−6.5860e+002	−1.9134e+002
1.3051e+006	Ty	4741	−1.0000e+001	7.1837e+002	−2.4718e+002
2.5673e+006	Tx	902	8.3966e-006	−1.9840e+002	−7.6130e+000
3.0745e+006	Tz	3683	−1.4199e+001	−3.4030e+002	−8.4574e+001
6.3546e+006	Tz	1952	−1.6011e+001	5.1664e+002	−1.5459e+002
9.3031e+006	Tz	1755	−9.0105e+000	9.0931e+002	−2.4098e+002
1.3706e+007	Tz	4166	1.2096e+001	−8.5146e+002	−2.6362e+002

TABLE 5.19 Translational pivot distribution.

Structural loadings	Value	%
Vertical	10.E6-> 10.E7	1.2079e−002
	10.E7-> 10.E8	8.0525e−003
	10.E8-> 10.E9	1.7079e+001
	10.E9-> 10.E10	8.2800e+001
	10.E10-> 10.E11	1.0066e−001
Single-sided	10.E5-> 10.E6	1.6725e−002
	10.E6-> 10.E7	5.5751e−003
	10.E7-> 10.E8	7.2476e−002
	10.E8-> 10.E9	8.8476e+000
	10.E9-> 10.E10	9.0377e+001
	10.E10-> 10.E11	6.8016e−001
Double-sided	10.E5-> 10.E6	1.2534e−002
	10.E6-> 10.E7	4.3868e−002
	10.E7-> 10.E8	1.5667e−001
	10.E8-> 10.E9	6.9123e+000
	10.E9-> 10.E10	8.8820e+001
	10.E10-> 10.E11	4.0421e+000
	10.E11-> 10.E12	1.2534e−002

Case 5.2: Tension or dead-end assembly.

The same span length of 35 m was taken by terminating the cable on both sides with tension fittings and providing a suitable suspension support. The setup was prepared to provide aeolian vibration, and the fibers were spliced together to make a single loop. At a frequency of 28 Hz for 1 million s, the cable was again vibrated. The attenuation of the loop was again measured after the test, and no physical damage on the cable or visual damage on hardware accessories was found at a frequency of 28 Hz. A complete result for this case is presented in Table 5.24, and respective variation is shown by Fig. 5.17.

6. Optimal power flow test systems vis-à-vis simulation results

The modified IEEE 30-bus test power system is undertaken for this study, and its SLD is referred to as Fig. 5C in the Appendix. The values of the control

TABLE 5.20 Equilibrium results.

Structural loadings	Components	Applied Forces	Reactions	Residual	Relative Magnitude error
Vertical	Fx (N)	1.1642e−009	1.3887e−008	1.5051e−008	4.7254e−011
	Fy (N)	−2.3194e−008	3.9158e−008	1.5964e−008	5.0120e−011
	Fz (N)	−2.3558e+003	2.3558e+003	−2.2957e−008	7.2075e−011
	Mx (Nxm)	2.2307e+002	−2.2307e+002	5.0993e−009	8.3717e−011
	My (Nxm)	1.8545e+001	−1.8545e+001	−1.4508e−009	2.3818e−011
	Mz (Nxm)	3.0052e−009	−5.5417e−010	2.4510e−009	4.0239e−011
Single-sided	Fx (N)	0.0000e+000	−3.5000e−008	−3.5000e−008	4.1607e−011
	Fy (N)	2.0000e+003	−2.0000e+003	−2.0453e−008	2.4314e−011
	Fz (N)	−1.0912e+003	1.0912e+003	1.0369e−007	1.2326e−010
	Mx (Nxm)	−3.2913e+002	3.2913e+002	3.2913e+002	1.0449e−010
	My (Nxm)	−1.8037e+000	1.8037e+000	1.1103e−008	1.4391e−011
	Mz (Nxm)	−3.6028e+000	3.6028e+000	2.8165e−008	3.6507e−011
Double-sided	Fx (N)	4.4703e−008	5.8506e−009	5.0554e−008	1.5957e−011
	Fy (N)	1.7166e−005	−1.7180e−005	−1.4074e−008	4.4425e−012
	Fz (N)	−2.0000e+003	2.0000e+003	−1.1152e−007	3.5202e−011
	Mx (Nxm)	7.5895e−006	−7.4246e−006	1.6497e−007	5.6777e−011
	My (Nxm)	−3.6028e+000	3.6028e+000	−1.3014e−008	4.4788e−012
	Mz (Nxm)	−1.3081e−007	1.4743e−007	1.6623e−008	5.7208e−012

TABLE 5.21 Global sensors (energy (J)).

Parameter	Structural loadings		
	Vertical	Single-sided	Double-sided
Sensor value	0.051	0.333	0.527

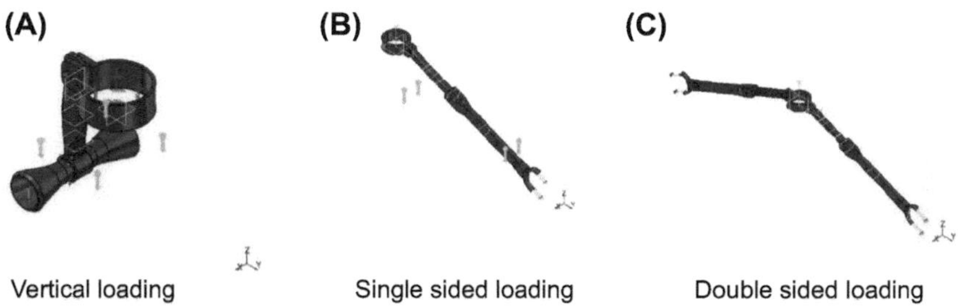

FIGURE 5.12 Boundary conditions.

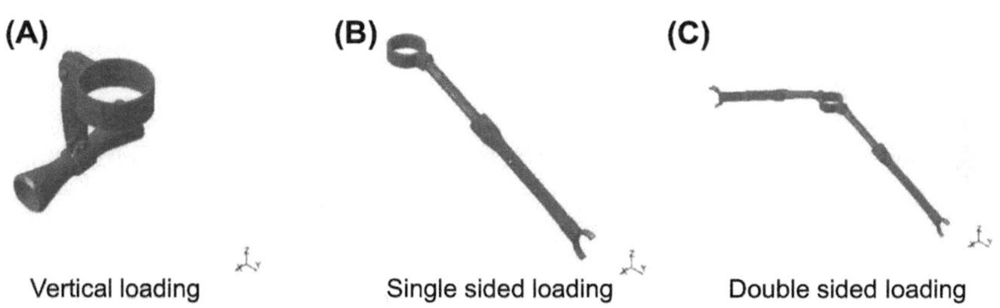

FIGURE 5.13 Graphical results of deformed mesh.

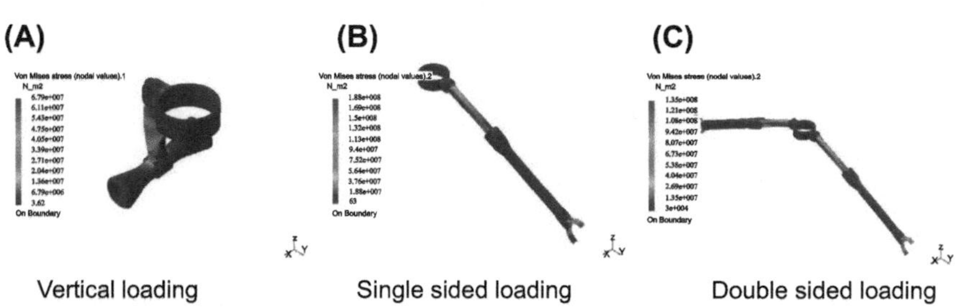

FIGURE 5.14 Graphical results of Von Mises stress.

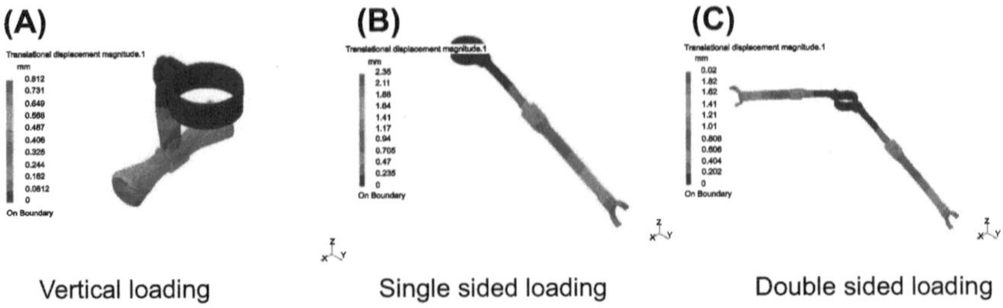

FIGURE 5.15 Graphical results of translational displacement.

TABLE 5.22 Tensile strength test results.

Components	Test condition for 1 min	Observation for 1 min
Terminating helix	31.2 kN load applied	No slip at 31.2 kN
Twisted eye link and suspension clipper	No visual damage at 70 kN	No damage at 70 kN
Hardware (except jumper cable clamp)	No visual damage at 70 kN	No damage at 70 kN

variables besides bus data, line data, and corresponding variable limits are taken from Ref. [58]. In this work, the CAB algorithm is tested to solve the OPF problem for the modified IEEE 30-bus test case system. The code is developed with MATLAB 2008a computing software tools on a 3.9 GHz Intel Turbo i5 Laptop with 8 GB RAM. Interesting results are **boldfaced** in tables to ascertain the potential of the CAB concept with 30 independent test runs.

6.1 Fuel cost minimization

The optimum control parameter tunings by DE [61], BBO [63], GSA [64], and the undertaken CAB are presented in Table 5.25. The *FC* value obtained from the CAB algorithm is **798.3339 $/h**, which shows a reduction of **0.043%** compared with the corresponding output (i.e., 798.675143 $/h) by the GSA method as reported in Ref. [64]. A statistical comparison of the CAB and a few other methods is presented in Table 5.26. The CAB-based convergence characteristics of *FC* ($/h) for this test case are presented in Fig. 5.18.

6.2 Minimization of transmission line losses

Curtailment of P_{Loss} is one of the objective functions for the undertaken test network to be executed by the proposed CAB method. The corresponding

TABLE 5.23 Galloping or fatigue test (Case 5.1).

Fiber No		Attenuation before test-A (dB)		Attenuation before test-B (dB)		Change in attenuation B-A (dB)	
		1310 nm (A1)	1550 nm (A2)	1310 nm (B1)	1550 nm (B2)	1310 nm (B-A1)	1550 nm (B-A2)
1	13	0.085	0.06	0.11	0.099	0.025	0.039
2	14	0.073	0.058	0.105	0.097	0.032	0.039
3	15	0.09	0.074	0.108	0.096	0.018	0.022
4	16	0.084	0.058	0.1	0.091	0.016	0.033
5	17	0.077	0.061	0.103	0.088	0.026	0.027
6	18	0.098	0.08	0.119	0.104	0.021	0.024
7	19	0.072	0.059	0.107	0.082	0.035	0.023
8	20	0.084	0.064	0.109	0.09	0.025	0.026
9	21	0.071	0.053	0.095	0.078	0.024	0.025
10	22	0.078	0.066	0.103	0.085	0.025	0.019
11	23	0.081	0.069	0.109	0.096	0.028	0.027
12	24	0.074	0.057	0.097	0.089	0.023	0.032

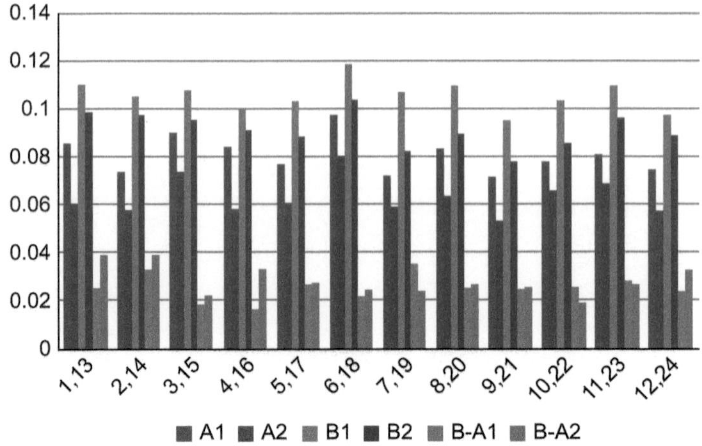

FIGURE 5.16 Fiber No vs Attenuation characteristics (Case 5.1).

outcomes along with other methods are presented in Table 5.27. Minimum P_{Loss} yielded by the adopted CAB method is found to be **3.202 MW**. The value of P_{Loss} (MW) by the CAB algorithm is **0.0012 MW** (i.e., **0.0374%**) less than the EGA-DQLF results [60]. The CAB technique-based convergence for a characteristic of P_{Loss} (MW) for the studied test system is shown in Fig. 5.19.

7. Summaries and future scope of work

The restructuring of the energy grid from a centralized traditional system to a demand-oriented decentralized SG is likely to entail substantial resilience of the communication infrastructure and OPF requirements. This article summarizes compatibility of hardware fittings for ADSS cable and the CAB algorithm application along with its testing. This outreach of ADSS hardware makes it more flexible with installation and brings innovation to the hardware required to mount it either on boundary walls or existing poles. The innovative hardware has helped overcome the barrier of RoW for new installations using the existing infrastructure with retrofit solutions. Concurrently, the low-cost hardware may leverage the spread of network diversification as well penetration into the congested area. In spite of being versatile in range, it also prevents damage over the minimum bending radius, maximum actual tension, vibration damping, etc. This will help us optimize cost and provide a contingency-free network. The direct method of FEA was applied to test structural resilience of the proposed hardware against deformation, and corresponding strain energy is given as follows—vertical loading: 5.097e−002 J; single-sided loading: 3.327e−001 J; double-sided loading: 5.275e−001 J. In this static-case solution, the structural mass is taken into account, and it is notable that on that date not a single piece of research work had been surfaced

TABLE 5.24 Aeolian vibration test (Case 5.2).

Fiber No		Attenuation before test-A (dB)		Attenuation before test-B (dB)		Change in attenuation B-A (dB)	
		1310 nm (A1)	1550 nm (A2)	1310 nm (B1)	1550 nm (B2)	1310 nm (B-A1)	1550 nm (B-A2)
1	13	0.085	0.06	0.104	0.082	0.019	0.022
2	14	0.073	0.058	0.099	0.087	0.026	0.029
3	15	0.09	0.074	0.114	0.095	0.024	0.021
4	16	0.084	0.058	0.098	0.081	0.014	0.023
5	17	0.077	0.061	0.108	0.089	0.031	0.028
6	18	0.098	0.08	0.109	0.102	0.011	0.022
7	19	0.072	0.059	0.095	0.079	0.023	0.02
8	20	0.084	0.064	0.11	0.091	0.026	0.027
9	21	0.071	0.053	0.088	0.082	0.017	0.029
10	22	0.078	0.066	0.097	0.085	0.019	0.019
11	23	0.081	0.069	0.107	0.097	0.026	0.028
12	24	0.074	0.057	0.094	0.081	0.02	0.024

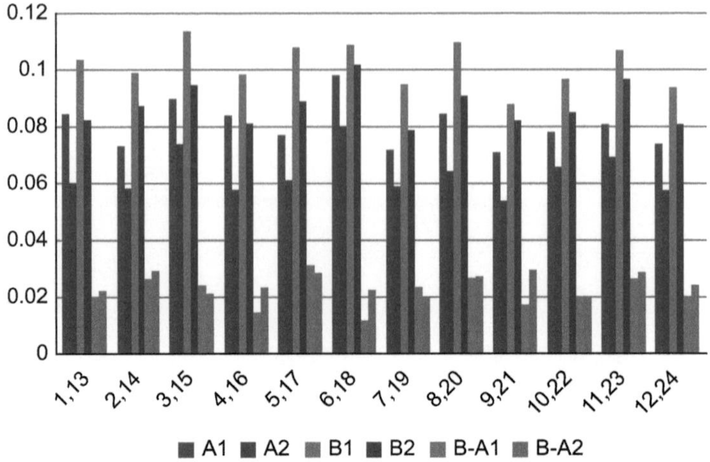

FIGURE 5.17 Fiber No vs Attenuation characteristics (Case 5.2).

TABLE 5.25 Best control variable settings for *FC* minimization objective.

Control variables	DE[1]	BBO[2]	GSA[3]	CAB[4]
Generator output				
$P_{G\text{-}1}$ (p.u.)	1.762592	1.770177	1.75749826	1.77699
$P_{G\text{-}2}$ (p.u.	0.485602	0.486410	0.48165537	0.46165
$P_{G\text{-}5}$ (p.u.)	0.213402	0.212390	0.21381724	0.22384
$P_{G\text{-}8}$ (p.u.)	0.220553	0.211360	0.21561405	0.19685
$P_{G\text{-}11}$ (p.u.)	0.117785	0.119440	0.12417360	0.13236
$P_{G\text{-}13}$ (p.u.)	0.120217	0.120540	0.12510199	0.12510
Generator voltage				
V_1 (p.u.)	1.0999	1.1000	1.086235	0.9984
V_2 (p.u.)	1.0890	1.0876	1.046685	1.0812
V_5 (p.u.)	1.0659	1.0614	1.035570	1.0328
V_8 (p.u.)	1.0697	1.0695	1.076962	1.0752
V_{11} (p.u.)	1.0965	1.0982	1.077452	0.9789
V_{13} (p.u.)	1.0996	1.0998	1.099999	1.0539
Transformer tap ratio				
$T_{6\text{-}9}$ (p.u.)	1.0429	1.05	0.939297	0.9784
$T_{6\text{-}10}$ (p.u.)	0.9179	0.90	1.006593	0.9932
$T_{4\text{-}12}$ (p.u.)	1.0190	0.99	0.907372	1.0218
$T_{28\text{-}27}$ (p.u.)	0.9836	0.97	0.921855	1.0352

TABLE 5.25 Best control variable settings for *FC* minimization objective.—cont'd

Control variables	DE[1]	BBO[2]	GSA[3]	CAB[4]
Capacitor bank				
$Q_{C\text{-}10}$ (p.u.)	0.045453	0.05	0.02190333	0.0190
$Q_{C\text{-}12}$ (p.u.)	0.044158	0.05	0.05000000	0.0159
$Q_{C\text{-}15}$ (p.u.)	0.041734	0.05	0.00000000	0.0016
$Q_{C\text{-}17}$ (p.u.)	0.025171	0.05	0.02715239	0.0281
$Q_{C\text{-}20}$ (p.u.)	0.020916	0.05	0.00000672	0.0060
$Q_{C\text{-}21}$ (p.u.)	0.041990	0.05	0.00000000	0.0428
$Q_{C\text{-}23}$ (p.u.)	0.025527	0.04	0.00000593	0.0190
$Q_{C\text{-}24}$ (p.u.)	0.043812	0.05	0.00000000	0.0251
$Q_{C\text{-}29}$ (p.u.)	0.027503	0.03	0.00000000	0.0500
FC ($/h)	799.2891	799.1116	798.675143	**798.3339**
P_{Loss} (MW)	8.6150	8.63	8.386049	8.279
TVD (p.u.)	NR*	NR*	0.872862	0.7861
CPU time (s)	NR*	11.02	10.7582	10.5262

*NR means not reported in the referred literature.
[1]*DE means DE [61],*
[2]*BBO means BBO [63],*
[3]*GSA means GSA [64],*
[4]*CAB means CAB (proposed).*

TABLE 5.26 Statistical analysis of simulation results for *FC* minimization objective for different techniques.

	FC ($/h)			
Methods	Minimum	Average	Maximum	CPU time (s)
MDE[1]	802.376	802.382	802.404	23.25
EGA[2]	802.06	NR*	802.14	76
IGA[3]	800.805	NR*	NR*	NR*
PSO[4]	800.41	NR*	NR*	NR*
EADDE[5]	800.2041	800.2412	800.2784	3.32
DE[6]	799.2891	NR*	NR*	NR*

Continued

TABLE 5.26 Statistical analysis of simulation results for *FC* minimization objective for different techniques.—cont'd

Methods	*FC* ($/h) Minimum	Average	Maximum	CPU time (s)
BBO[7]	799.1116	799.1985	799.2042	11.02
GSA[8]	798.675143	798.913128	799.028419	10.7582
CAB[9]	798.3339	798.2356	798.9689	10.5262

*NR means not reported in the referred literature.
[1]*MDE means MDE [59],*
[2]*EGA means EGA [57],*
[3]*IGA means IGA [56],*
[4]*PSO means PSO [58],*
[5]*EADDE means EADDE [62],*
[6]*DE means DE [61],*
[7]*BBO means BBO [63],*
[8]*GSA means GSA [64],*
[9]*CAB means CAB (proposed).*

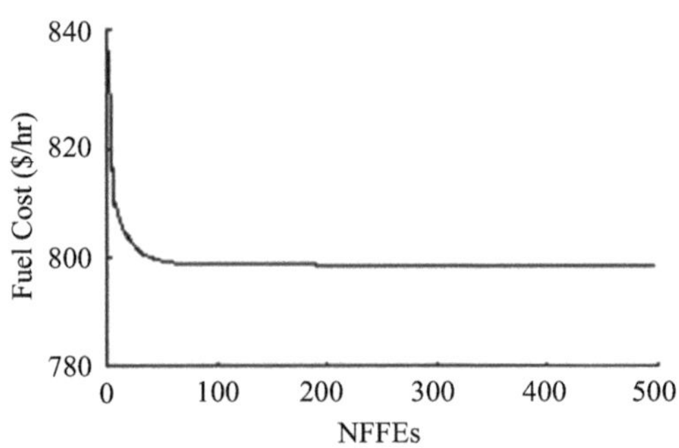

FIGURE 5.18 CAB based convergence profiles for fuel cost minimization objective.

TABLE 5.27 Best control variable settings for P_{Loss} minimization objective.

Control variables	PSO[1]	EGA-DQLF[2]	CAB[3]
Generator output			
$P_{G\text{-}1}$ (p.u.)	NR*	NR*	0.525540
$P_{G\text{-}2}$ (p.u.)	0.7906	0.80	0.789893
$P_{G\text{-}5}$ (p.u.)	0.50	0.50	0.500000
$P_{G\text{-}8}$ (p.u.)	0.35	0.35	0.341090
$P_{G\text{-}11}$ (p.u.)	0.2953	0.30	0.312300
$P_{G\text{-}13}$ (p.u.)	0.36134	0.40	0.397201

TABLE 5.27 Best control variable settings for P_{Loss} minimization objective.—cont'd

Control variables	PSO[1]	EGA-DQLF[2]	CAB[3]
Generator voltage			
V_1 (p.u.)	1.0	1.0435	1.0000
V_2 (p.u.)	0.996	1.04353	1.0424
V_5 (p.u.)	0.978	1.02470	1.0221
V_8 (p.u.)	0.98	1.03470	1.0357
V_{11} (p.u.)	1.032	1.07	1.0474
V_{13} (p.u.)	1.0415	1.043	1.0487
Transformer tap ratio			
$T_{6\text{-}9}$ (p.u.)	0.9	1.0375	0.9758
$T_{6\text{-}10}$ (p.u.)	1.0	0.925	0.9590
$T_{4\text{-}12}$ (p.u.)	0.95	0.975	0.9852
$T_{28\text{-}27}$ (p.u.)	0.9375	0.975	0.9710
Capacitor bank			
$Q_{C\text{-}10}$ (p.u.)	0.05	0.05	0.0497
$Q_{C\text{-}12}$ (p.u.)	0.05	0.03	0.0391
$Q_{C\text{-}15}$ (p.u.)	0.03	0.0	0.0367
$Q_{C\text{-}17}$ (p.u.)	0.04	0.01	0.0015
$Q_{C\text{-}20}$ (p.u.)	0.05	0.04	0.0489
$Q_{C\text{-}21}$ (p.u.)	0.02	0.02	0.0193
$Q_{C\text{-}23}$ (p.u.)	0.02	0.05	0.0409
$Q_{C\text{-}24}$ (p.u.)	0.06	0.05	0.0500
$Q_{C\text{-}29}$ (p.u.)	0.04	0.05	0.02910
FC ($/h)	956.45	967.86	966.3669
P_{Loss} (MW)	3.6294	3.2008	**3.202**
TVD (p.u.)	NR*	NR*	0.7872
CPU time (s)	NR*	NR*	9.4378

*NR means not reported in the referred literature.
[1]*PSO means PSO [60],*
[2]*EGA-DQLF means EGA-DQLF [60],*
[3]*CAB means CAB (Proposed) and.*

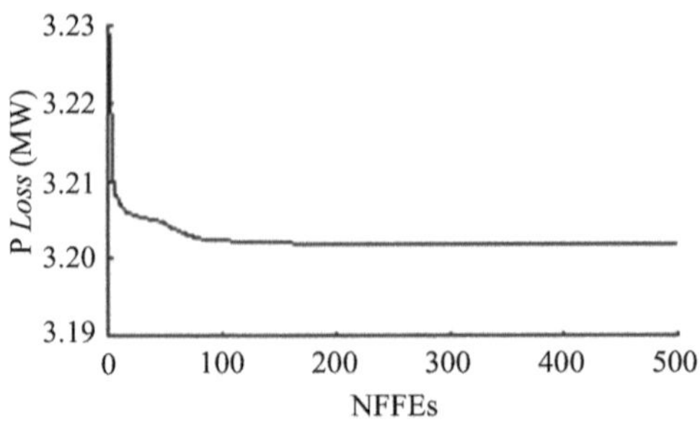

FIGURE 5.19 CAB based convergence profiles for P_{Loss} minimization objective.

on ADSS hardware validation to meet the robustness requirements of the SG. Even this hardware will be cost-effective in rapid deployment of high-speed Internet connectivity to end users. In this work, the OPF problem is expressed as a nonlinear optimization problem with equality and inequality constraints of the power network. One recently developed supervised algorithm like CAB has been successfully implemented to solve the complicated OPF problem of the modified IEEE 30-bus test system. As future work, we aim to validate more communication infrastructure hardware as well as examine new learning-based algorithms for effective planning, operation, and security of SGs.

Appendix

In this section, two drawings of smart grid networking hardware accessories (Figs. 5A and 5B) have been presented for assembly of suspension and tension fittings, respectively. The SLD of the undertaken test system for the OPF study is portrayed in Fig. 5C.

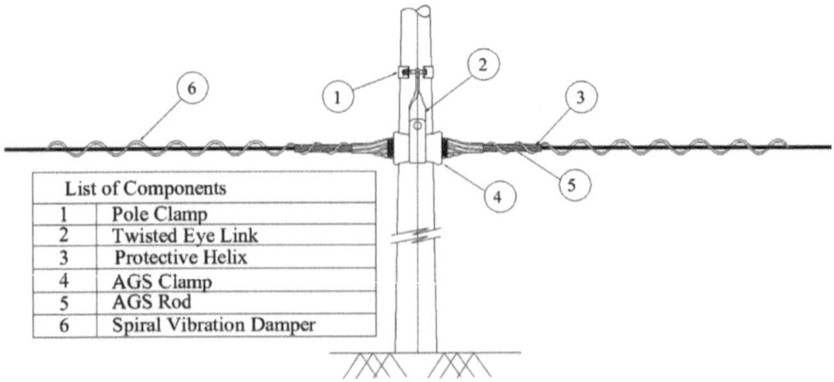

FIGURE 5A Schematic representation of ADSS suspension assembly.

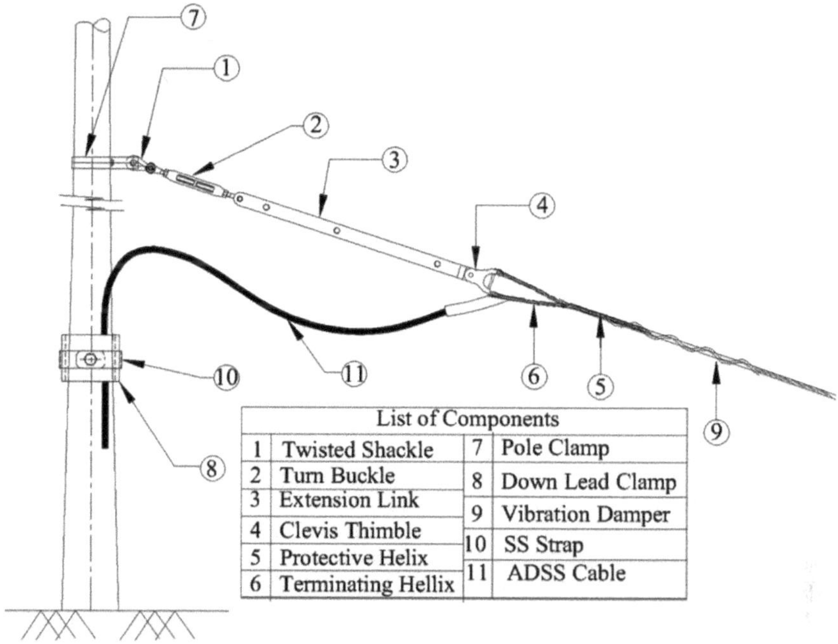

FIGURE 5B Schematic representation of ADSS tension assembly.

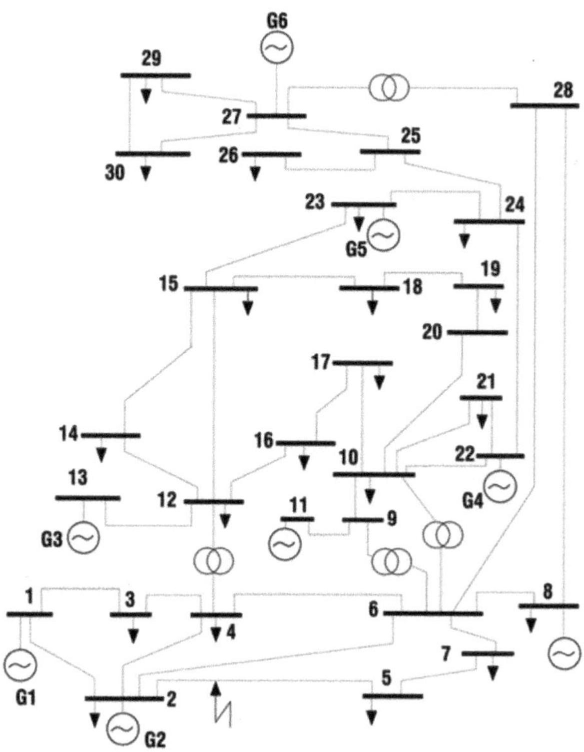

FIGURE 5C Single-line diagram of the modified IEEE 30-bus test power system.

Acknowledgments

The apparent part of the current research, mainly validation and compatibility study of ADSS cable networking infrastructure hardware, was based on the PGCIL work order. Any opinions and recommendations expressed in this article are those of the authors and do not necessarily reflect the views of the PGCIL authority.

References

[1] E. Ancillotti, R. Bruno, M. Conti, The role of communication systems in smart grids: architectures, technical solutions and research challenges, Comput. Commun. 36 (17–18) (2013) 1665–1697.

[2] H. Habibzadeh, T. Soyata, B. Kantarci, A. Boukerche, C. Kaptan, Sensing, communication and security planes: a new challenge for a smart city system design, Comput. Network. 144 (2018) 163–200.

[3] V. Glass, E. Woychik, E. Glassa, Software defined network communications: the likely standard for smart grids, Electr. J. 32 (9) (2019) 106639.

[4] S. Döbelt, M. Jung, M. Busch, M. Tscheligi, Consumers' privacy concerns and implications for a privacy preserving Smart Grid architecture—results of an Austrian study, Energy Res. Soc. Sci. 9 (2015) 137–145.

[5] E. Vahedi, M. Bayat, M.R. Pakravan, M. Reza Aref, A secure ECC-based privacy preserving data aggregation scheme for smart grids, Comput. Network. 129 (2017) 28–36.

[6] U.B. Baloglu, Y. Demir, Lightweight privacy-preserving data aggregation scheme for smart grid metering infrastructure protection, Int. J. Crit. Infrastruct. Protect. 22 (2018) 16–24.

[7] X. Jiajia, L. Liu, Z. Bao, Smart grid privacy protection using solar energy and rechargeable battery, Energy Proc. 158 (2019) 6170–6175.

[8] T. Dimitriou, T. Giannetsos, L. Chen, REWARDS: privacy-preserving rewarding and incentive schemes for the smart electricity grid and other loyalty systems, Comput. Commun. 137 (2019) 1–14.

[9] S. Sultan, Privacy-preserving metering in smart grid for billing, operational metering, and incentive-based schemes: a survey, Comput. Secur. 84 (2019) 148–165.

[10] J.D. Hmielowski, A.D. Boyd, G. Harvey, J. Joo, The social dimensions of smart meters in the United States: demographics, privacy, and technology readiness, Energy Res. Soc. Sci. 55 (2019) 189–197.

[11] L. Zhang, L. Zhao, S. Yin, C.-H. Chi, R. Liu, Y. Zhang, A lightweight authentication scheme with privacy protection for smart grid communications, Future Generat. Comput. Syst. 100 (2019) 770–778.

[12] W. Kong, J. Shen, P. Vijayakumar, Y. Cho, V. Chang, A practical group blind signature scheme for privacy protection in smart grid, J. Parallel Distr. Comput. 136 (2020) 29–39.

[13] P. Shakarian, J. Shakarian, A. Ruef, Cyber attacks by nonstate hacking groups: the case of anonymous and its affiliates, in: P. Shakarian, J. Shakarian, A. Ruef (Eds.), Introduction to Cyber-Warfare Syngress, Boston, 2013, pp. 67–110.

[14] E.D. Knapp, J.T. Langill, Industrial Network Security: Securing Critical Infrastructure Networks for Smart Grid, SCADA, and Other Industrial Control Systems, Syngress, 2014.

[15] N. Kilic, V. Cagri Gungor, Analysis of low power wireless links in smart grid environments, Comput. Network. 57 (5) (2013) 1192–1203.

[16] D. Sahin, V.C. Gungor, T. Kocak, G. Tuna, Quality-of-service differentiation in single-path and multi-path routing for wireless sensor network-based smart grid applications, Ad Hoc Netw. 22 (2014) 43–60.

[17] H. Elkhorchani, K. Grayaa, Novel home energy management system using wireless communication technologies for carbon emission reduction within a smart grid, J. Clean. Prod. 135 (2016) 950–962.

[18] M. Faheem, V.C. Gungor, MQRP: mobile sinks-based QoS-aware data gathering protocol for wireless sensor networks-based smart grid applications in the context of industry 4.0-based on internet of things, Future Generat. Comput. Syst. 82 (2018) 358–374.

[19] F. Muhammad, S.B.H. Shah, R.A. Butt, B. Raza, M. Anwar, M. Waqar Ashraf, Md A. Ngadi, V.C. Gungor, Smart grid communication and information technologies in the perspective of industry 4.0: opportunities and challenges, Comput. Sci. Rev. 30 (2018) 1–30.

[20] F. Al-Turjman, M. Abujubbeh, IoT-enabled smart grid via SM: an overview, Future Generat. Comput. Syst. 96 (2019) 579–590.

[21] A. Aydeger, N. Saputro, K. Akkaya, S. Uluagac, SDN-enabled recovery for smart grid teleprotection applications in post-disaster scenarios, J. Netw. Comput. Appl. 138 (2019) 39–50.

[22] E.P.K. Gilbert, M. Lydia, K. Baskaran, E.B. Rajsingh, Trust aware fault tolerant prediction model for wireless sensor network based measurements in smart grid environment, Sustain. Comput. Inf. Sys. 23 (2019) 29–37.

[23] H. Hui, Y. Ding, Q. Shi, F. Li, Y. Song, J. Yan, 5G network-based Internet of Things for demand response in smart grid: a survey on application potential, Appl. Energy 257 (2020) 113972.

[24] Z.A. Khan, Y. Faheem, Cognitive radio sensor networks: smart communication for smart grids-A case study of Pakistan, Renew. Sustain. Energy Rev. 40 (2014) 463–474.

[25] U.S. Premarathne, I. Khalil, M. Atiquzzaman, Secure and reliable surveillance over cognitive radio sensor networks in smart grid, Pervasive Mob. Comput. 22 (2015) 3–15.

[26] O. Ergul, A.O. Bicen, O.B. Akan, Opportunistic reliability for cognitive radio sensor actor networks in smart grid, Ad Hoc Netw. 41 (2016) 5–14.

[27] S. Althunibat, Q. Wang, F. Granelli, Flexible channel selection mechanism for cognitive radio based last mile smart grid communications, Ad Hoc Netw. 41 (2016) 47–56.

[28] A. Abuadbba, I. Khalil, A. Ibaida, M. Atiquzzaman, Resilient to shared spectrum noise scheme for protecting cognitive radio smart grid readings– BCH based steganographic approach, Ad Hoc Netw. 41 (2016) 30–46.

[29] E. Fadel, M. Faheem, V.C. Gungor, L. Nassef, N. Akkari, M.G.A. Malik, S. Almasri, I.F. Akyildiz, Spectrum-aware bio-inspired routing in cognitive radio sensor networks for smart grid applications, Comput. Commun. 101 (2017) 106–120.

[30] M.W. Khan, M. Zeeshan, QoS-based dynamic channel selection algorithm for cognitive radio based smart grid communication network, Ad Hoc Netw. 87 (2019) 61–75.

[31] K.A. Darabkh, O. Amro, H. Bany Salameh, R.T. Al-Zubi, AZ overview of the in-band full-duplex cognitive radio networks, Comput. Commun. 145 (2019) 66–95.

[32] Z. Guan, S. Guanlin, Achieving privacy-preserving big data aggregation with fault tolerance in smart grid, Digital Commun. Netw. 3 (4) (2017) 242–249.

[33] S. Marksteiner, H. Vallant, K. Nahrgang, Cyber security requirements engineering for low-voltage distribution smart grid architectures using threat modeling, J. Inf. Sec. Appl. 49 (2019) 102389.

[34] I.A. Kamil, S.O. Ogundoyin, A big data anonymous batch verification scheme with conditional privacy preservation for power injection over vehicular network and 5G smart grid slice, Sustain. Energy Grids Netw. 20 (2019) 100260.

[35] M.A.P. Chamikara, P. Bertok, D. Liu, S. Camtepe, I. Khalil, An efficient and scalable privacy preserving algorithm for big data and data streams, Comput. Secur. 87 (2019) 101570.

[36] P. Eder-Neuhauser, T. Zseby, J. Fabini, G. Vormayr, Cyber attack models for smart grid environments, Sustain. Energy Grids Netw. 12 (2017) 10–29.

[37] M. Attia, S. Mohammed Senouci, H. Sedjelmaci, E.-H. Aglzim, D. Chrenko, An efficient intrusion detection system against cyber-physical attacks in the smart grid, Comput. Electr. Eng. 68 (2018) 499–512.

[38] D. Velusamy, G. Kumar Pugalendhi, Fuzzy integrated Bayesian Dempster–Shafer theory to defend cross-layer heterogeneity attacks in communication network of smart grid, Inf. Sci. 479 (2019) 542–566.

[39] X. Wang, X. Luo, M. Zhang, X. Guan, Distributed detection and isolation of false data injection attacks in smart grids via nonlinear unknown input observers, Int. J. Electr. Power Energy Syst. 110 (2019) 208–222.

[40] L. Lee, P. Hu, Vulnerability analysis of cascading dynamics in smart grids under load redistribution attacks, Int. J. Electr. Power Energy Syst. 111 (2019) 182–190.

[41] N.Z. Bawany, J.A. Shamsi, SEAL: SDN based secure and agile framework for protecting smart city applications from DDoS attacks, J. Netw. Comput. Appl. 145 (2019) 102381.

[42] X. Luo, X. Wang, M. Zhang, X. Guan, Distributed detection and isolation of bias injection attack in smart energy grid via interval observer, Appl. Energy 256 (2019) 113703.

[43] X.G. Shan, J. Zhuang, A game-theoretic approach to modeling attacks and defenses of smart grids at three levels, Reliab. Eng. Syst. Saf. 195 (2020) 106683.

[44] J.-S. Chou, N.-T. Ngo, Smart grid data analytics framework for increasing energy savings in residential buildings, Autom. ConStruct. 72 (2016) 247–257.

[45] S. Wallace, X. Zhao, D. Nguyen, K.-T. Lu, R. Buyya, R.N. Calheiros, A.V. Dastjerdi, Big data analytics on smart grid: mining pmu data for event and anomaly detection, Big Data Princ. Paradigms 17 (2016) 417–429.

[46] A.A. Munshi, A.-R.I. Mohamed Yasser, Big data framework for analytics in smart grids, Elec. Power Syst. Res. 151 (2017) 369–380.

[47] T. Wilcox, N. Jin, P. Flach, J. Thumim, A Big Data platform for smart meter data analytics, Comput. Ind. 105 (2019) 250–259.

[48] C. Shang, F. You, Data analytics and machine learning for smart process manufacturing: recent advances and perspectives in the big data era, Engineering 5 (2019) 1010–1016.

[49] J. Holweger, M. Dorokhova, L. Bloch, C. Ballif, N. Wyrsch, Unsupervised algorithm for disaggregating low-sampling-rate electricity consumption of households, Sustain. Energy Grids Netw. 19 (2019) 100244.

[50] S. Rathore, J.H. Park, Semi-supervised learning based distributed attack detection framework for IoT, Appl. Soft Comput. 72 (2018) 79–89.

[51] T. Ahmad, H. Chen, Potential of three variant machine-learning models for forecasting district level medium-term and long-term energy demand in smart grid environment, Energy 160 (2018) 1008–1020.

[52] Y. Li, Y. Wang, L. Qi, B. Cheng, X. Jiang, S. Sun, Incremental semi-supervised learning on streaming data, Pattern Recogn. 88 (2019) 383–396.

[53] J.P. Seevers, J. Johst, T. Weiß, H. Meschede, J. Hesselbach, Automatic time series segmentation as the basis for unsupervised, non-intrusive load monitoring of machine tools, Proc. CIRP 81 (2019) 695–700.

[54] A. Arlandis, S. Ciriani, How firms interact and perform in the ICT ecosystem? Commun. Strat. 79 (2010) 121–141.

[55] J. Carpentier, Contribution a l'etude du dispatching economique, Bull. Soc. Fr. Elect. 3 (1) (1962) 431–447.

[56] L.L. Lai, J.T. Ma, R. Yokoyama, M. Zhao, Improved genetic algorithms for optimal power flow under both normal and contingent operation states, Int. J. Electr. Power Energy Syst. 19 (5) (1997) 287–292.

[57] A.G. Bakirtzis, P.N. Biskas, C.E. Zoumas, P. Vasilios, Optimal power flow by enhanced genetic algorithm, IEEE Trans. Power Syst. 17 (2) (2002) 229–236.

[58] M.A. Abido, Optimal power flow using particle swarm optimization, Int. J. Electr. Power Energy Syst. 24 (7) (2002) 563–571.

[59] S. Sayah, K. Zehar, Modified differential evolution algorithm for optimal power flow with non-smooth cost functions, Energy Convers. Manag. 49 (11) (2008) 3036–3042.

[60] M.S. Kumari, S. Maheswarapu, Enhanced genetic algorithm based computation technique for multi-objective optimal power flow solution, Int. J. Electr. Power Energy Syst. 32 (6) (2010) 736–742.

[61] El Ela, A.A. Abou, M.A. Abido, S.R. Spea, Optimal power flow using differential evolution algorithm, Elec. Power Syst. Res. 80 (7) (2010) 878–885.

[62] K. Vaisakh, L.R. Srinivas, Evolving ant direction differential evolution for OPF with non-smooth cost functions, Eng. Appl. Artif. Intell. 24 (3) (2011) 426–436.

[63] A. Bhattacharya, P.K. Chattopadhyay, Solution of optimal reactive power flow using biogeography-based optimization, Int. J. Electr. Electron. Eng. 4 (8) (2010) 568–576.

[64] S. Duman, U. Güvenç, Y. Sönmez, N. Yörükeren, Optimal power flow using gravitational search algorithm, Energy Convers. Manag. 59 (2012) 86–95.

[65] E. Cuevas, M. Gonzalez, D. Zaldivar, M. Perez-Cisneros, G. García, An algorithm for global optimization inspired by collective animal behavior, Discrete Dynam Nat. Soc. 2012 (2012).

[66] D.J.T. Sumpter, The principles of collective animal behaviour, Phil. Trans. Biol. Sci. 361 (1465) (2006) 5–22.

[67] K. Lee, Y. Park, D. Lee, Measuring efficiency and ICT ecosystem impact: hardware vs. software industry, Telecommun. Pol. 42 (2) (2018) 107–115.

[68] R.C. Basole, H. Park, B.C. Barnett, Coopetition and convergence in the ICT ecosystem, Telecommun. Pol. 39 (7) (2015) 537–552.

[69] Z. Ghassemlooy, S. Arnon, M. Uysal, Z. Xu, J. Cheng, Emerging optical wireless communications-advances and challenges, IEEE J. Select. Areas Commun. 33 (9) (2015) 1738–1749.

[70] R.W. Tkach, Scaling optical communications for the next decade and beyond, Bell Labs Tech. J. 14 (4) (2010) 3–9.

[71] National Telecommunications & Information Administration, http://www.ntia.doc.gov/blog/2013/household-broadband-adoption-climbs-724-percent.

[72] F. Travostino, M. Joe, G. Karmous-Edwards (Eds.), Grid Networks: Enabling Grids with Advanced Communication Technology, John Wiley & Sons, 2006.

[73] A. Nespola, S. Abrate, R. Gaudino, C. Zerna, B. Offenbeck, N. Weber, High-speed communications over polymer optical fibers for in-building cabling and home networking, IEEE Photon. J. 2 (3) (2010) 347–358.

[74] R.G. Olsen, Laboratory simulation of dry band arcing on all-dielectric self-supporting fiber optic cable near high voltage power lines, in: 1999 IEEE International Symposium on Electromagnetic Compatability. Symposium Record (Cat. No. 99CH36261), 2, IEEE, 1999, pp. 993–997.

[75] D.C. Sunkle, D.F. Christian, The effects of tension on vibration of all-dielectric self-supporting fiber optic cable, in: Proceedings of ESMO'95-1995 IEEE 7th International Conference on Transmission and Distribution Construction, Operation and Live-Line Maintenance, IEEE, 1995, pp. 197–203.

[76] D. Srinivasan, G.G. Karady, Development of the dry-band arc on all-dielectric self-supporting cables during laboratory tests, IEEE Trans. Power Deliv. 19 (4) (2004) 1746–1750.

[77] B. Mukherjee, WDM optical communication networks: progress and challenges, IEEE J. Sel. Area. Commun. 18 (10) (2000) 1810–1824.

[78] M. Gibbens, C. Gniady, B. Zhang, Towards eco-friendly home networking, Sustain. Comput. Inf. Syst. 11 (2016) 16–25.

[79] C.A. Roa-Sepulveda, B.J. Pavez-Lazo, A solution to the optimal power flow using simulated annealing, Int. J. Electr. Power Energy Syst. 25 (1) (2003) 47–57.

[80] Z.-L. Gaing, Constrained optimal power flow by mixed-integer particle swarm optimization, in: IEEE Power Engineering Society General Meeting, 2005, IEEE, 2005, pp. 243–250.

[81] R. Cozzens, CATIA V5 Workbook Release V5-6R2013, Sdc Publications, 2013.

[82] J.J. Connor, F. Susan, Fundamentals of Structural Engineering, Springer, 2016.

[83] C.J. Wong, M.D. Miller (Eds.), Guidelines for Electrical Transmission Line Structural Loading, American Society of Civil Engineers, 2009.

Chapter 6

Internet of Things for smart grid applications

P. Manoj, Y. Bhuvan Kumar, M. Gowtham, D.B. Vishwas, A.V. Ajay
Department of Computer Science and Engineering, NIE Institute of Technology, Mysuru, Karnataka, India

1. Introduction to the Internet of Things

The Internet of Things (IoT) is a system consisting of interconnected devices (computing devices), which provides connectivity with other systems such as mechanical and digital machines and other entities in the real world that are encapsulated with unique identification numbers capable of transferring data over the network using some protocols without human interactions. A number of smart gadgets and machines are now monitored and controlled using IoT protocols.

The enhanced IoT concept has developed as a result of the convergence of advanced technologies consisting of real-time analytics, emulations, machine learning, embedded systems, wireless networks, and automation. All of these should be combined to make the description accurate. Such IoTs integrate many of the sensors that contribute internally to contact and conduct other operations (data transfer).

1.1 Deployments in the real world

Intel's assembly manufacturing plant utilizes IoT to constantly catch enormous amounts of information among devices and frameworks for quicker basic leadership, directed reaction, and prescient results for upgrading hardware execution at lower cost while opening up new.

2. Internet of Things protocols

IoT communication protocols are communication modes that secure and safeguard the data sent or received by connected devices. Regularly, IoT gadgets are associated by means of the Internet Protocol (IP) system to the Internet. Gadgets—for example, Bluetooth and RFID—additionally permit nearby system administration of IoT gadgets. There is a distinction in power,

Advances in Smart Grid Power System. https://doi.org/10.1016/B978-0-12-824337-4.00006-0

range (domain), and measure of memory utilized in such a situation. In any case, the correspondence utilizing IP systems is relatively mind-boggling and requires more memory and force from IoT gadgets, whereas range is not an issue. Non-IP systems, on the other hand, need less force and memory yet have a range impediment. Human life is fully dependent on these innovations—for instance, electricity is one of the basic needs of life. To the extent IoT correspondence conventions or frameworks are concerned, it is conceivable to think about relying on a mix of both IP and non-IP systems. IoT convention types commonly can be isolated into two separate classes.

2.1 Internet of Things network protocols

To connect gadgets over the system, IoT arrangement conventions are utilized, typically over the Internet as a progression of correspondence conventions. Start-to-finish information correspondence is allowed throughout the system by utilizing the IoT arrangement convention hypertext transfer protocol (HTTP).

HTTP is the best example of an IoT network protocol. This protocol forms the basis of web-based data communication. When there much data to be written, it is the most common protocol used by IoT devices. Nevertheless, due to its size and effects on battery life, the HTTP protocol is not preferred or energy saving, and it has additional constraints.

Additive production/3-D printing is one of the HTTP use cases. It allows computers to connect network 3-D printers and print 3-D objects and predetermined prototypes of processes.

- Long-range wide-area network

The long-range wide-area network (LoRaWan) is a low-power convention. In both private and worldwide systems, LoRaWan dynamically joins battery-powered devices to the Internet. Savvy urban areas primarily utilize this correspondence convention where a huge number of gadgets work with less force and memory.

Shrewd road lighting is a LoRaWan IoT functional use case. Utilizing this convention, road lights can be connected to a LoRaWan portal. The entryway, as a result, interfaces with the cloud system to naturally screens the light force contingent upon the surrounding lighting, which decreases daytime power use.

- Bluetooth

One of the most regularly utilized short-distance correspondence conventions is Bluetooth, a standard remote information-moving IoT convention. This correspondence convention is secure and perfect for transmission between electronic gadgets for short-range, low-power, minimal efforts remotely. Bluetooth Low Energy is a Bluetooth low-vitality variation that reduces power utilization and assumes a significant role in interfacing IoT gadgets.

Bluetooth is generally utilized in savvy, wearable and other cell phones to share little sections of information without using high levels of energy and memory. Bluetooth offers usability and tops the IoT framework list of available conventions.

- ZigBee

ZigBee is an IoT convention that permits correspondence between brilliant items. It is generally utilized for home mechanization. ZigBee is utilized with applications that permit low-rate information to move across short separations progressively in regular modern settings.

The ZigBee communication protocol is used for street lighting and electric meters in urban areas and provides low power consumption. It is also used in smart homes and security systems.

The table below compares various protocols (Table 6.1).

2.2 Internet of Things data protocols

In order to connect low-power IoT devices, IoT data protocols are used. Such protocols enable user-side point-to-point contact with the hardware without any connection to the Internet. IoT data protocol communication can be via a wired or wireless network when we need to transfer data to other systems such as an IoT network between our devices. Some devices are equipped with sensors that regularly push data. Other sensors work with a pull model where information first needs to be requested. We can use several data protocols in IoT.

- Message queue telemetry transport (MQTT)

MQTT accumulates information from different electronic gadgets and encourages remote framework control, one of the most famous conventions for IoT applications. It is a buy-in/distribute convention running over transmission control protocol (TCP), which means it underpins occasion-driven trade of messages by means of remote systems.

MQTT is for the most part utilized in gadgets that are practical and require less energy and memory—for example, fire locators, vehicle sensors, shrewd watches, and applications for content-based information.

- Constrained application protocol (CoAP)

CoAP is a convention for Internet utility for devices that are compelled. The customer can send a solicitation to the server utilizing this convention, and the server can restore the reaction in HTTP to the customer. It utilizes the user datagram protocol for lightweight execution and reduces space utilization. The convention utilizes the efficient extensible markup language (XML) interchange's twofold information position.

TABLE 6.1 Comparison of various Internet of Things network protocols.

Sl. No	Standard/Protocol	Frequency (MHz)	Range	Data rates	Power usage	Cost
1	2G/3G	Cellular bands	Several miles	10 Mbps	High	High
2	LTE Cat 0/1	Cellular bands	Several miles	1–10 Mbps	Medium	High
3	NB-IoT	Cellular bands	Several miles	0.1–1 Mbps	Medium	High
4	Bluetooth 4.2 core specification	2400 MHz (ISM)	50.0–150.0 m (Smart/BLE)	1 Mbps	Low	Low
5	ZigBee 3.0 based on IEEE 802.15.4	2400 MHz	10.0–100.0 m	0.25 Mbps	Low	Medium
6	Z-Wave Alliance ZAD12837/ ITU-T G.9959	900 MHz (ISM)	30.0 m	0.0096/0.04/0.1 Mbps	Low	Medium
7	6LoWPAN/RFC6282	Bluetooth smart (2400 MHz) or ZigBee (sub-1000 MHz)	NA	NA	Low	Low
8	Thread, based on IEEE 802.15.4 and 6LoWPAN	2400 MHz (ISM)	NA	NA	Low	Low
9	Wi-Fi based on 802.11n	2400 and 5000 MHz bands	~50.0 m	600 Mbps (MAX)/ 150–200 Mbps	Medium	Low
10	NFC (ISO/IEC 18000-3)	13.56 MHz (ISM)	0.10 m	0.1–0.42 Mbps	High	Low

The CoAP convention is principally utilized in robotization, versatile, and small-scale controller applications. The convention sends a solicitation to the endpoints of the application—for example, home apparatuses—and returns the reaction of the application's administrations and assets.

- Advanced message queuing protocol (AMQP)

AMQP is a center product condition programming layer convention that gives direction and lining. It is utilized for highlight-point solid association and supports consistent and secure information trading between associated gadgets and the cloud. AMQP comprises three distinctive exchange, message queue, and binding components. These three components guarantee that messages are shared and put away safely and productively. It likewise sets up the connection between one message and another.

The AMQP convention is utilized primarily in the financial industry. The convention records the message each time a message is sent by a server until each message is sent to the expected clients/goals without disappointment.

- Machine-to-machine (M2M) communication protocol

M2M is an open industry convention intended to give remote control to an IoT framework application. The conventions of M2M correspondence are financially savvy and utilize open systems. This makes for a domain where two machines associate and trade information. This convention encourages PC self-observation and causes gadgets to adjust to their evolving atmosphere.

M2M contact conventions are utilized in smart homes, mechanized vehicle security, candy machines, and ATMs.

- Extensible messaging and presence protocol (XMPP)

XMPP is designed to be exclusive. It uses a push method for real-time exchange of messages. XMPP is versatile and can be implemented smoothly with changes. Developed using free XML, XMPP acts as an indication of the availability status of servers or computers that send or receive messages.

An addition to instant messaging applications like Google Talk and WhatsApp, XMPP is also used in online gaming, news websites, and Voice over Internet Protocol (Table 6.2).

3. Implementation

Renewable energy (RE) generation is extremely volatile and variable, unlike conventional energy sources. Large-scale renewable generation integration requires a special balancing system to resolve instability and volatility to maintain grid stability and safety. In recent decades, electricity demand has increased exponentially with advancements in technology and new scientific concepts. Certain losses and fatal accidents have occurred in this field, and moreover while the majority of income has come from the electricity department, the

TABLE 6.2 Comparison of various IoT data protocols.

Feature	HTTP	MQTT	CoAP	Modbus TCP
Transport layer	TCP	UDP	TCP	TCP
Transport port	80,443	5683	1833, 8883	502, 802
Network layer	IPv6/IPv4	IPv6	IPv6/IPv4	IPv6/IPv4
Infrastructure	Ethernet, wireless fidelity	6LoWPAN	Ethernet, wireless fidelity	Ethernet, wireless fidelity
Model	Synchronous	Asynchronous	Asynchronous	Synchronous
Quality level	One level	Two—CON or NON	Three—QoS 0,1,5	One level
Encoding	ASCII text	RESTful	UTF-8	Binary
Security	SSL,TLS	DTLS	SSL,TLS	TLS
Mechanism	One-to-one	One-to-one	One-to-many	One-to-one

CoAP, constrained application protocol; *HTTP*, hypertext transfer protocol; *MQTT*, message queue telemetry transport; *TCP*, transmission control protocol; *UDP*, user datagram protocol.

sector has not been upgraded to an efficient extent. In addition to the introduction of capacity and electrification in remote areas, microgrid technology also needs attention. Successful integration of such large-scale RE projects requires the active participation of all stakeholders—i.e., from government agencies to NGOs, from producers to R&D organizations, from financial institutions to developers—and of course, a new breed of energy businesses.

The seamless integration of emerging technologies in the fields of monitoring, automation, control, communication, and IT systems with the active involvement of all stakeholders is inevitable for addressing these issues and delivering performance. In this direction, a smart grid that combines 21st-century technologies and the 20th-century electrical grid has the ability to revolutionize the network of electricity supply and provide greater access to electricity.

Each global smart grid driver applies to India, but it has additional drivers as follows:

For utilities:

- Reduction of transmission-and-distribution losses in all utilities as well as increased collection capacity
- Peak load management—multiple options from direct load monitoring to consumer price discounts
- Reduction of power purchase costs
- Better asset management

- Increased system reliability
- Self-healing grid
- Renewable incorporation

For customers:

- Extend access to energy—“energy for all”
- Improve supply efficiency for all customers—less or no power cuts, less or no more distributed-generation sets and inverters
- Improve supply quality
- User-friendly and clear utility interface
- Increased consumer options, including green power
- “Prosumer”(producers and consumers) help
- Save money by moving loads

For government and regulators:

- Satisfied consumers
- Financially sound utilities
- Grid redesign and improvement
- Reduction in pollution intensity

Significant obstacles occur in implementation of the transition from a current grid with its one-way power flows from central generation, to dispersed loads, to a modern two-way power flow infrastructure, two-way and peer-to-peer customer interactions, and distributed generation thermal power generation.

- **Interoperability:** As the smart grid covers the globe and many participants are interested in the worldwide development and deployment of smart grids, interoperable specifications need to be implemented by the network.
- **Cybersecurity:** Information and communication infrastructures must play a key role in linking and leveraging the network layers available. Smart-grid architecture will help to withstand unwanted physical and cyberintrusions and protect customer privacy.
- **Communication technology**: For the smart-grid data communication network, various technologies are being implemented. Issues of concern include costs, contact success rate, and technical issues, viz. electromagnetic interference, signal-to-noise ratio, baud rate, and parity check.
- **Social Barriers:** Before end users participate in the program, no program will be effective. Consumers are concerned about the issue of the smart grid. Awareness programs are required and they must provide clear benefits.
- **Political Barriers:** Because initial implementation of the system entails high costs, managers, policy-makers, and implementers must be persuaded of returns on investments and end results, including interruptions of supply to customers during program implementation. Approximately “32,000 smart homes and businesses are at risk of leaked data” according to an article published by *Forbes*. Thus, exploring the potential of IoT protocols

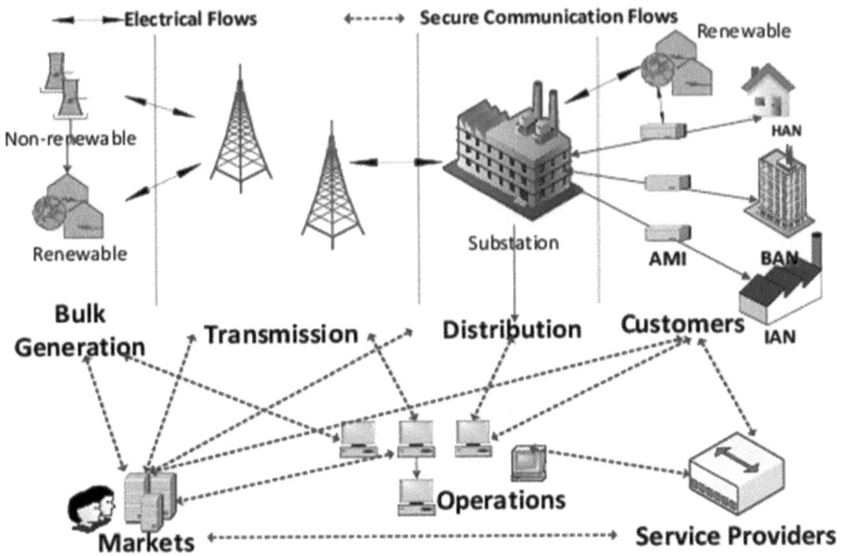

FIGURE 6.1 Map of the power grid from the place of the generation to place of deployment.

and standards, which provide a secure environment, is essential. These protocols can communicate and exchange data with the cloud via local gateways and other connected devices (Fig. 6.1).

This is a simple diagram of the power grid from the generation to distribution site. Due to advances in technology, the use of smart systems in every platform is increasing rapidly. Such advanced systems provide technicians, supervisors, and managers with access to track performance and control of smart devices. Since these systems are clubbed together with multiple devices, all of them can be operated by single users on their own in an efficient manner.

This chapter's main objective is to provide an effective and important way to detect disconnected areas of the grid's main supply. This proposed system is called intelligent detection. Smart detection is designed to track the current and voltage ratio of three-phase supply and send the status to respective substations where the error is detected and corrected if there is any voltage drop in the grid. Currently we come across two types of solutions for the above problem:

- pole-to-pole detection
- spur line monitoring

3.1 Pole-to-pole detection

The essential components used in this solution are

1. **Arduino Nano**: This is the intermediate component used for programming and is the place where all other components are interlinked and the actual task or specific functional performance takes place (Fig. 6.2).
2. **Voltage sensors:** This module is used to detect incoming voltage and sends data for further processing to the Arduino (Fig. 6.3).

FIGURE 6.2 Arduino nano.

FIGURE 6.3 Voltage sensor.

FIGURE 6.4 Current sensor.

3. **Current sensors**: This module is used to detect the voltage input and sends data to the Arduino for further processing (Fig. 6.4).
4. **Global system for mobile communications (GSM) module:**

This is used to send the message over the network to the receiver and receive the data/message from the sender, and can serve as both the sender and

FIGURE 6.5 Global system for mobile communications (GSM) module.

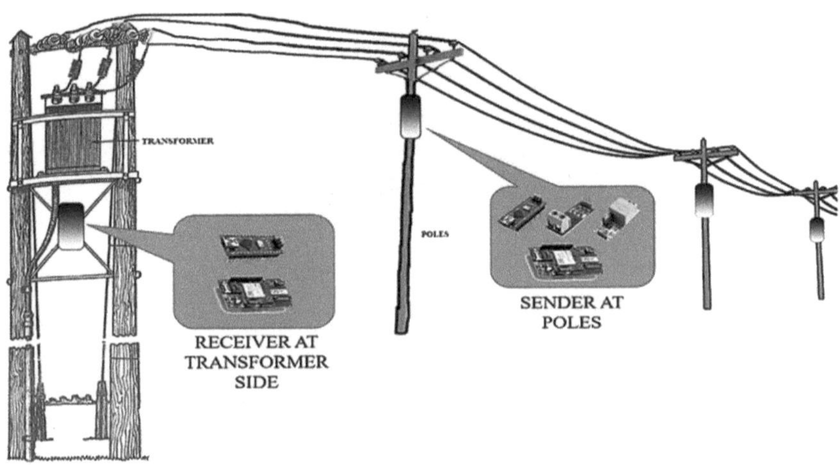

FIGURE 6.6 Model of pole-to-pole detection.

receiver. Pole-to-pole detection with this device is primarily suggested for detecting shortcomings that may have been triggered by certain environmental or human actions between the poles. There are two sections of this device:

- receiver
- sender

The sender component is the IoT installed within the poles consisting of voltage and current sensors that track the voltage and current ratios and send data back to the receiver, which is the GSM module mounted onto the transformer that communicates with the servers at the substations (Figs. 6.5 and 6.6).

Initially, the transformer is connected to the first pole and then followed by another *n* poles. Each pole gives a certain reading for the voltage and current sensor values, poles are connected in series, and the previous pole's voltage and current readings are transferred to the successive poles, and so on.

With any discontinuity among poles, the pole affected sends its current V (voltage) and A (current) readings to the previous poles until they reach the transformer. It sends a notification to the substation through the GSM module, but this causes a lengthy delay in sending back readings as they must travel across each pole. So we came up with an advancement to the above technique, where all the poles are connected to the centralized transformer—like star topology in a computer network. Any distortion in the flow of supply for that particular pole will pass on the readings directly to the transformer, which intern notifies the substation.

3.1.1 Tension and conductor icing

The mechanical loadings on an overhead line result from conductor strain, wind, ice, and temperature variations. Extreme (or breaking-point) loads can be assessed by the quality of parts. Conductor icing can be determined by the results of micrometeorology and tension sensors. The data analysis system at the monitoring center analyzes the information collected by the sensors and makes early-warning decisions, allowing for ice flashover to be alleviated or avoided.

3.2 Transformer protection

The transformer comprises temperature, voltage, and current sensors as well as GSM modules. The temperature sensor measures the temperature of transformer oil (mineral oil), which when below the threshold level harms the transformer. When the level of the transformer oil is reduced, the sensor sends a notification to the substation through the GSM module.

Normally the transformer receives about 480 V, 60 A of power from a power station which is then stepped down to 240 V, 50 Hz, 6.6 A. The range of acceptance for the transformer is between 460 and 500 V (approximation based on the type of transformer used).

For sensing voltage, current, and temperature sensor values, the respective sensors are driven by the microcontroller (Arduino or Raspberry Pi), and with any distortions from the standard acceptance range, the power supply is automatically shut down and a notification is sent to the respective power substations through the GSM module.

3.3 Spur line monitoring

As we know, after reaching the substation, the supply lines emerge as a single path called the trunk line. Whenever a short-circuit problem arises in the trunk line, the substation can be notified, but this trunk line is further divided into multiple spur lines (this is the most common model used in rural areas), which are further subdivided into residual lines (Fig. 6.7).

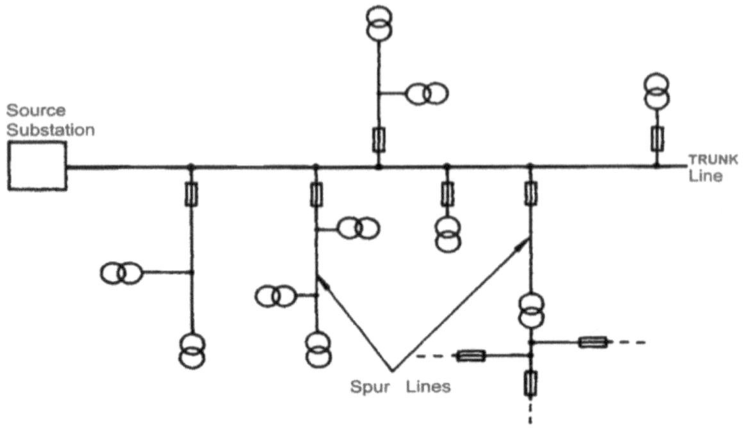

FIGURE 6.7 Spur line monitoring.

In this approach, we insert a voltage monitoring system in the spur line that tests the range of incoming voltage and compares it with the preset value and sends an error message (if the voltage is less or greater than the actual value) to the substation. This method includes only one component that reads and tests the voltage. The proposed system consists of the Arduino, voltage sensor, and GSM module to transmit the message to respective servers at the substation.

In this system, we will set up a voltage range so that any distortion due to wind should not be considered a defective line, and the received message will be checked three times over to determine whether the lines are actually faulty (Figs. 6.8–6.13).

4. Smart grid applications

The sender comprises smart meters, and in the sense of power network and communication system upgrades, developments in the smart grid have also accelerated development and improvements in smart metering. Such advances have been prompted by criteria for tracking and measuring improved technology for sensors. It is recognized that the criteria for connectivity and measurement play a vital role in developing the smart grid as well as the novelties of the power network. Smart meters are based on voltage and current measurements—i.e., watt-hour meters. Because of overlapping networks of sensors, the measured and conditioned signals are converted to digital signals to generate the measurement data used for storage and delivery to remote monitoring and operating centers. Therefore, by using two-way communication and power transfer, remote monitoring, calculation, control, and decision-making can be accomplished. Using a smart meter user interface, consumers can track and control their smart household appliances.

Advanced smart meters enable users to handle recloses, switches, and relays for controlling or running devices for efficient energy consumption. Smart

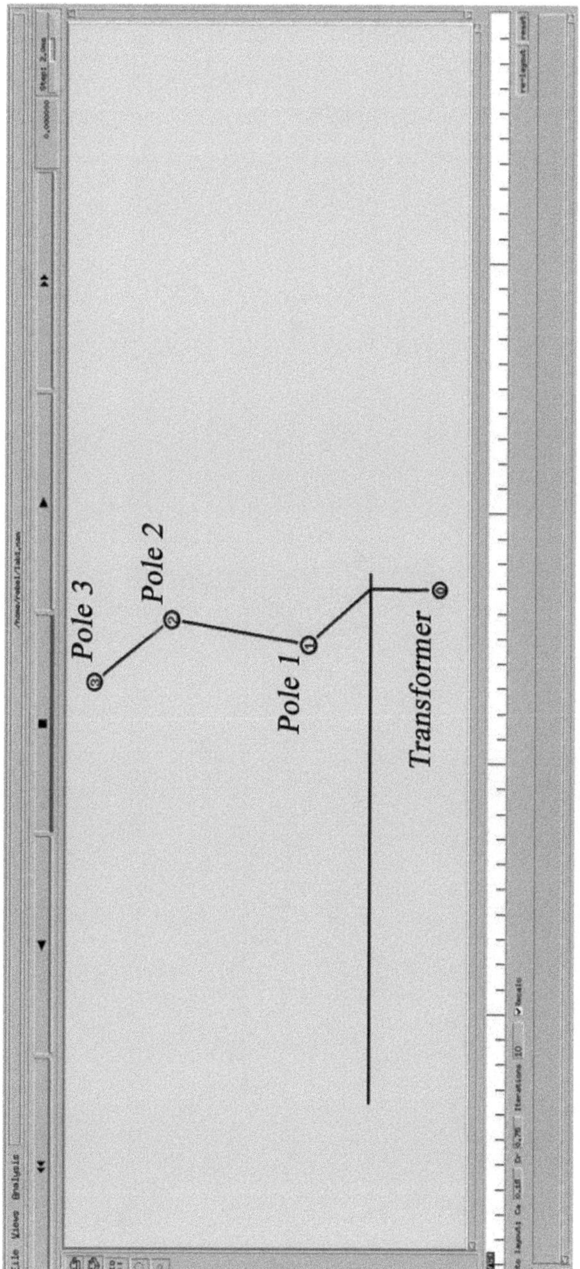

FIGURE 6.8 Setup of a typical smart grid.

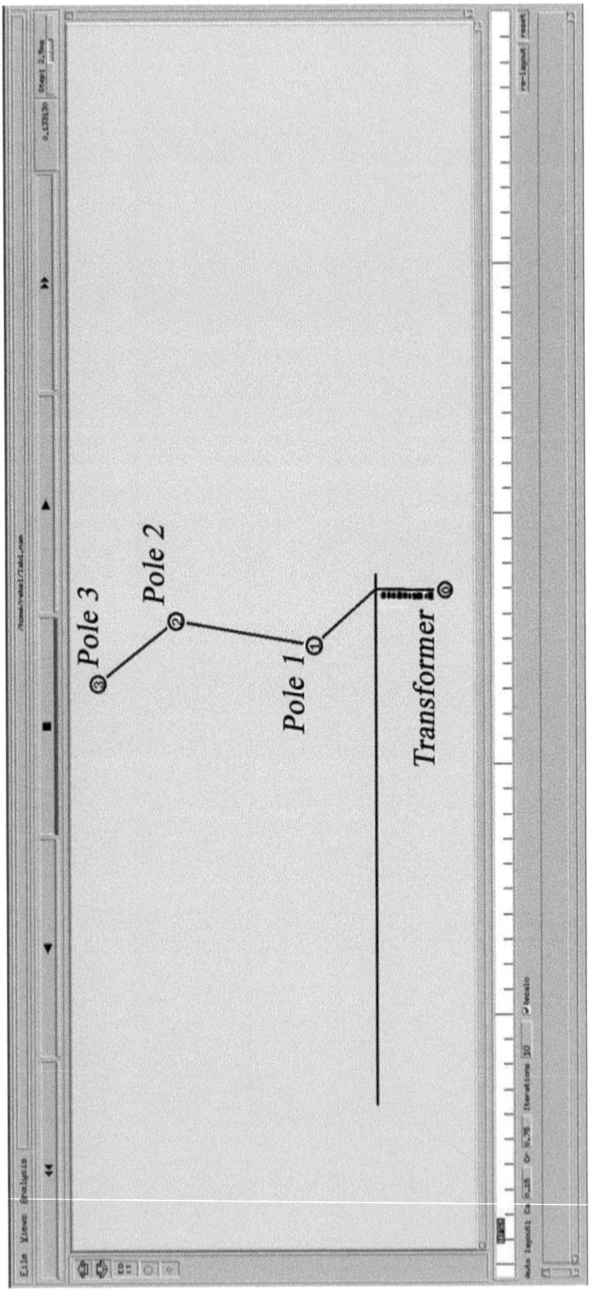

FIGURE 6.9 Supply from the transformer to respective poles.

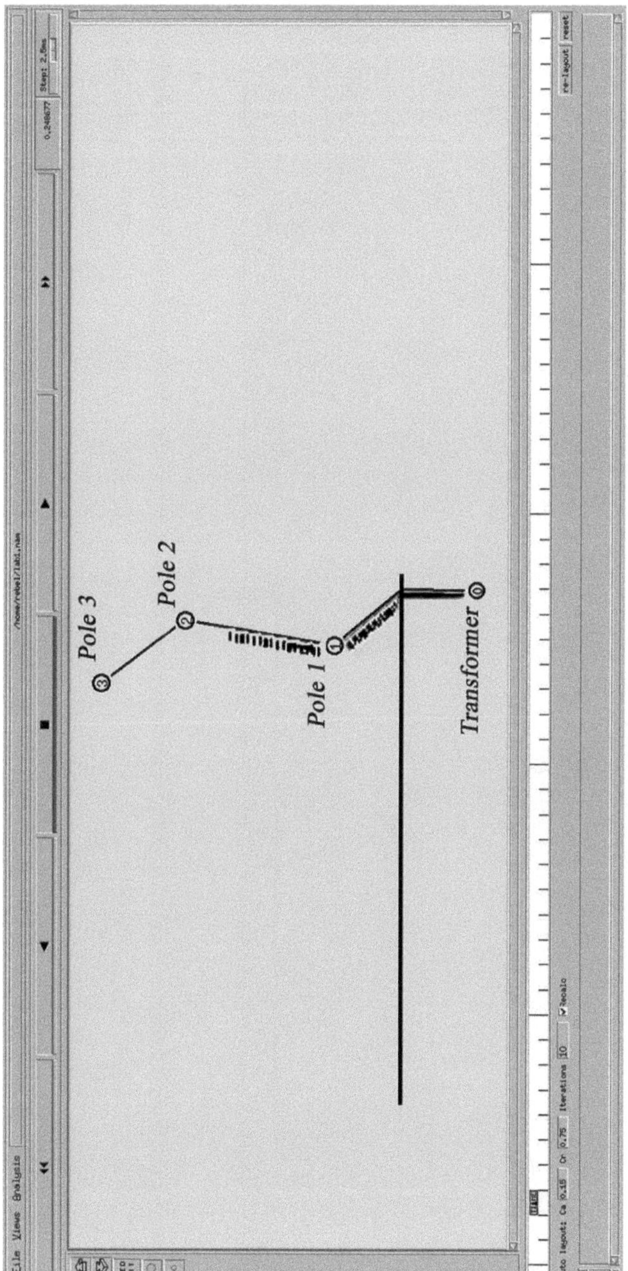

FIGURE 6.10 Successive supply from the transformer to Pole 1 without leakage.

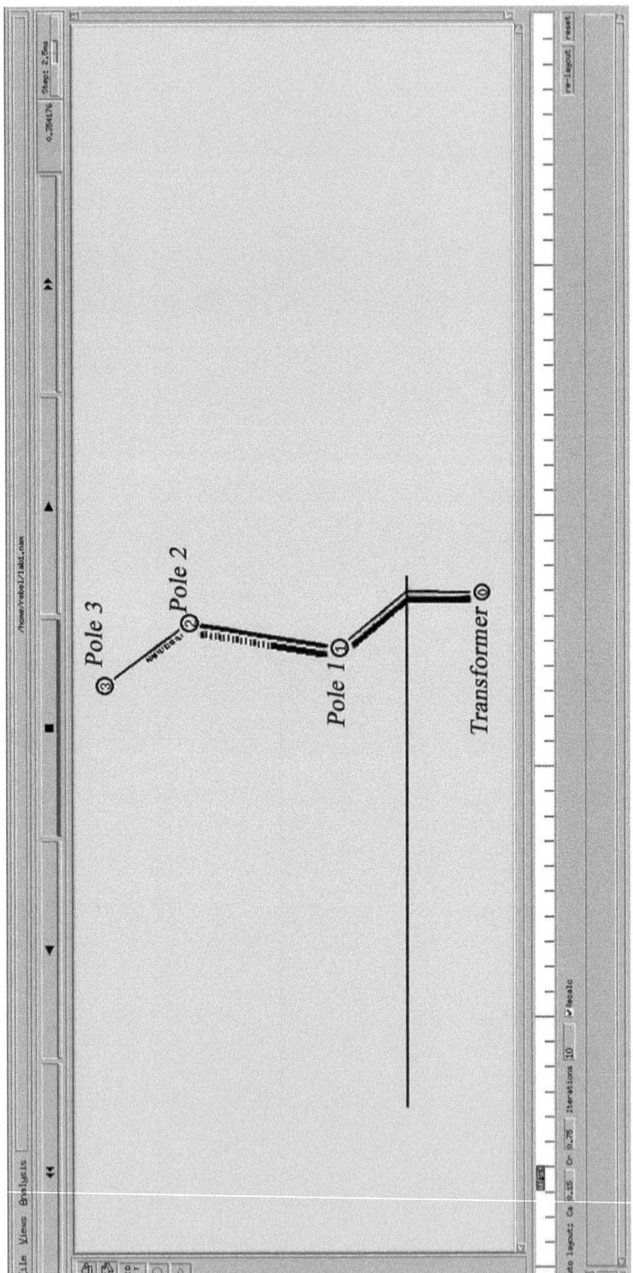

FIGURE 6.11 Successive supply from Pole 1 to Pole 2 without leakage and followed by successive poles.

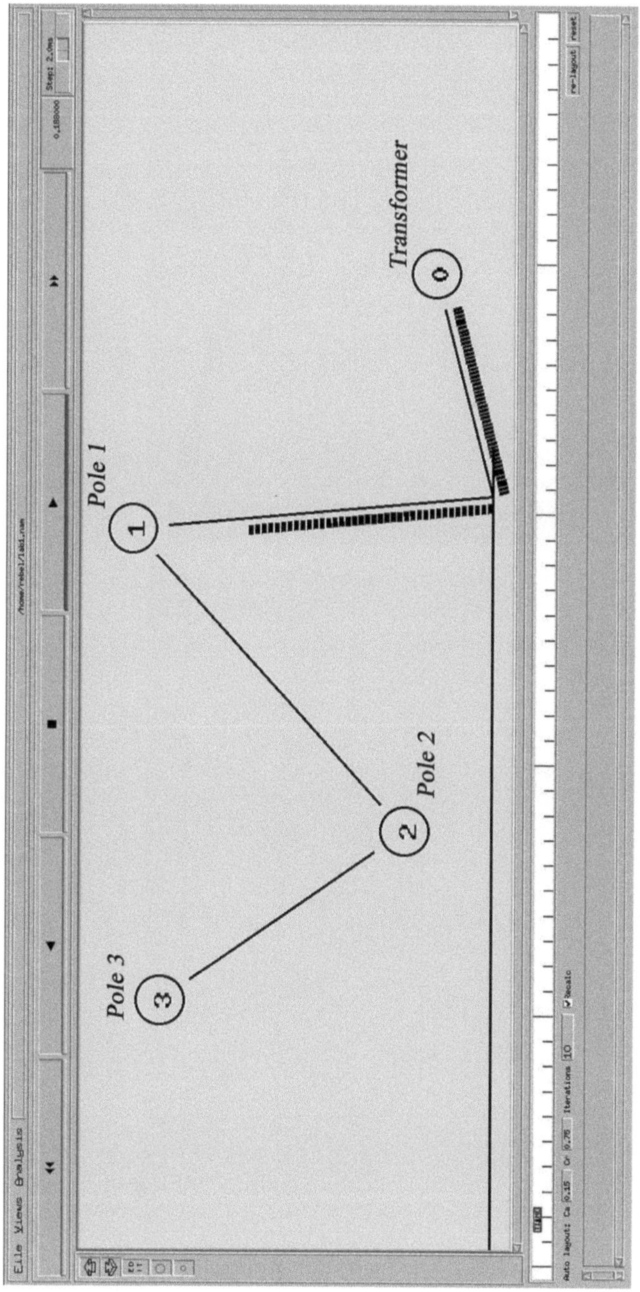

FIGURE 6.12 Successive supply from transformer to Pole 1.

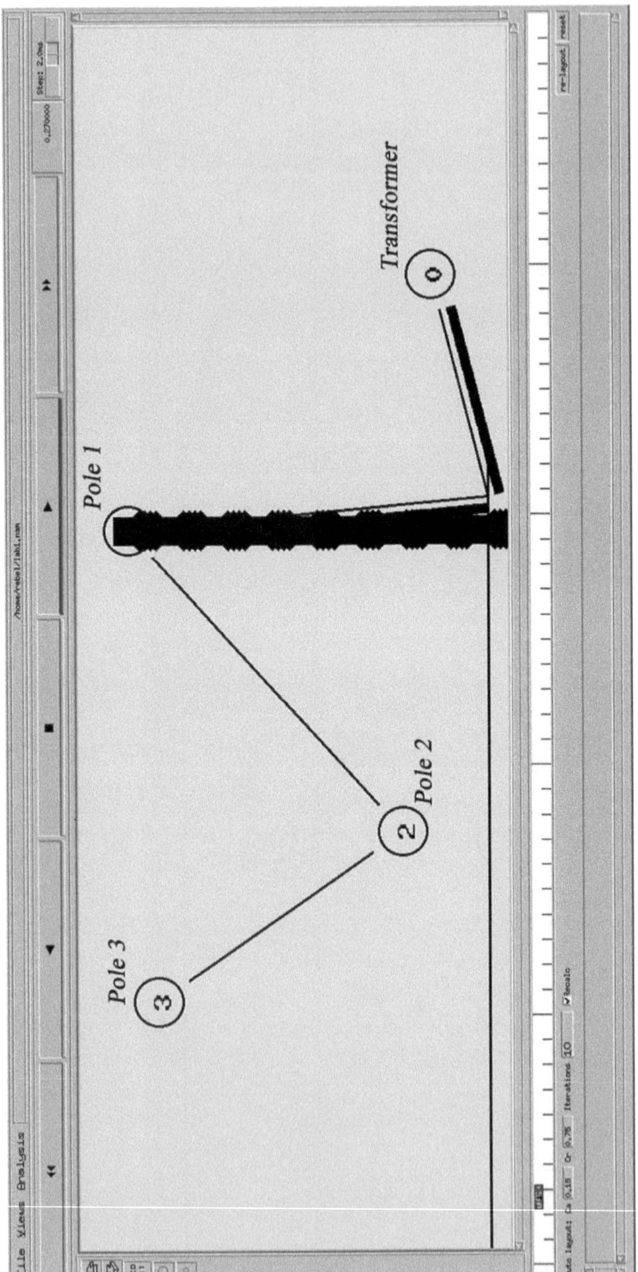

FIGURE 6.13 Defective Pole 1 causes current leakage.

meters are primarily responsible for calculating energy consumption for the entire system with which they are associated, storing data on consumption, and continuously transmitting the stored data to memory. Smart meters provide customers with a number of benefits and management incentives. For example, smart meters can easily detect or forecast outage situations and provide operators and customers with restore services. On the other hand, smart meters identify and organize priorities for everyday use and provide customers with cost savings. Smart meters provide secure and reliable communication in terms of integrity, authenticity, and privacy for transmission links between customers and electrical boards.

Smart meters provide every node with completely efficient control of the entire distribution system. Smart meters can provide daily and 10- to 15-minute intervals of consumption logs. Then again, for handing off or changing the smart grid at the hub, moment access to shrewd meters is likewise accessible. Almost all ongoing savvy meters are fitted with in—home showcases and web-based UIs to enable purchasers to control immediate vitality utilization. In this way, by decreasing wasted utilization, customers have an incentive to save money.

The gadgets produce yield signals that are perfect for any chip's ADC association. In the latest smart meters, stage and recurrence location capacities are also included.

The chip procures estimation information through its ADC ports and afterward stores that information in the inner database. A microchip with sequential and parallel correspondence interfaces likewise handles the information interfaces. Notwithstanding Ethernet and CAN transport remote correspondence capacities, most shrewd meter microchips are outfitted with infrared and RFID correspondence interfaces.

The systems are utilized for charging, following, and DSM purposes by the electrical board, while the home territory organizer offers access to brilliant family apparatuses, Internet passages, and home vitality control frameworks. It is centered on the transmission of information utilizing existing electrical cables as methods for transmission. Notwithstanding the two-way power transmission, electrical cables are fit for giving two-way information transmission.

For checking and control focus information, the control interface is essential for the electrical board to convey information about charging, the board, and control tasks. The control framework makes up the metering framework for information controls (Fig. 6.14).

Demand response in the literature involves numerous studies and analysis that have been published on scalable and robust utility networks. There has also been much research on improving the versatility of the smart grid. This applies to controlling and improving consumption habits to improve grid efficiency. There is no motivation or payback potential for traditional grid applications. Under these conditions, consumers are not aware of efficient energy use, and to maintain the balance between generation and usage, the

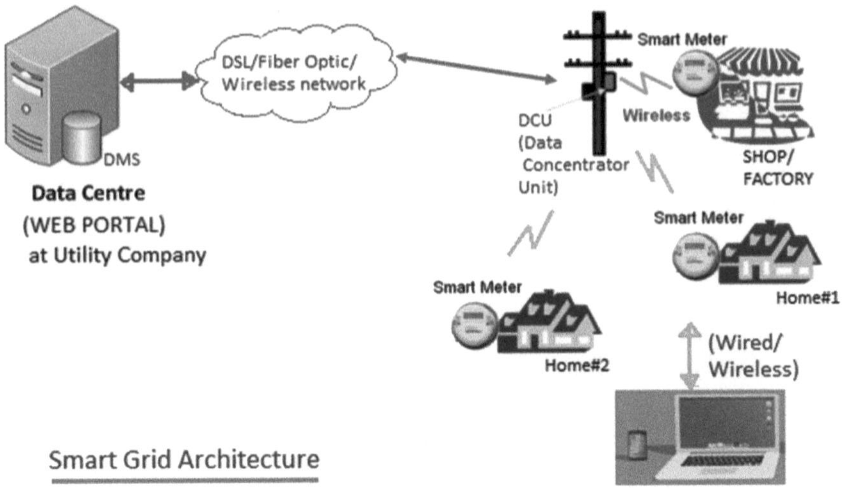

FIGURE 6.14 Smart grid architecture.

electrical board is required to control the power grid. Both services, however, contribute to price-based grid management in the smart grid.

- During peak hours or other periods, the first group of consumers will not change their behavior.
- The second group shifts their peak demand times to off-peak hours when energy costs during peak-load periods become quite high.
- The third group includes a few customers who, at peak-load times, change their energy usage times.

DR incentive programs are capable of controlling customer loads in peak hours or emergencies, and by doing so, consumers benefit. The profiles of load controls are direct load controls.

The three rule operations are based on rice-based demand programs. On the other hand, through incidents or causative conditions, all forms of incentive-based strategies are triggered and can be disabled at any time when necessary. DR methods based on prices are triggered on a regular basis.

The upgraded appropriation arrangement incorporates various sustainable plants and loads with new innovations—for example, electric vehicles, savvy gadgets, and interfaces for correspondence. In modern and local locations, a high number of IEDs are coordinated into the smart network framework.

Improving the savvy network requires various changes in substation structure and arrangements to incorporate numerous hubs—for example, transmission lines, transformers, generators, compensators, and loads. The substations assume an essential job in broad region following, cooperating, directing, and keeping up the energy organizing. The robotization framework is in this way vigorously associated with the substations in substations (Fig. 6.15).

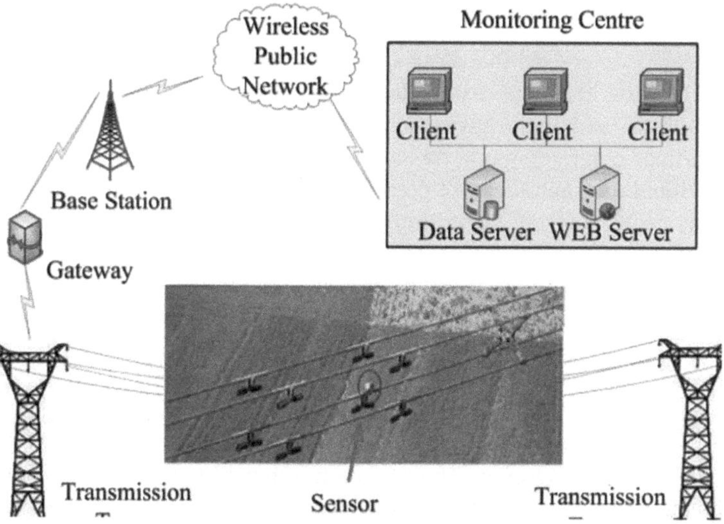

FIGURE 6.15 Data-transmission network.

TABLE 6.3 Overall sensors and the hardware used in this project.

Component	Hardware
Voltage sensor	Single-phase AC voltage sensor circuit
Current sensor	Single-phase AC current sensor circuit
Wireless communication	HC-5 Bluetooth
Microcontroller board	Arduino nano

The data-transmission network includes the meteorological sensor, vibration sensor, ultrasonic sensor, tower-leaning sensor, infrared sensor, leakage current sensor, camera, and backbone node of power transmission line and tower-monitoring system.

In this case, for the sake of connectivity between the sensors and GSM module, we can have several protocols such as Bluetooth, Wi−Max, ZigBee, etc. We can pick ZigBee protocols for communication, as it ensures a better rate of transmission.

Similarly, the same process takes place in the receiver part of the transformer, which sends the message to the server whenever it receives the "0" status (Table 6.3).

5. Cybersecurity

Cybersecurity is one of the most significant themes that ought to be incorporated because the majority of IoT gadgets are hacked in light of minor vulnerabilities, and if the programmer becomes acquainted with any of the weak parts of our framework, he can break into our system and could get all the individual data and likewise control the information on the framework. To avert the occurrence of these things, it is smarter to have security in the framework itself.

Offering security to organization is not as natural as it looks, yet there is a way to utilize the idea of blockchain, which involves hashing the information in its individual calculation. Initially, individuals utilize blockchain for exchanging digital forms of money. It is important as a record where singular individuals make sections with the assent of different individuals. The information gets put away in various squares over the chain. In this record, it is difficult to make the squares. There is a ton of handling and time to make even one square. There is a reference to the past passages in each square and along these lines, it shapes a chain.

On the off chance that a programmer attempts to get into the system, the blockchain security stops him by hashing the information into one structure that is encoded by some other structure, and this goes about as squares of chains that are entomb-connected to each other, and if the programmer attempts to split the first hashing, he would be propositioned by another, and this proceeds until a specific key is utilized. So this is the most precise approach to shield our framework from cyberattacks.

5.1 Blockchain-based secure mechanism for security with smart grid

Remote frameworks are often powerless against various ambushes, and along these lines, necessities rise to confirm their general circumstances (Biswas, Kamanashis, and Vallipuram Muthukkumarasamy). The blockchain innovation is one of the indisputable and first-class strategies for use in consolidating security with remote frameworks to approve a more conspicuous degree of security and by-and-large execution. Blockchain implies world-class security and careful innovation wherein a propelled record is maintained. The automated record is extremely clear, and there is no degree of control in the records by the intermediates or any executive. The records of the significant number of trades are marked in the blockchain record, and the exercises are submitted last with different displays and computations that can't be hacked by the outcast interferences [1].

In front-line remote circumstances, general execution is noteworthy and required with the objective that the general framework condition can be checked. The work shows the use of blockchain-based security with remote

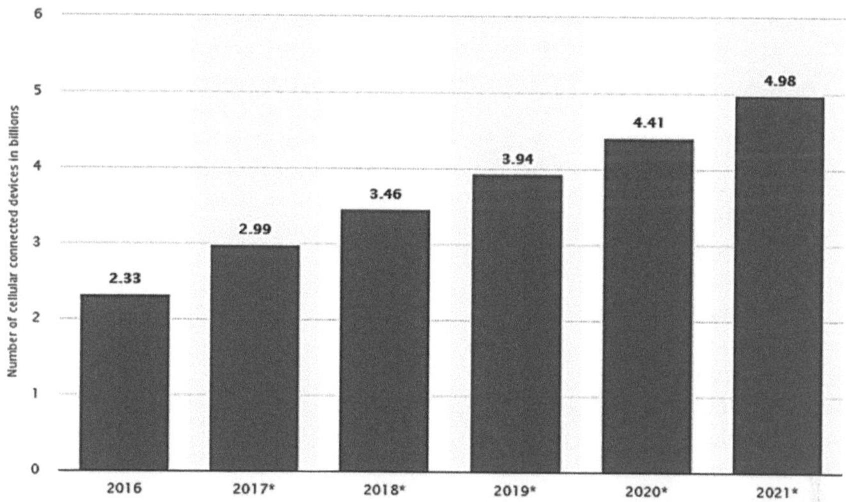

FIGURE 6.16 Number of cellular-connected devices in billions.

frameworks using Python-based libraries that can be fused with Raspberry Pi, Arduino, or some other open-source board for the coordination and approval of circumstances [2] (Fig. 6.16).

5.2 Key aspects of blockchain

Blockchain is a cutting-edge innovation that is always associated with security and has progressively gained acceptance as it has sifted up through applications. Recently, blockchain innovation has not been kept only to motorized sorts of cash but in addition is under execution for various social and corporate uses. These pieces join e-affiliation, individual to particular correspondence, online business, transportation, composed endeavors, capable correspondence, and distinctive others [3].

Blockchain is a bleeding-edge innovation that is continually associated with security and has logically grown in share for security in built applications. Blockchain innovation has moved beyond cryptographic sorts of cash to various social and corporate locales. These parts include e-affiliation, individual to solitary correspondence, online business, transportation, joint endeavors and gifted trades, and diverse others [4].

Following are some of the examples of Blockchain Implementations

- **Entertainment**
  - KickCity
  - Guts
  - Spotify
  - B2Expand
  - Veredictum

- **Social networks**
  - Matchpool
  - Steepshot
  - MeWe
  - Minds
  - Mastodon
  - DTube
  - Sola
- **Cryptocurrency**
  - Namecoin
  - Bitcoin
  - Dogecoin
  - Primecoin
  - Ripple
  - Litecoin
  - Nxt
  - Ethereum
- **Retail**
  - Blockpoint
  - Warranteer
  - Loyyal
  - Shopin
  - Opskins
  - Spl.yt
  - Portion
  - Fluz Fluz
  - Ecoinmerce.io
  - Every.Shop
  - Buying.com

In a blockchain arrangement, there exists the squares of various information components and records. Each square partakes in the blockchain system and is unchanging. The term changeless here alludes to verified and nonfragile (Fig. 6.17).

Consequently, it shapes the blockchain of a verified chain of squares with no probabilities of purposeful or coincidental altering or spillage in the information. The main square in the chain of system is known as the Genesis Block from where the blockchain starts the exchanges.

With expansion by embedding various squares with encryption of each square with past squares, it becomes verified, and in this manner, it is hard to break past states in view of such a significant number of encryptions.

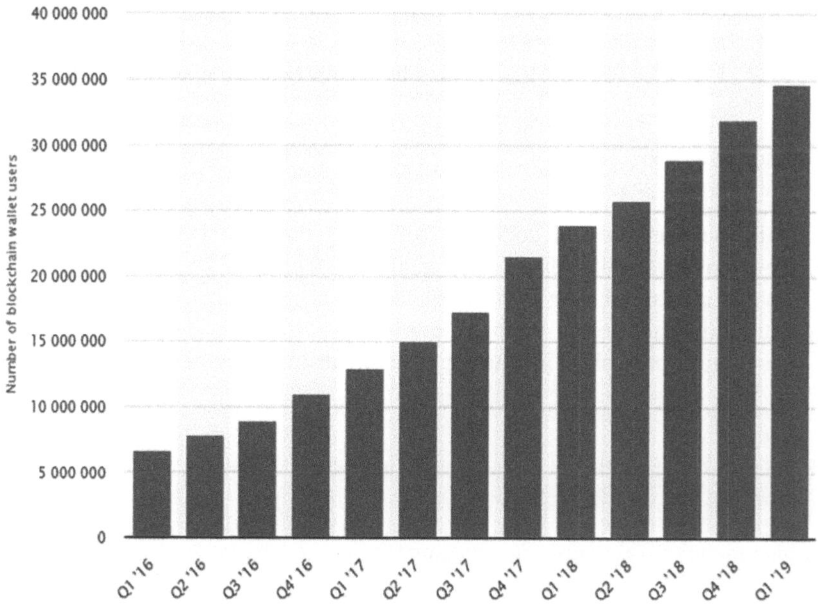

FIGURE 6.17 Number of blockchain wallet users.

```
class Block:
def __init__(this, myidx, ts, mydata, backsecuredhash):
this.myidx = myidx
this.ts = ts
this.mydata = mydata
this.backsecuredhash = backsecuredhash
this.securedhash = this.securedhashop()
def securedhashop(this):
shasecuredhash = securedhasher.sha256()
shasecuredhash.update(mystr(this.myidx) + mystr(this.ts) +
mystr(this.mydata) + mystr(this.backsecuredhash))
return shasecuredhash.hexdigest()
def keyblock():
return Block(0, date.datetime.now(), "Keyblock Block", "0")
def next_block(last_block):
this_myidx = last_block.myidx + 1
this_ts = date.datetime.now()
this_mydata = "Block" + mystr(this_myidx)
this_securedhash = last_block.securedhash
return Block(this_myidx, this_ts, this_mydata, this_securedhash)
blockchain = [keyblock()]
IoT_Secured_Block = blockchain[0]
```

```
maxblocks = 20
for i in range(0, maxblocks):
block_to_add = next_block(IoT_Secured_Block)
blockchain.append(block_to_add)
IoT_Secured_Block = block_to_add
echo "Block #{} inserted in Blockchain".format(block_to_add.myidx)
echo "Securedhash Value: {}\n".format(block_to_add.securedhash)
```

With the execution of code, the following outcome is obtained that has different hash values and provides a higher degree of security using cryptography functions. Using these hash values, attempts at hacking or sniffing the transaction are nearly impossible.

Generation of hash values using blockchain implementation

```
Block #1 Gen in Blockchain
Node:
60afeff7f57bc04dc1c411763778f661d9088c77d7789b4881a67d46557154ac
Block #2 Gen in Blockchain
Node:
8084d8e09b74f295082ea38cc3bbd892f27719b0fa9a732295fac34ed03debb9
Block #3 Gen in Blockchain
Node:
247086d83f38210185aed834e1efe218822c6e1223e465321a80d3d78f7ee803
Block #4 Gen in Blockchain
Node:
e7af556dd479088f4fb8cfb11f8648e192c8bb66835cdc748b039e0fc5fac646
Block #5 Gen in Blockchain
Node:
439ed4470f39e44475852dee4d42e4474d535431b61807d02892af38e4a4395e
Block #6 Gen in Blockchain
Node:
452d166db9c397cac1870afab08b462b00e15d7f5450485b03f7243a1a684b36
Block #7 Gen in Blockchain
Node:
301fabc673d1d29b0c18e0e8cf3c59d85051cf2a1eb4bd11c92abe0c0e531441
Block #8 Gen in Blockchain
Node:
bb876789e44fbd8d8d108d161cbde1b5d5d71cd5de388cdf2c0815c0e66a8635
Block #9 Gen in Blockchain
Node:
0fa214cf91613631d59514eecab2e9bf74ce13fba4ba1582bcf2b40dae2510db
Block #10 Gen in Blockchain
Node:
900139625df05d0dab34eab2e0921e4055cd2358dbbf64fd643c03d364745ad8
Block #11 Gen in Blockchain
```

```
Node:
86d641c12979caaec401ba3a8966df562bb8439183be3e41361f1eb9a14f8fbb
Block #12 Gen in Blockchain
Node:
fcaee1e98c7719df6769323fa0aaa79dc068c1c000380b454dd0e1cbb45c6e3a
Block #13 Gen in Blockchain
Node:
9429c9d0ecdfb284f189dc0c5359d565eeda405f5970fda0a6ce962942088d20
Block #14 Gen in Blockchain
Node:
fc9d9357c58c46ea1ec067cbb5e305aac6ef54295da7792f2c6732a6c7c93ee5
Block #15 Gen in Blockchain
Node:
aa7692f3f300e8178953328d99f9194c3c601fecc23b81319b739e6c801a4b6c
Block #16 Gen in Blockchain
Node:
9d145ad7051eaf167c227277497a8eb5418ad17925d24c88276c412f99026cc1
Block #17 Gen in Blockchain
Node:
297e1510bc43890077d36a65ed13fab28f1b8dacaada1b31f4e247de77bed878
Block #18 Gen in Blockchain
Node:
092eee32f96382135723d9ebe849b8e3b6c52c432f7930264d5565defb6e4f6c
Block #19 Gen in Blockchain
Node:
0ea694cea5e535107d840fd15d150f505269d6922ec348fdf8d212007cce2f83
Block #20 Gen in Blockchain
Node:
52b567064c45680ee110a2d660f2c0ff061451510bf9b2848649940c0f032589
Block #19980 inserted in Blockchain
HashValue:
58d86f4e14068cd1f56c5b27b7088a9894aceca01bd51bb304569b57c4e2a37e
Block #19981 inserted in Blockchain
HashValue:
d8d86d541ed3abb30bddd98d35b6f34d57958cb266aa70774251e5491e07d8ac
Block #19982 inserted in Blockchain
HashValue:
baf04029f374008683118504391e9a85f0af1593cad9b4d6e32088c5dcf4814d
Block #19983 inserted in Blockchain
HashValue:
43fa6ddebcc9b433711b3e613152223aea4ced529515a61b7b497a7f82fc0a86
Block #19984 inserted in Blockchain
HashValue:
b124ea343e9608f48b98cdc20835d6af14d492cbc60a2aff2809a12741f506a8
Block #19985 inserted in Blockchain
```

```
HashValue:
149a0ebe5ba5cde548e24531f8097dc7ea0b5b2b88307597f6775bfa55bd9686
Block #19986 inserted in Blockchain
HashValue:
e7569db4f1353d1f462ef467cccf479cd45f143e3a711b319d3d0ba59df7550d
Block #19987 inserted in Blockchain
HashValue:
a1ae0818dd75ac1edc82af2e6b6eacc549cbed4e6216cbe2aba5d44467ed4639
Block #19988 inserted in Blockchain
HashValue:
231ffb45f1bab279cd1e89d3fb44d47d900e1cc396b693bd2ea719bc8c83397b
Block #19989 inserted in Blockchain
HashValue:
53816e89cb4b8b5c3725f27b2cc466dc7d1d32da72c6ce7549d65fa357ac2789
Block #19990 inserted in Blockchain
HashValue:
1f8a7ac887af2741c3efdbdef04dc785b2a415f607f11ca346cdf9e4b7680777
Block #19991 inserted in Blockchain
HashValue:
63cb49f0159d5389b35bdf7d38d1abc09132d561574913d30a46caee5243b9d0
Block #19992 inserted in Blockchain
HashValue:
e25d51c6e2124653e73ae6ea69feb93f9fd28688afa824345bee481a5dd97874
Block #19993 inserted in Blockchain
HashValue:
7cadfde4d902eef011a88f7bc370d36165521e9bab96515254042fd9b7f0f094
Block #19994 inserted in Blockchain
HashValue:
9f8e6a12738a00fca346832860e2d2c4a5ac9f58a552ea642514689297eb957d
Block #19995 inserted in Blockchain
HashValue:
ec85bd1133893a6178eedc3e7832c350c40f217418a9d363926caa44fe71108d
Block #19996 inserted in Blockchain
HashValue:
d5384c12997497c193c791971423c7ed619bad2ce3b202d14175c3995182f0a1
Block #19997 inserted in Blockchain
HashValue:
e95d733459d11e25ea1de53d987e2acb6350e547315e124bbf5028146f6c6528
Block #19998 inserted in Blockchain
HashValue:
b96522fce352e01351e25e40affc43f6708d4fb7844c7cfa6485b45356d5274a
Block #19999 inserted in Blockchain
HashValue:
b88e5bfbc8c8853fc82676875f9b5f277ea8966a9ca60c0c597f1623a654e675
Block #20000 inserted in Blockchain
HashValue:
fdd851eafca41027cd3ffb3d0a535f0f53ad6f70ff1f87ce091295eb9d53fa79
```

5.3 Internet of Things—enabled blockchain for secure scenarios

As in the past model, the execution of hash work with squares is done on an independent system. In the case of a legitimate event on the blockchain, it is required to be passed on with the objective that different customers can begin their trades and squares. For spread and electronic executions, there are different structures in Python. In blockchain programming, proof-of-work (PoW) is one of the critical estimations. It is used to proclaim and bolster trades with the objective that the new squares are combined in the blockchain. It is proposed as the key understanding to checking and authenticating trades. In blockchains, different participants show enthusiasm for endorsing and completing trades. For productive backings, the diggers are repaid with progressively pushed sorts of cash as their remuneration.

This procedure also keeps up a key segment of the twofold encountering issue with the objective that moved cash or trades are executed in an avowed manner—for example, if A transmits a record or remote message to B. For this circumstance, that specific report or cash receipt in the records of A specifically needs to be destroyed and after that should be reflected only in the records of B. This is done by the remote controller as a comprehensive capture. On the off chance that there ought to be an event of the blockchain plan, it is executed with no extensive capturing and is grasped normally using express counts. In case there are events of nondeleting the trade from the sender, it will de-evaluate the remote message regardless of the kind.

As portrayed in Fig. 6.18, the execution of code, and generally speaking, usage with every record and exchange, can be broken down so that straightforward tasks occur with no attempts at hacking. Utilizing proof-of-work, the respectability of exchanges is logged and submitted. With this methodology, brilliant urban areas and smart homes, or related parts of the IoT, can be verified utilizing blockchain-based advances.

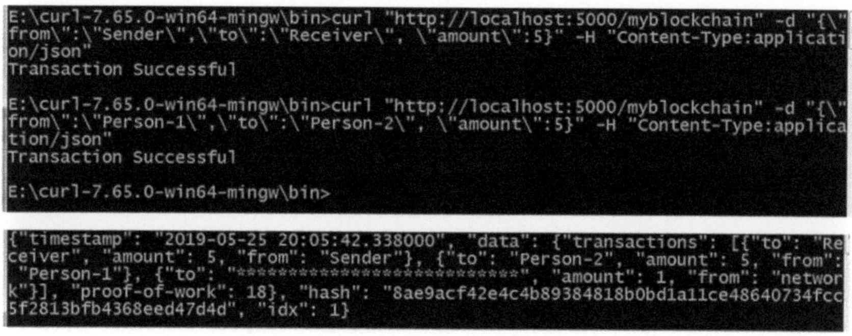

FIGURE 6.18 Exaction of code.

5.3.1 Protection to the device

The framework comprises numerous electronic contraptions like Arduino, voltage sensors, and comparable gadgets that may be damaged by nature. So we need to deal with the framework by measuring all security estimations and afterward placing it in a proper spot. This should be possible by setting the sensors inside fiberglass, which is firmly secured by a plastic box as an external covering. So this gives two-layered security to the sensors, and in the situation of precipitation, water cannot get to it, ensuring its safety from the elements. Superfluous holes are closed except for wiring reasons with scarcely any gaps at the base of the crate.

6. Conclusion

IoT is considered the third transformation in computerized innovation after the PC and the Internet. The necessities of computerization and intellectualizing prompt thoughtful reconciliations of IoT. The continuous observation and early admonition of disaster can be acknowledged, which will effectively oppose or lessen the harm of significant disastrous events to the force network. This necessity that has caused the advancement of brilliant network ideas was urgent for burdening and sourcing the executives, effective metering, and checking all phases of intensity frameworks. This new idea is acknowledged to introduce an information-correspondence condition for shipping different signs that contain significant data based on estimation, security necessities, observation, the board, and control forms. In our future work, the expense of our capacity-transmission-line-checking framework should be diminished, and the unwavering quality of our framework likewise should be improved. Inquiries about innovative enhancements identified with the physical structure and components of the traditional matrix have demonstrated that dynamic usage of correspondence frameworks in utilities should be incorporated to productively perform metering, checking, and controlling procedures. The correspondence framework of the smart network is one of the most significant parts of this idea and needs to assure the information transmission process in a safe, solid, and proficient way.

References

[1] J. Sun, J. Yan, K.Z. Zhang, Blockchain-based sharing services: What blockchain technology can contribute to smart cities, Financial Innovation 2 (1) (2016) 26.

[2] P.K. Sharma, S.Y. Moon, J.H. Park, Block-VN: A distributed blockchain based vehicular network architecture in smart City, JIPS 13 (1) (2017) 184–195.

[3] D. Puthal, N. Malik, S.P. Mohanty, E. Kougianos, C. Yang, The blockchain as a decentralized security framework [future directions], IEEE Consumer Electronics Magazine 7 (2) (2018) 18–21.

[4] C. Pop, T. Cioara, M. Antal, I. Anghel, I. Salomie, M. Bertoncini, Blockchain based decentralized management of demand response programs in smart energy grids, Sensors 18 (1) (2018) 162.

Further reading

[1] W.K. Edwards, Discovery systems in ubiquitous computing, IEEE Pervasive Computing, 2006, p. 7077.

[2] E. Fleisch, F. Mattem, Das Internet der Dinge, 1 edition, Springer, July 2005.

[3] F. Jammes, H. Smit, Service-oriented paradigms in industrial automation, IEEE Trans. Ind. Inf. (February 2005) 62–70.

[4] M. Bowman, S.K. Debray, L.L. Peterson, Reasoning About Naming Systems, 1993.

[5] W. Ding, G. Marchionini, A Study on Video Browsing Strategies, University of Maryland at College Park, 1997. Technical Report.

[6] B. Fröhlich, J. Plate, The cubic mouse: a new device for three-dimensional input, in: Proceedings of the SIGCHI Conference on Human Factors in Computing Systems Tavel, P. 2007 Modelling and Simulation Design, AK Peters Ltd, 2000.

[7] L.D. Brown, H. Hua, C. Gao, A Widget Framework for Augmented Interaction in SCAPE, 2003.

[8] Y.T. Yu, M.F. Lau, A comparison of MC/DC, MUMCUT and several other coverage criteria for logical decisions, J. Syst. Softw. 79 (5) (2006) 577–590.

[9] A.Z. Spector, Achieving application requirements, in: S. Mullender (Ed.), Distributed Systems, 1989.

[10] J. Pacheco, D. Ibarra, A. Vijay, S. Hariri, IOT security framework for smart water system, 2017 IEEE/ACS 14th Int. Conf. Comput. Syst. Appl. IEEE 9 (3) (2017) 11–30.

[11] S.-R. Oh, Y.-G. Kim, Development of IOT security component for interoperability, IEEE 12 (4) (2017) 67–89.

[12] U.M. Mbanaso, G.A. Chukwudebe, Requirement analysis of IOT security in distributed systems, 2017 IEEE 3rd Int. Conf. Electro. Technol. Natl. Dev. (NIGERCON) IEEE 5 (7) (2017) 20–30.

[13] J. Shi, J. Wan, H. Yan, H. Suo, A survey of cyber-physical systems, in: Proc. Int. Conf. Wireless Commun. Signal Process, November 2011, pp. 1–6.

[14] M. Weiser, The computer for the 21st century, SIGMOBILE Mob. Comput. Commun. Rev. 3 (July 1999) 3–11.

[15] M. Green, S. Hohenberger, B. Waters, Outsourcing the decryption of ABE ciphertexts, in: *Proc. 20th USENIX Secur. Symp.* San, USENIX Assoc., Francisco, CA, USA, August 2011, pp. 1–16. J. Lai, R. H. Deng, C. Guan, J.Weng, Attribute-based encryption with verifiable outsourced decryption,' IEEE Trans. Inf. Forensics Secur., 8, 1343–1354, August 2013.

[16] J. Li, X. Huang, J. Li, X. Chen, Y. Xiang, Securely outsourcing attribute- based encryption with checkability, IEEE Trans. Parallel Distrib. Syst. 25 (8) (August 2014) 2201–2210.

[17] B. Waters, Ciphertext-policy attribute-based encryption: an expressive, efficient, 19. and provably secure realization, in: Public Key Cryptography PKC (Lecture Notes in Computer Science), Springer, Taormina, Italy, 2011, pp. 53–70.

[18] B. Qin, R.H. Deng, S. Liu, S. Ma, Attribute-based encryption with efficient verifiable outsourced decryption, IEEE Trans. Inf. Forensics Secur. 10 (7) (Jul. 2015) 1384_1393.

[19] Y. Yan, Y. Qian, H. Sharif, D. Tipper, A survey on smart grid communication infrastructures: motivations, requirements and challenges, IEEE Commun. Surv. Tutor. 15 (1) (2013) 5–20.

[20] W. Wang, Y. Xu, M. Khanna, A survey on the communication architectures in smart grid, Comput. Netw. 55 (15) (2011) 3604–3629.

[21] A. Usman, S.H. Shami, Evolution of communication technologies for smart grid applications, Renew. Sustain. Energy Rev. 19 (2013) 191–199.

[22] X. Lu, W. Wang, J. Ma, An empirical study of communication infrastructures towards the smart grid: design, implementation, and evaluation, IEEE Trans. Smart Grid 4 (1) (2013) 170–183.

[23] M. Kuzlu, M. Pipattanasomporn, S. Rahman, Communication network requirements for major smart grid applications in HAN, NAN and WAN, Comput. Netw. 67 (2014) 74–88.

[24] R.H. Khan, J.Y. Khan, A comprehensive review of the application characteristics and traffic requirements of a smart grid communications network, Comput. Netw. 57 (3) (2013) 825–845.

[25] E. Ancillotti, R. Bruno, M. Conti, The role of communication systems in smart grids: architectures, technical solutions and research challenges, Comput. Commun. 36 (17) (2013) 1665–1697.

Chapter 7

Smart grid modernization using Internet of Things technology

Mamata Rath[1], Anuradha Tomar[2,3]
[1]*School of Management (Information Technology), Birla Global University, Bhubaneswar, Odisha, India;* [2]*Electrical Engineering Department, Eindhoven University of Technology, Eindhoven, The Netherlands;* [3]*Electrical Engineering Department, JSS Academy of Technical Education, Noida, India*

1. Smart grid design challenges

Modern power systems are spread over a large geographical area with multiple interconnection points because of issues related to the economy, the environment, safety and reliability, and stability. As demand for electricity grows daily, there is a need for numerous network grids as high sources of such services. More investment is required to activate power system components to maximum utilization, which is a challenge. There is strong competition among distributors in electricity markets to fulfill the requirements of growing populations.

Fig. 7.1 demonstrates outdated power network grid with one mode of electricity flow. In the traditional mode, there is centralized generation of electricity in bulk, which leads to higher dependency on coal and oil with few automated processes. Situational awareness is minimal because consumers have lack the data to manage for all types of use. A customary power network grid comprises countless loosely organized synchronous network grids. It works on the basis of three primary capacities—age, distribution, and dissemination of electrical energy—for a power stream with just a single mode as a specialist organization for purchasers. Right off the bat in power age, countless power plants create electrical energy mostly by consuming CO_2 (carbon dioxide) fills. Furthermore, in power transmission, power is distributed from power plant furnaces to distant locations, requiring communication lines with tremendous current. Next, because of this power dispersion, these electrical frameworks disseminate electronic energy to the last purchasers at diminished voltages. Every network grid is controlled and checked to guarantee that power plant

Advances in Smart Grid Power System. https://doi.org/10.1016/B978-0-12-824337-4.00007-2

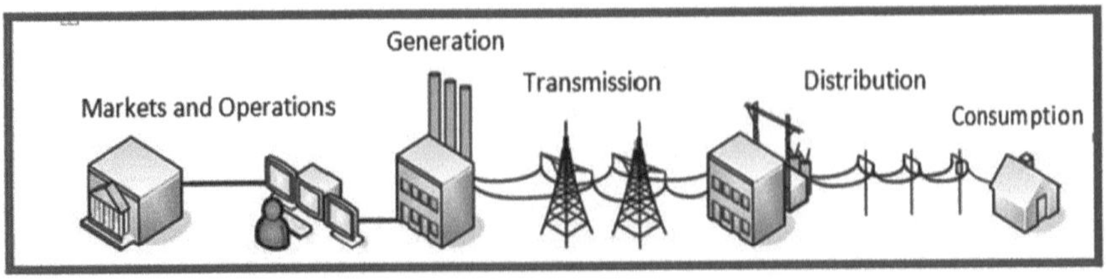

FIGURE 7.1 Electricity mode in traditional power network grid.

furnaces produce electrical energy to meet purchaser requirements within the limitations of their power frameworks. These are the age, broadcast, and dissemination characteristics of electromechanical power controlled by service organizations that provide energy to purchasers and charge them likewise to recoup expenses and generate a profit.

The customary power network grid has performed brilliantly since its commencement [1]. Despite the fact that buyers' interest in energy developed extraordinarily, its delivery was immobile and unsurprising. Notwithstanding this, incredible modifications have been made with regard to digital electromechanical energy utilization since 1970, as the heap of electrical gadgets has turned into the quickest developing component of total power demand with the growth of new wellsprings of high power utilization—for example, electric vehicles. Control network grids persevere through noteworthy energy expenditures caused by various elements: purchasers' wasteful apparatuses, the absence of shrewd innovation, inefficient directing and regulation of electrical energy, problematic correspondence and observation, and in particular the absence of a system to accumulate the electromechanical energy produced [2–4]. Moreover, power network grids encounter different difficulties, including developing energy to meet new demand, maintaining quality, security, increasing allocations to sustainable power sources, and the maintenance and replacement of aging facilities, to give some examples.

To understand these difficulties, the smart network grid worldview has appeared as an auspicious arrangement through assorted data as well as correspondence innovations. Such innovations can achieve the viability, proficiency, unwavering quality, security, supportability, stability, and adaptability of the conventional power network grid [5]. The smart grid (SG) contrasts with conventional power network grids in numerous ways. For example, the SG offers a dual-way transmission stream among specialist organizations as well as buyers, while a conventional power network grid has only single-directional correspondence with the specialist co-op to said customers. An SG provides unconventional metering organization, smart metering systems, adaptation for noncritical failures, unauthorized-use identification, and burden adjustment [6–8], such as self-mending—i.e., location and recuperation after deficiencies [9].

The SG [2,3] takes care of the issue of electricity-based energy expenditure by creating electrical energy [4] and intensely coordinating its use among competing interests. SG settles significant choices in the best interests of overall energy provisioning—for example, ongoing evaluation, self-recuperating, power-utilization planning, and enhanced electrical energy use. Such choices can essentially improve power quality as well as the proficiency of the network grid [6] by maintaining harmony between power age and use.

SGs utilize different devices for detecting, analyzing, and monitoring the network grid. These observational devices are set up at power plants, communication lines, communication towers, transmission junctions [1], and customer premises with costs totaling millions or even billions. Some principal problems for SGs are the network and the mechanization and successful operation of a gigantic number of appliances, which need observation, investigation, and switching by rapid, universal and two-mode advanced communications. SGs need dispersed robotization for such equipment or "things." The concept has been authenticated through Internet of Things (IoT) modernization. The IoT is characterized as a system that can interface any article with the Internet dependent on a convention for trading data and correspondence between different smart gadgets as a way to accomplish observation and monitoring of transmissions with board and area ID destinations [10, 41]. In recent years, IoT innovation has received increased critical consideration for different applications and has accounted for the interconnection of the Internet for various system-installed gadgets utilized in day-to-day life [11].

From the outset, the Internet has provided accessibility for people-to-people and people-to-things interactions. By 2008, the number of things related to the Internet had become greater than the number of people in the world [1], and thus the impact of IoT expertise keeps rising. IoT is an arrangement of physical things or things related to the Internet. Such things are equipped with embedded development to interface with internal and outside circumstances. These articles sense, analyze, control, and pick, alone or jointly, various things through quick and two-mode propelled transactions in a scattered, autonomous, and inescapable mode. This is really what is required for the SG. Going forward, IoT development can help SGs support diverse frameworks in managing power, accumulation, transmission, distribution, and use by joining IoT devices (for instance, sensors, actuators, and smart meters) and similarly by providing networking, computerization, and monitoring for such devices.

When communication architecture and information technology are implemented in traditional power network grids to enhance their capability, the system is referred to as a "smart network grid." Fig. 7.2 illustrates a smart network grid architecture with communication and information technology devices attached. Generally, smart network grid denotes a group of

FIGURE 7.2 Smart network grid with communication infrastructure.

technologies that individuals use to establish effective electricity-release schemes in the modern age via automated remote access and control systems with high automation levels. Such schemes benefit from two-mode digital-message transfer innovations and automated system dispensation that have been utilized for some time by companies in other industries. Now they are also being used in electromechanical devices—based systems ranging from generators such as power mines and windmills to users of electrical energy in places such as homes and commercial sectors [42]. They propose many advantages to clients and consumers, with better energy proficiency and reliability on safety-based electricity network grids.

Fig. 7.3 validates application areas of the smart network grid. These grids are used in thermal plants, solar power generation, wind power plants, nuclear power plants, electric vehicles, smart homes, cities and offices, industrial plants, etc. The advantages of using smart network grids include improved power consistency and quality, enhanced monitoring of sensors in systems and

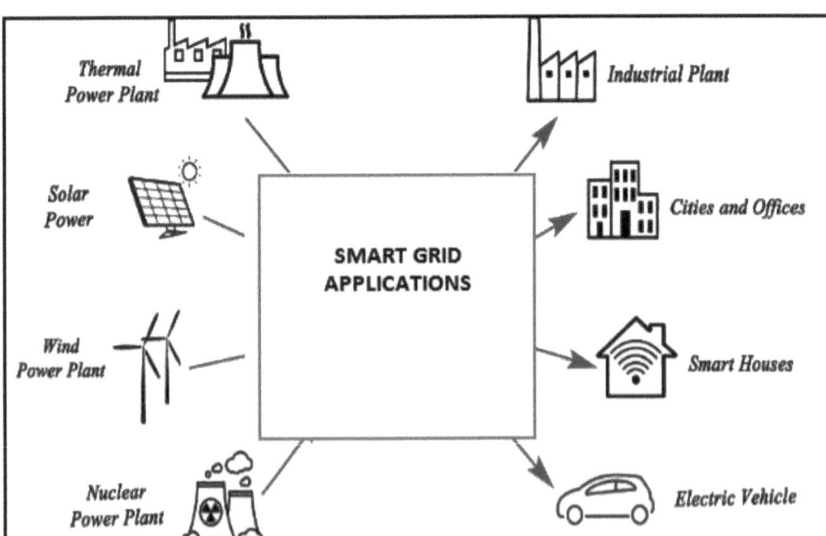

FIGURE 7.3 Smart network grid applications.

transmission, improved, quick matching of supply and demand, diminished need for backups to the main load power plant furnaces, better demand management from the consumer side, and the use of advanced metering infrastructures.

1.1 Chapter organization

The chapter is organized in parts. Section 1 offered the introduction to SG design challenges. Section 2 presents the SG control infrastructure, Section 3 illustrates the road map from the SG to the IoT concept, and lastly, Section 4 concludes the chapter.

2. Smart grid control infrastructure

In recent years, IoT innovation has gained tremendously in being considered for different applications and has accounted for the interconnection of the Internet to various system-inserted gadgets utilized in day-to-day life. It has mechanized the activity of different frameworks: for example, medical services, transportation, military, home machines, security, reconnaissance, horticulture, and force network grids. IoT gadgets are ordinary items outfitted with handsets, and smaller-scale controller and convention stacks that empower their communication with different gadgets just as with outside elements (e.g., people), thus allowing for totally computerized frameworks that make them a fundamental piece of the Internet [11,38,39].

Various endeavors in the literature have reviewed different viewpoints of IoT in smart network grid frameworks. The creators focused on as well as overviewed advanced metering infrastructure (AMI) and shrewd meter checking advances aimed at checking dependability and power excellence in IoT-associated SG frameworks [14]. Be that as it may, the primary spotlight of that research is only on AMI and shrewd metering [24,26] instead of wrapping in the alternate significant parts of IoT-helped SG frameworks: applications, models, and so forth. There were two recent reviews on safety in IoT-supported SG frameworks, for the most part centered around safety in those frameworks [15] but also reviewing susceptibilities and safety strings and their effect on IoT-supported SG frameworks. The creators concentrated on profound bundle examination as a security device for SG and mechanical IoT [16]. The creators considered the current literature as well as suggestions from the scholarly community and administrative system and assessed profound parcel investigation as a security instrument for SG. In any case, their focus was for the most part on profound bundle assessment. Both reviews presented in Refs. [15,16] predominantly investigated security angles in IoT-aided SG frameworks. The authors also addressed how IoT should be connected to electric power and energy frameworks—for example, the worth and significance of IoT in electric power organization and the monetary, ecological, and cultural effects of IoT-aided electric power frameworks as well as the difficulties and suggestions identified with IoT [17]. It provides just one utilization of IoT in rechargeable power and energy frameworks. Additionally, the researchers in Ref. [18] concentrated on giving mindfulness to analysts on the ground of IoT and SG frameworks.

Following are the basic contributions of IoT-based applications to the smart network grid for the automation and modernization of different components as discussed in this chapter:

- smart solicitations of smart IoT in the smart network grid [1]
- IoT-based smart choice low-power wide area network (LPWAN) [2]
- condition-based maintenance and SG life-cycle management [3]
- IoT-centered smart network grid automation scheme [4]
- design of IoT-centered smart dynamism metering [5]
- protection of insolent network grid using IoT systems [6]
- transmission tower protection system using IoT [7]
- employment of a basic meter smart network grid surrounded in IoT platform [8]
- IoT-based smart load enabler device [9]
- safety and issues in energy-based IoT [11]
- smart plug with IoT for power monitoring [12]

2.1 Smart applications of smart internet of things in smart network grid [1]

The technology of IoT innovation has extraordinary potential for acquiring, breaking down, handling, controlling, and sustaining the data of element properties (area, status, conduct, trademark, and so forth.), which will assume a functioning job in advancing improvements of the smart network grid. One research article [1] centers on uses of IoT innovation in the smart network grid field. A sort of framework construction of the smart network grid is managed by IoT framework architecture. The framework architecture of the smart network grid is partitioned into three strata—the discernment, correspondence, and application layers. The utilization of IoT innovation in each stratum (layer) of the smart network grid's framework architecture is broken down and examined in detail. Finally, the application prospects and issues to be tackled are gauged. It is discovered that IoT innovation will make the smart network grid increasingly productive, progressively canny, increasingly secure, and progressively advantageous.

2.2 Internet of things–based smart choice low-power wide area network [2]

The short power wide area system, which is presently progressing well with improvements in the Web of Things (WoT) may suggest a technically new explanation for existing smart network grid communications because of impressive highlights of little control, long collection, and great limit. Purposeful smart network grid interchanges need protected and dependable associations among the benefits and gadgets along with a great capacity for administration. This is hard to accomplish for unrestrained LPWAN advances because of the jam-packed permit-free band. Short-band IoT (SB-IoT), by way of an authorized SPWAN innovation, was created dependent upon current large haul development determinations and offices. In this manner, it can give cell-level QoS and from now on could be seen as a gifted possibility for smart network grid interchanges. Research [2] acquaints SB-IoT with smart network grid and contrasts it and existing agent correspondence innovations with regards to smart network grid interchanges as far as information rate, inertness, extent, and so on. The general necessities of interchanges in the smart network grid from both measurable and subjective viewpoints are extensively researched, and every one of these is painstakingly analyzed for SB-IoT. It further investigates delegated submissions in the smart network grid and breaks down the related plausibility of SB-IoT. In addition, the exhibition of SB-IoT in run-of-the-mill smart network grid correspondence conditions—for example, urban and rustic regions—is painstakingly assessed by means of Monte Carlo recreations.

2.3 Condition oriented preservation and smart grid life-cycle administration [3]

IoT-based innovation is exceptionally incorporated and exhaustively utilized in the new age of data innovation and has been advanced for national improvement procedures by government committees. The smart network grid is a significant submission field in IoT innovation. So as to quicken the union of IoT innovation and the smart network grid, the State Grid Corporation of China sorted out IoT innovation ventures [3], choosing two 10 kV dissemination lines for dispersion organizing gear with on-line observing, condition-oriented support, and life span of the board dependent on IoT innovation to determine operational administration, lack of condition-centered upkeep implications, and so on. The task achieved great outcomes in functional applications.

2.4 Internet of things–related smart network grid automation system [4]

The smart network grid is essentially "reconciliation of development of electrical data with the capacity to speak with clients and the utility." This works faster and more efficiently than the customary communication and conveyance framework. Construction robotization is an outstanding idea for the programmed and smart exercises essential to any structure/shopping center—for example, heat control, entry mode resistor, and weight regulator. The IoT-based system is staged to build a system of various devices/actuators over the Internet. Correspondence and regulation over secure and quick modes are conceivable from anyplace; this sphere principally utilizes the Internet of Energy (IoE). IoT also helps in information stockpiling and investigation and outsider impedance. Another research article [4] fundamentally focuses on (1) assembly and innovation of the smart network grid; (2) IoT construction essentials with preferences; (3) submission and mixing of "smart " and IoE explicitly in the structured computerization area; and (4) requests for administration. The research in Ref. [4] illustrates "energy protection by sustainable power source and by observing and governing electromechanical energy along with the utilization of IoT-based applications in existing structures to gain financial advantage" in the context of smart network grid modernization.

2.5 Design of internet of things–based smart energy metering [5]

IoE can be utilized to outfit logical administration of the power supply and its utilization under diverse conditions. Over the years, with the development of IoE and other advances, the smart network grid has continually become smoother than before. The upcoming power network grid should be actualized in a disseminated network structure that could powerfully retain distinctive energy resources. IoE can be used for different uses of the smart network grid comprising disseminated control plant observing, control age and utilization expectation, control utilization checking, energy stockpile observation, smart digital checker, electric automobile charging, control request part administration, and different territories of energy creation [5]. Utilizing IoT capacities, a smart energy metering stage has been structured comprising smart fittings, entry modes, and cloud servers [29].

3. Road map from smart grid to Internet of Energy concept

The IoT is a system that can interface with any item on the Internet, dependent on a convention for trading data and correspondence among different smart gadgets to accomplish checking, following, and board and area ID destinations [10]. The IoT centers around the acknowledgment of three primary ideas, in particular things-situated, Internet-arranged and semantic-arranged. The things location idea includes smart gadgets: for example, RFID labels, devices, actuating devices, video cameras, scanning machines, the worldwide positioning system, and smart platforms using 5G [33]. The Internet-situated idea empowers correspondence among smart gadgets through different communication advances—for example, ZigBee,

Wi-Fi, Bluetooth, and cell interchanges—and furthermore interfaces them to the Internet. The semantic-arranged idea understands an assortment of applications with the assistance of savoir-faire gadgets.

During our study on IoT protocols and identifying intelligent IoT applications, we have categorized the various applications of IoT in the smart network grid into five prominent categories as follows:

- IoT conventions in smart network grid applications/research issues
- smart network grid/metering consuming IoE
- cost-effective/smart micronetwork grid using IoT
- IoT plug-ins and disaster prevention
- security provision and issues in IoT-based smart network grid applications

For every category, we have identified innovations/basic purpose, creativity, associated challenges, and solution methods. At last, we conclude with our strict observations on how smart network grid automation using IoT protocols has really benefitted mankind, what related research is ongoing for better development, and how recent trends present a realistic view of an advanced smart network grid with a favorable vision.

Table 7.1 presents IoT conventions in smart network grid applications/ research issues, Table 7.2 presents smart network grid/metering using IoT, Table 7.3 illustrates a cost-effective/smart micronetwork grid using IoT, Table 7.4 demonstrates IoT plug-ins and disaster prevention and, Table 7.5 depicts security provisions and issues in IoT-centered smart network grid applications.

3.1 Superior metering infrastructure and smart metering using internet of energy

Power superiority and consistency create conflicts in mutually serving service suppliers and customers in conservative control network grids. The fragmentary technical development in the IoE period delivered enhanced explanations to improve the administration of many experiments and implement the process of a smart network grid. Superior meter-checking organization and smart metering-based knowledge are proficient knowledge bases from which to transform the conformist-influenced network grid to expose the secreted particulars of electric department authority by setting up of two-mode message schemes during power operation, leading to progressive advantages for clients.

The energy necessities of the cutting-edge century are expanding quickly because of the inhabited developments in keen urban areas and a high standard of living. Extensive endeavors are being made to make the power network grid all the more sharp to make it progressively agreeable to the energy needs of customers and supply improved proficiency and consistency of high-intensity frameworks. WoT has developed as one of the empowering innovations for a

TABLE 7.1 Internet of Things (IoT) conventions in smart grid applications/research issues.

Internet of Things (IoT) conventions in smart network grid applications/research issues				
Ref. no	**Authors**	**Year**	**IoT and smart network grid implementation/research theme**	**Associated issues/factors and challenges discussed**
[16]	A. Kanyilmaz, A. Cetin	2019	Fog regulated architecture proposal for IoT with secluded nodes: a shrewd home application	Smart devices, fog systems, fog computing architecture, IoT, sequestered nodes, smart home system, data confidentiality, Internet connection usage, architecture design, smart home application, cloud computing systems, computational power constraint problem, fog nodes, internet of things, internet service quality, data processing
[4]	S. Sidid, S. Gaur	2017	Smart network grid building automation based on IoT	Transmission system, smart network grid, smart network grid construction automation, Web of Things, Internet, building automation domain, third party interference, pressure control, door control, temperature control, smart activities, distribution system
[1]	Y.E. Song et al.	2015	Research on solicitations of Internet-based things in the smart network grid	IoE, smart network grid, IoT system architecture, communication stratum (layer), perception stratum (layer), application stratum (layer)
[21]	S. Borkar et al.	2016	Submission of 5G next group network to internet of energy things	Heterogeneous network, coordinated multipoint networking, small cell, adaptable resource management, flexible duplex method, control plane, user plane, millimeter frequency range, dynamic spectrum sharing, carrier aggregation utilization, spectral efficiency, radio and antenna system, moving robot, connected car
[22]	Song and Zeng et al.	2018	Reliable sensing mechanism using IoT	IoT devices, trustworthy sensing scheme, critical smart city applications, smart network grid, smart building, IoT data, wireless network intelligent transportation, mobile healthcare, fake data generation, RealAlert scheme
[15]	Zhen and Chen et al.	2011	Incorporation of IoT through smart network grid	Smart network grid, IoT, power transmission line monitoring, smart patrol, smart home, electric vehicle management
[13]	W. Wen et al.	2011	Investigation on the key machineries of IoT applied on smart network grid	IoT, smart network grid, power system stability, power system security

TABLE 7.2 Smart network grid/metering using internet of things.

Smart network grid/metering using internet of things				
Ref. no	**Authors**	**Year**	**IoT and smart network grid implementation/research theme**	**Associated issues/factors and challenges discussed**
[17]	S.E. Collier	2017	Conjunction of the smart network grid with IoE	Control plants, Ethernet, IoT, smart network grid convergence, emerging Enernet
[10]	Sharjat and Safkhani	2017	Regulator over smart metering is disconnected in BAN gate mode	Lightweight AKA scheme, mutual authentication, smart network grid, multicast key generation phase, smart meters, IoE—IoT ecosystem, building area network gate mode, denial of service attack
[5]	Yaghmaee and Hejazi et al.	2018	Strategy and employment of IoT-based smart dynamism metering	Smart plugs, intelligent management, energy distribution, distributed topology, power generation, energy sources, smart network grid, IoT, smart energy metering, distributed power plant monitoring
[20]	P. Ramanathan	2018	Confidentiality assurances in the IoE world	Sensitive information, associated users, extensive research, privacy challenge, privacy assurances, smart network grids, smart homes, smart city projects, IoT technologies, security, IoT applications, IoT application areas, IoT world
[19]	H. Geng	2017	IoT and standardization of smart network grid	Protocols, IoT conventions, power efficiency, security
[8]	SpanA and Pascoli et al.	2015	Metering with smart network grid surrounded in an IoT platform	User interface, distributed IoT server, gate modes, ZigBee smart meters, abstraction stratum (layer), actuator mapping, univocal sensor, data access, heterogeneous sensor communication protocols, customer-centric, distributor-centric, smart building systems, smart home, IoT, last-meter smart network grid

TABLE 7.3 Cost-effective/smart micronetwork grid using internet of things (IoT).

Cost effective/smart micronetwork grid using internet of things (IoT)				
Ref. no	**Authors**	**Year**	**IoT and smart network grid implementation/research theme**	**Associated issues/factors and challenges discussed**
[9]	Satti et al.	2018	Low-cost smart load enabler device	Commercial electricity infrastructure, client-server communication mechanism, wireless communication, electrical wiring, smart load enabler, domestic electricity infrastructure
[7]	Zhen et al.	2012	Communication tower defense system based on Web of Things in smart network grid	Power network grid infrastructure management, smart network grid, early warning, threat location, threat classification, high-voltage backbone transmission tower, high-voltage backbone transmission facilities
[9]	M.B. Satti, N. Ali, M.M.A. Asif, A.H. Abid, A. Hasan	2018	Developed device with smart and less cost	IoT, smart network grids
[3]	Jinhua et al.	2016	IoT in Henan smart distribution network demonstration project	Operational management, life-cycle management, condition-based maintenance, on-line monitoring, distribution network equipment, distribution lines
[2]	Yang et al.	2018	Innovative system with narrow bandwidth for energy-based metering	LPWAN technologies, licensed LPWAN technology, smart network grid communication environments, quality of service, cellular-level QoS, Monte Carlo simulations

TABLE 7.4 Security provision and issues in internet of things (IoT)-based smart network grid applications.

Security provision and issues in internet of things (IoT)-based smart network grid applications				
Ref. no	**Authors**	**Year**	**IoT and smart network grid implementation/research theme**	**Associated security issues/factors/related domains and challenges discussed**
[11]	J. Cao et al.	2014	The safety for power IoT (PIoT): outline, policies, and prevention	Power Internet of Energy, IoT technology, smart network grid, PIoT, sensing applications, security framework, security policies, complex cyber-physical network, perception stratum (layer), intelligent devices, application stratum (layer), security measures
[23]	C. Jiang, Y. Lu	2018	IoT submission and safety risk analysis for smart network grid	IoT application, security risk analysis, smart network grid, intelligent equipment, wireless communication, intelligent access terminals, information security protection, IoT security
[14]	G.S. Sharvani et al.	2016	Standardized security measures for unified IoT submissions of smart network grids	Industrial control systems, security requirements, smart network grid application, end-to-end protection system, computer threats, cyber security, smart network grid management, backend communication network, multiple heterogeneous systems, physical systems, cyber systems, backbone network, industrial applications, remote location, internet communication, centralized IoT applications, calibrated security measures
[18]	K. Chae et al.	2015	Protected authentication for organized smart network grid system	Utility systems, smart meters
[6]	Y.Y. Ma et al.	2012	Investigation on application and safety protection of IoT in smart network grid	Power consumer interactions, industrial sector

TABLE 7.5 Smart internet of things protocols designed for smart network grid applications.

Internet of things protocol	Mechanism	Application field in smart grid
Energy-efficient and dependable data gathering steering protocol [34]	Software-defined federal regulator and various mobile sources for dynamism competent and consistent data collection	Wireless sensor networking system based smart network grid operations
Pairing-based verification protocol for V2G systems in shrewd network grid [35]	Guarantees privacy of transportations, protects characteristics of users, and restricts attackers from hacking the automobile	Prevention of signaling congestion and minimized bandwidth utilization
Minimally aggressive attack mitigation via recognition segregation and localization [36]	Online and anomaly discovery algorithm [36] which is accessible and competent of timely discovery	Prevention of circulated (D-DoS) attack on the global network using IoT-based botnets
Energy resourceful and QoS-based steering protocol [EQRP] for smart network grid industry 4.0 [37]	Encouraged by the genuine behavior of the bird breeding optimization method	Develops system reliability considerably and lacks extreme data frames distributions for wireless network-oriented smart network grid implementations
Combination-based anonymous and protected significant agreement protocol for shrewd network grid edge calculating [41]	Permits the smart metering and checking to get associated with effectiveness organization secretly to benefit.	Secured private key is used in the nonappearance of reliable authority, useful for high security alert need

smart network grid arrangement. As IoT-associated gadgets carry on to deliver at a quick pace, one of the biggest challenges is security because the gadgets are online, thus making the savvy network grid vulnerable to significant assaults. Since an IoT-based brilliant network grid would involve a great many hubs, it is the chief external assault risk for an IoT-centered digital assault.

Presently, the circulated age (DG) has become quite basic because of the developing widespread consideration of dependable and practical electric force conveyance to coordinate increasingly sustainable and elective energy sources and decrease the weight and disappointment in existing transmission frameworks [7]. In DG, disparate energy assets can be incorporated to shape an energy association to meet the cases of neighborhood clients. The accentuation in a flowed age is expanding, as it can likewise helpfully continue electrical energy prerequisites in remote and peaceful regions [3] where no significant utility force arrangement exists or that which does is deceitful. A smaller-scale network grid, right now, to a restrained arrangement for many demands and dispersed miniaturized-scale energy sources that can supply electrical capacity to its neighboring zones [9] can gainfully synchronize modified sorts of scattered energy stockpiling through nearby force providers.

3.2 Internet of things plug-ins and disaster prevention

Technically developed energy networks in current use comprise apparatus capable of fulfilling their utility in an energy-efficient manner and with message and remote control functions. Therefore, such devices, such as smart energy meters, become smart for use in power production and distribution engineering, achieving the visualization of smart network grids. However, there are many problems to be solved in demand to attain a completely functional and security-based aware smart network grid. Supporting correct measurement, control, communication, control, reporting, and synchronization management are big challenges in this effort.

3.3 Security provision in smart network grid

The insolent network grid has initiated new ideas in the sectors of power and energy with a real-time pricing system, load balancing, management of electricity demand, and unification of renewable power sources. Also, the smart network grid has produced many additional access points for customer use. From a security standpoint, any admittance points can be influenced by assailants to hack a system and have access to direct a core program and modify the distribution channel to undermine the network grid in impulsive modes [11]. Therefore, a smart network grid safety method may be compulsory at numerous strata (layers) counting substantial and reasonable strata (layers) [23]. Physically, smart network grid systems and constituents should be protected from damage, mitigation, theft, destruction, and interruption.

Examples of corporeal stratum (layer) security include connection of boundary marker, movie inspection, and alert arrangement [18]. Sanctuary in the logical stratum (layer) deals with defending numerical data. Some logical stratum (layer) defense mechanisms in smart network grid implementation are presented here:

- coding–decoding
- verification
- submission safety controls
- safety patches
- malware elimination

3.4 Internet of things in micronetwork grid-distributed energy sources

The smart network grid is fundamentally the "incorporation of developed electrical evidence that has the capability of corresponding with customers as well as effectiveness." It is faster and more efficient than outmoded broadcast and allocation organizations. It incorporates a cluster of unified work distribution and a distributed power source surrounded by obviously distinct electromechanical surroundings that acts as a solitary convenient object in the form of network grid. A micronetwork grid can unite and separate from the network grid to enable it to control in both network grid-connected and unconnected modes.

Smaller-scale microgrids (Fig. 7.4) are small-scale power supply organizations intended to provide additional capacity to little networks. They

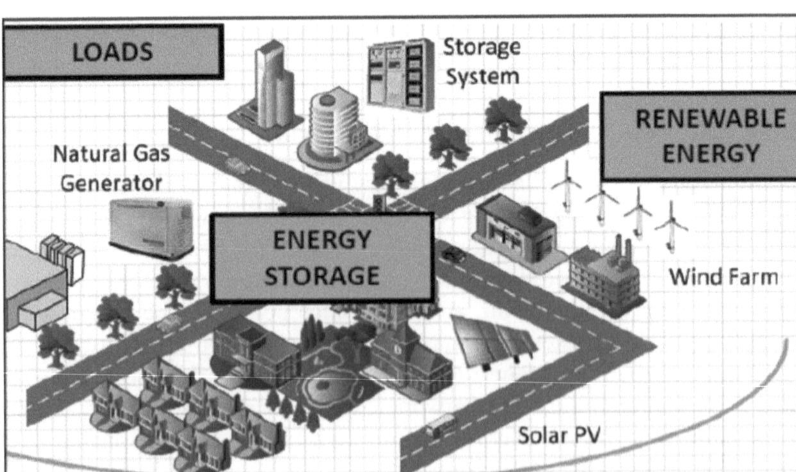

FIGURE 7.4 A microgrid with energy storage, renewal energy, and loads.

empower neighborhood power for nearby loads and contain various, small force-producing sources that makes them profoundly adaptable and proficient. They are associated with both the neighborhood-producing units and the utility network grid, consequently forestalling forced blackouts. Overabundant power can be offered to the utility network grid or can be stored away in the framework. The size of the micronetwork grid may extend from a lodging bequest (barely any kW) to a civil locale (a few MW).

Smaller-scale network grids satisfy the fundamental prerequisites of the keen network grid. By applying IoT advancements [30,31], different smart administrations can be made for miniaturized-scale network grids. The advancement of most parts of the small-scale network grid would be improved by applying IoT. Three-stratum (layer) engineering of IoT for the brilliant network grid [32] has been proposed in various applications. Different data and correspondence strategies of IoT are also being applied to the brilliant network grid. Especially, IoT application arrangements are given in detail [2] to control transmission-line observation, smart monitoring, brilliant home, and electric vehicle executions.

Fig. 7.5 depicts implementation of a network grid for an assortment of uses. Micronetwork grids provide the most encouraging methods for incorporating many conveyed sources into the force network grid, which is especially significant for sustainable power sources. Smaller-scale network grids can give higher unwavering quality, provide energy security and surety, and open the entryway to critical framework effectiveness enhancements that utilize combined heating and power. Key functions and characteristics of smart network grids are self-mending ability, unwavering quality of the power framework, energy-stockpiling frameworks, security and protection, sustainable power source incorporation, and force quality.

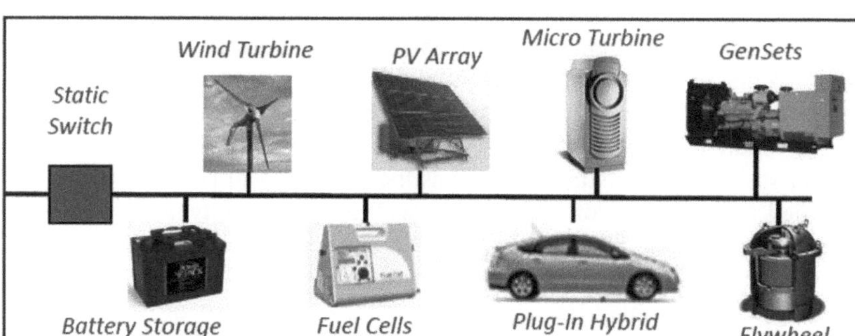

FIGURE 7.5 Implementation of micronetwork grid in a variety of applications.

3.5 Information and Communication Technology in the smart network grid

This area is about the necessity for progress in the smart network grid and execution of the basic benchmarks for information and communication technology (ICT) support in anticipated smart network grids. ICT is a prerequisite for information substitution between the distinctive market strata (layers) in the power production network and the safe, financial, and naturally delicate activity of smart network grids [6]. Currently, ICT is applied at transmission and subtransmission levels of smart network grids and ends at the transport bars of the subtransmission (110 kV)/medium voltage substations. Assorted standard conventions [25] at various voltage levels and for different sorts of hardware are utilized. Though cumbersome, medium and low voltage levels are considered by constrained ICT if monetarily suitable. Institutionalized models that depend on open data models and correspondence administration for all information transmission inside the whole power inventory network and electric force supply framework are required. Basic, extreme, sheltered, and adaptable correspondence foundations are profoundly basic to permit checking, association, control, and dispatching exercises at all levels down to dispersion and clients. Normal data and information models for all data-building barriers, to guarantee a predictable database for the board [28], should be characterized at all degrees of the force framework and power store network. Well-working ICT arrangements are a fundamental principle for maintaining supply security and the productive connection of market players.

3.6 Internet of things protocols designed for smart network grid applications

The techniques of IoE are the subsequently emerging development of our Web era, where several corporeal objects and smart things equipped with calculation and message-passing capabilities could be effortlessly included, at diverse levels, to the Web and Internet [27,40]. Table 7.5 shows smart IoT protocols designed for smart network grid applications.

The smart network grid, measured as an important part of the most significant structures, is distinctly stated as the conventional power network grid amplified through a major-scale ICT as well as generous dynamism incorporation. It can be taken to be prominent in the principal IoE networking system. The smart network grid would occupy billions of smart devices such as smart metering systems, smart devices, smart instruments, smart projecting appliances, and smart vehicles in accumulating a number of message setups that may be community or confidential. Nevertheless, protection is perceived as one of the key points causing the speedy and huge scaling and consumption of IoE revelations and the technology of the smart network grid.

4. Conclusion

The smart network grid is a promising conception that supports pioneering methods to resolve power eminence and consistency issues for both service sectors and consumers. The prominent aim of the smart network grid is to conserve a certain level of residents' lifestyle values while sustaining the complete spectrum of cost-effective activities. Conventions and protocols that support IoT technology are used for better performance of the smart network grid power distribution mechanism. This chapter provides a rational description of the challenges and major applications of IoT-based methods in different applications along with challenging protocols developed for better management of smart network grid operations in an optimized mode.

References

[1] Y.E. Song, Y. Liu, S. Fang, S. Zhang, Research on applications of the internet of things in the smart network_grid, in: 2015 7th International Conference on Intelligent Human-Machine Systems and Cybernetics, Hangzhou, 2015, pp. 178–181.

[2] Y. Li, X. Cheng, Y. Cao, D. Wang, L. Yang, Smart choice for the smart network_grid: narrowband internet of things (NB-IoT), IEEE Internet Things J. 5 (3) (June 2018) 1505–1515.

[3] Z. Jingchao, H. Jinhua, M. Linchang, S. Ningxi, L. Xiaolei, The implementation of the internet of things technology in Henan smart distribution network demonstration project, in: 2016 China International Conference on Electricity Distribution (CICED), Xi'an, 2016, pp. 1–5.

[4] S. Sidid, S. Gaur, Smart network_grid building automation based on internet of things, in: 2017 Innovations in Power and Advanced Computing Technologies (i-PACT), Vellore, 2017, pp. 1–4.

[5] M.H. Yaghmaee, H. Hejazi, Design and implementation of an internet of things based smart energy metering, in: 2018 IEEE International Conference on Smart Energy Network_grid Engineering (SEGE), Oshawa, ON, 2018, pp. 191–194.

[6] Y.F. Wang, W.M. Lin, T. Zhang, Y.Y. Ma, Research on application and security protection of internet of things in smart network_grid, in: IET International Conference on Information Science and Control Engineering 2012 (ICISCE 2012), Shenzhen, 2012, pp. 1–5.

[7] Y. Zhen, X. Li, Y. Zhang, L. Zeng, Q. Ou, X. Yin, Transmission tower protection system based on internet of things in smart network_grid, in: 2012 7th International Conference on Computer Science & Education (ICCSE), Melbourne, VIC, 2012, pp. 863–867.

[8] E. Spanò, L. Niccolini, S.D. Pascoli, G. Iannacconeluca, Last-meter smart network_grid embedded in an internet-of-things platform, IEEE Trans. Smart Network_grid 6 (1) (January 2015) 468–476.

[9] M.B. Satti, N. Ali, M.M.A. Asif, A.H. Abid, A. Hasan, A low-cost smart load enabler device, in: 2018 15th International Conference on Smart Cities: Improving Quality of Life Using ICT & IoT (HONET-ICT), Islamabad, 2018, pp. 61–64.

[10] M. Shariat, M. Safkhani, How the control over smart meters is lost in the Yan et al. lightweight AKA scheme for smart network_grids, in: 2017 9th International Conference on Information and Knowledge Technology (IKT), Tehran, 2017, pp. 82–84.

[11] Y. Zhang, W. Zou, X. Chen, C. Yang, J. Cao, The security for power internet of things: framework, policies, and countermeasures, in: 2014 International Conference on Cyber-Enabled Distributed Computing and Knowledge Discovery, Shanghai, 2014, pp. 139–142.

[12] A.S. Musleh, M. Debouza, M. Farook, Design and implementation of smart plug: an internet of things (IoT) approach, in: 2017 International Conference on Electrical and Computing Technologies and Applications (ICECTA), Ras Al Khaimah, 2017, pp. 1–4.

[13] W. Shu-wen, Research on the key technologies of IOT applied on Smart Network_grid, in: 2011 International Conference on Electronics, Communications and Control (ICECC), Ningbo, 2011, pp. 2809–2812.

[14] A.K. Koundinya, G.S. Sharvani, K.U. Rao, Calibrated security measures for centralized IoT applications of smart network_grids, in: 2016 International Conference on Computation System and Information Technology for Sustainable Solutions (CSITSS), Bangalore, 2016, pp. 153–157.

[15] X. Chen, L. Sun, H. Zhu, Y. Zhen, H. Chen, Application of internet of things in power-line monitoring, in: 2012 International Conference on Cyber-Enabled Distributed Computing and Knowledge Discovery, Sanya, 2012, pp. 423–426.

[16] A. Kanyilmaz, A. Cetin, Fog based architecture design for IoT with private nodes: a smart home application, in: 2019 7th International Istanbul Smart Network_grids and Cities Congress and Fair (ICSG), Istanbul, Turkey, 2019, pp. 194–198.

[17] S.E. Collier, The emerging enernet: convergence of the smart network_grid with the internet of things, IEEE Ind. Appl. Mag. 23 (2) (March–April 2017) 12–16.

[18] I. Doh, J. Lim, K. Chae, Secure authentication for structured smart network_grid system, in: 2015 9th International Conference on Innovative Mobile and Internet Services in Ubiquitous Computing, Blumenau, 2015, pp. 200–204, https://doi.org/10.1109/IMIS.2015.32.

[19] H. Geng, Internet of things and smart network_grid standardization, in: Internet of Things and Data Analytics Handbook, Wiley, 2017.

[20] P. Ramanathan, Privacy assurances in the internet of things (IoT) world, in: 2018 31st International Conference on VLSI Design and 2018 17th International Conference on Embedded Systems (VLSID), Pune, 2018 (xlii-xlii).

[21] S. Borkar, H. Pande, Application of 5G next generation network to internet of things, in: 2016 International Conference on Internet of Things and Applications (IOTA), Pune, 2016, pp. 443–447.

[22] W. Li, H. Song, F. Zeng, Policy-based secure and trustworthy sensing for internet of things in smart cities, IEEE Internet Things J. 5 (2) (April 2018) 716–723.

[23] C. Jiang, Y. Lu, Research on IoT application and security risk analysis for smart network_grid, in: 2018 5th IEEE International Conference on Cloud Computing and Intelligence Systems (CCIS), Nanjing, China, 2018, pp. 1003–1007.

[24] F. Abate, M. Carratù, C. Liguori, V. Paciello, A low cost smart power meter for IoT, Measurement 136 (2019) 59–66, https://doi.org/10.1016/j.measurement.2018.12.069. ISSN 0263-2241.

[25] M. Saadeh, A. Sleit, K. Eddin Sabri, W. Almobaideen, Hierarchical architecture and protocol for mobile object authentication in the context of IoT smart cities, J. Netw. Comput. Appl. 121 (2018) 1–19, https://doi.org/10.1016/j.jnca.2018.07.009. ISSN 1084-8045.

[26] C. Bekara, Security issues and challenges for the IoT-based smart Network_grid, Procedia Computer Science 34 (2014) 532–537, https://doi.org/10.1016/j.procs.2014.07.064. ISSN 1877-0509.

[27] D. Bruneo, S. Distefano, M. Giacobbe, A.L. Minnolo, F. Longo, G. Merlino, D. Mulfari, A. Panarello, G. Patanè, A. Puliafito, C. Puliafito, N. Tapas, An IoT service ecosystem for smart cities: the #SmartME project, Internet Things 5 (2019) 12–33. ISSN 2542-6605.

[28] E. Kabalci, Y. Kabalci, Chapter 7 - Internet of things for smart network_grid applications, in: E. Kabalci, Y. Kabalci (Eds.), From Smart Network_grid to Internet of Energy, Academic Press, 2019, pp. 249–307. ISBN 9780128197103.

[29] A. Yassine, S. Singh, M. Shamim Hossain, G. Muhammad, IoT big data analytics for smart homes with fog and cloud computing, Future Generat. Comput. Syst. 91 (2019) 563–573, https://doi.org/10.1016/j.future. 2018.08.040. ISSN 0167-739X.

[30] M. Sanwar Hossain, M. Rahman, M. Tuhin Sarker, M. Ershadul Haque, J. Abu, A smart IoT based system for monitoring and controlling the sub-station equipment, Internet Things 7 (2019) 100085, https://doi.org/10.1016/j.iot. 2019.100085. ISSN 2542-6605.

[31] O. Bello, S. Zeadally, Toward efficient smartification of the Internet of Things (IoT) services, Future Generat. Comput. Syst. 92 (2019) 663–673, https://doi.org/10.1016/j.future.2017.09.083. ISSN 0167-739X.

[32] J.P. Astudillo León, L.J. de la Cruz Llopis, Emergency aware congestion control for smart network_grid neighborhood area networks, Ad Hoc Netw. 93 (2019) 101898, https://doi.org/10.1016/j.adhoc.2019.101898. ISSN 1570-8705.

[33] F. Al-Turjman, 5G-enabled devices and smart-spaces in social-IoT: an overview, Future Generat. Comput. Syst. 92 (2019) 732–744. ISSN 0167-739X.

[34] M. Faheem, R. Aslam Butt, B. Raza, M. Waqar Ashraf, MdA. Ngadi, V.C. Gungor, Energy efficient and reliable data gathering using internet of software-defined mobile sinks for WSNs-based smart network_grid applications, Comput. Stand. Interfac. 66 (2019) 103341, https://doi.org/10.1016/j.csi.2019.03.009. ISSN 0920-5489.

[35] L.F.A. Roman, P.R.L. Gondim, J. Lloret, Pairing-based authentication protocol for V2G networks in smart network_grid, Ad Hoc Netw. 90 (2019) 101745. ISSN 1570-8705.

[36] Y. Yılmaz, S. Uludag, Timely detection and mitigation of IoT-based cyberattacks in the smart network_grid, J. Franklin Inst. (2019). ISSN 0016-0032, https://doi.org/10.1016/j.jfranklin.2019.02.011.

[37] M. Faheem, V.C. Gungor, Energy efficient and QoS-aware routing protocol for wireless sensor network-based smart network_grid applications in the context of industry 4.0, Pplied Soft Computing 68 (2018) 910–922. ISSN 1568-4946.

[38] M. Rath, Security in IoT and big data with technical challenges, handbook of research on big data and the IoT, IGI Global pg (2019) 197–210, https://doi.org/10.4018/978-1-5225-7432-3.ch011.

[39] M. Rath, V.K. Solanki, Contribution of IoT and big data in modern health care applications in smart city, in: Handbook of IoT and Big Data, CRC Press, 2019, pp. 109–122. ISBN 9780429053290 - CAT# KE71438.

[40] M. Rath, B. Pati, Appraisal of soft computing methods in collaboration with smart city applications and wireless network, Int. J. e-Collaboration 14 (1) (2018) 27–42, https://doi.org/10.4018/IJeC.2018010102.

[41] K. Mahmood, L. Xiong, S. Ashraf Chaudhry, H. Naqvi, S. Kumari, A.K. Sangaiah, J.J.P.C. Rodrigues, Pairing based anonymous and secure key agreement protocol for smart network_grid edge computing infrastructure, Future Generat. Comput. Syst. 88 (2018) 491–500. ISSN 0167-739X.

[42] F. Al-Turjman, M. Abujubbeh, IoT-enabled smart network_grid via SM: an overview, Future Generat. Comput. Syst. 96 (2019) 579–590. ISSN 0167-739X.

Further reading

[1] S.E. Collier, The emerging enernet: convergence of the smart network_grid with the internet of things, in: 2015 IEEE Rural Electric Power Conference, Asheville, NC, 2015, pp. 65–68.

[2] X. Chen, J. Liu, X. Li, L. Sun, Z. Yan, Integration of IoT with smart network_grid, in: IET International Conference on Communication Technology and Application (ICCTA 2011), Beijing, 2011, pp. 723–726.

[3] V. Araujo, K. Mitra, S. Saguna, C. Åhlund, Performance evaluation of FIWARE: a cloud-based IoT platform for smart cities, J. Parallel Distr. Comput. 132 (2019) 250–261. ISSN 0743-7315.

[4] M. Rath, B. Pati, Security assertion of IoT devices using cloud of things perception, Int. J. Interdis. Telecommun. Netw. IGI Global 11 (4) (2019) 17–31, https://doi.org/10.4018/IJITN.201910010.

[5] M. Rath, B. Pati, B.K. Pattanayak, Design and development of secured framework for efficient routing in vehicular ad-hoc network, Int. J. Bus. Data Commun. Netw. IGI Global 15 (2) (2019) 55–72, https://doi.org/10.4018/IJBDCN.2019070104.

[6] E. Commission. Standardization Mandate to European Standardisation Organisations (ESOs) to Support European Smart Network_grid Deployment. Last accessed [Online]. Available from: ftp://ftp.cencenelec.eu/CENELEC/Smartnetwork_grid/M490.pdf.

[7] Smart Grids - European Standardisation. Available from: www.cencenelec.eu/standards/Sectors/SustainableEnergy/SmartNetwork_grids/Pages/default.aspx. (Accessed 30 December 2019).

[8] Smart Grids - European Standardisation. www.cenelec.eu/aboutcenelec/whatwedo/technology sectors/smartnetwork_grids.html. (Accessed 30 December 2019).

Chapter 8

Grid integration of single stage solar photovoltaic system using the exponential-based variable-step-size least mean square filtering control technique

Pushpendra, Rashmi Agarwal
Department of Electrical Engineering, J.C.Bose University of Science & Technology, YMCA, Faridabad, Haryana, India

1. Introduction

Today's vision of the future is that fuels such as coal, petrol, diesel, and others may not be accessible a half-century from now. Hence as an engineer, it is my responsibility to take an interest in renewable sources and develop improvements so that upcoming generations have alternatives. The electricity produced from the aforementioned sources results in significant air pollution in the form of increased levels of harmful gases like NO_2 for ozone depletion, CO_2 for climate change, and SO_2 as the cause of acid rain and other small particulates [1]. These effects result in harmful impacts on human health and the environment. Unconventional sources like solar, hydro, and wind are clean, create less pollution, and are environmentally friendly given their reduced overall carbon footprints. Solar provides a solution to the problems of conventional sources. Solar is a dominant energy source because its time investment for installation is less than that of others and because of its availability, maintenance, and pollution-free production. It requires no rotational parts for energy extraction.

A vital concern with renewable power sources is their intermittent, varying power availability, which requires a huge buffer energy source to provide a smooth flow of required load power. The solar photovoltaic (PV) system may be a standalone or grid-connected system. Grid integration plays an important role

Advances in Smart Grid Power System. https://doi.org/10.1016/B978-0-12-824337-4.00008-4

nowadays. As batteries are eliminated in a grid-connected solar system, the cost of the whole system is reduced greatly because 40% of total system cost is the cost of batteries connected to the system, and normally batteries must be replaced after 3 years. Another advantage is that the extra amount of power generated (after satisfying the load requirement) can be fed to the grid, which reduces required power generation at the generating end with conventional sources of energy. Therefore, grid-connected solar PV energy is gaining more popularity each day, indicating its market superiority to a standalone system. Weather conditions and solar irradiance vary continuously, hence the system requires a maximum power point tracking (MPPT) controller. More than 25 MPPT techniques are known; among them, incremental conductance (INC) is simple and easy to apply and has efficient tracking efficiency, whereas others are complex and less efficient [2,3]. Floating-size INC can self-adjust its step-size range to grasp the maximum power point (MPP), thus improving system behavior. An INC-based MPPT algorithm can manage the nonlinearity of the solar PV array. It results in the delivery of peak power under steady-state, dynamic, and variable solar irradiance conditions. Hence, the INC-based technique is used to obtain the peak amount of energy from the solar PV array. The use of power electronic devices makes the load nonlinear and may result in harmonics being injected into the system. Hence, different algorithms are used to mitigate the harmonics so that power quality is maintained under the standards conforming to IEEE-519 for nonlinear loads at the point of common coupling (PCC).

Adaptive filtering started in the early 1990s, T. Bose gave the concept of adaptive bilinear and nonlinear filtering to accelerate convergence [4,5]. In 1991 M. Givens from the USA explained the application of medium least mean square (LMS) algorithms to ADPCM systems. This is done to protect adaptive filters from impulsive noise [6]. M. D. Macleod proposed a robust LMS adaptive filter algorithm to reduce maladjustments. When noise is added to the input signal, it gives distorted output that is not desirable [7]. The realistic approach shown in this paper is to tackle unwanted noise entering in the system. The adaptive technique is used in different areas of research such as in CMOS technology [8], distribution estimation [9], adaptive graph filtering [10], mining applications [11], and spare space identification [12].

It can be seen and concluded from above that the LMS-based adaptive filter algorithm is encouraging for researchers because of its computational simplicity and ability to minimize harmonics from the output quantity.

In LMS, the error is calculated in a squared term as the cost function. The LMS technique was first developed by Widrow and Hoff in 1960 [13]. The LMS has a good convergence rate but is less effective for low signal-to-noise ratios (SNRs). To obtain the best characteristics of the adaptive filtering-based algorithm with a high convergence rate, it should be robust in a low SNR environment. An intelligent algorithm capable of maintaining both parameters, known as an exponential-based variable-step-size least mean square (EVSSLMS) algorithm, can be used [14]. The EVSSLMS algorithm is the simplest and most efficient method from the family of LMS adaptive filters. Hence, the EVSSLMS algorithm is a solution for the mitigation of power-quality problems in a grid-connected solar PV system, which is a challenging task under nonlinear load at PCC, as nonlinear loads are subject to inject harmonics into the system that pollute the supply load current. This is the concern about other equipment connected to the system that may lead to damage. The Simulink results of the proposed EVSSLMS algorithm extract fundamental current components and harmonics free from the square-shaped load current component as well as retain grid current total harmonic distortion (THD) below 5%, which is well under IEEE-519 standards [15].

2. Proposed three-phase three-wire solar photovoltaic system

Fig. 8.1 shows the proposed three-phase three-wire (3P3W) solar PV system with the EVSSLMS technique, which subsists of a solar PV array connected to a 3-phase voltage-source converter (VSC), coupling inductor (L_f), ripple filter (RC), three-phase AC mains, and nonlinear load. An insulated-gate bipolar transistor (IGBT)-based three-phase VSC is used for the conversion of output from DC to AC. The coupling inductor (L_f) is used for the elimination of ripples from grid currents. An RC is essentially needed for mitigating switching ripples from VSC. $S_{w1}-S_{w6}$ are the six switches of the VSC. Load unbalancing can be achieved by disconnecting phase "a" of the load with the help of a switch.

3. Control technique

The control technique is subdivided into two parts, MPPT control for maximum power extraction and the EVSSLMS control technique for VSC switch control. This proposed technique provides various functions like harmonics current mitigation, load balancing, active power injection, and reactive power compensation. A detailed description is given in the following.

3.1 Incremental conductance–based maximum power point tracking technique

Due to changing atmospheric conditions, solar PV arrays produce variable power. Therefore, to improve the efficiency of the system, the extraction of maximum output is necessary [2]. In most instances, the INC-based MPPT technique is used to achieve maximum power, which helps in estimating DC loss components of the proportional-integral (PI) controller by using solar PV

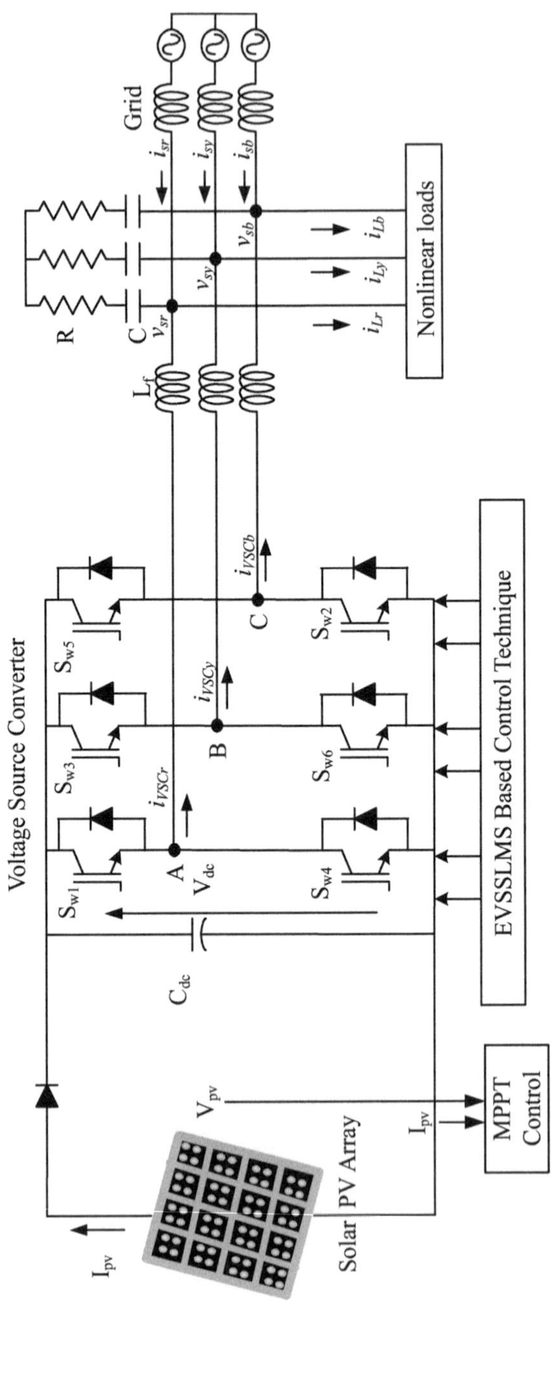

FIGURE 8.1 Three-phase three-wire solar photovoltaic system.

parameters. The INC technique depends on the steepness of the P_{pv}-V_{pv} curve and is explained below.

$$-\frac{I_{pv}}{V_{pv}} = \frac{\partial P_{pv}}{\partial V_{pv}} = 0, \text{atMPP} \tag{8.1}$$

Here conductance is 0 at the MPP.

$$-\frac{I_{pv}}{V_{pv}} < \frac{\partial P_{pv}}{\partial V_{pv}}, -\frac{I_{pv}}{V_{pv}} > 0, \text{left side of MPP} \tag{8.2}$$

When conductance is greater than zero, MPP is shifted to the left side.

$$-\frac{I_{pv}}{V_{pv}} > \frac{\partial P_{pv}}{\partial V_{pv}}, -\frac{I_{pv}}{V_{pv}} < 0, \text{right side of MPP} \tag{8.3}$$

When conductance is less than zero, MPP is shifted to the right side. This can be clearly seen from the preceding equation. Fig. 8.2 shows a flowchart of the INC control technique.

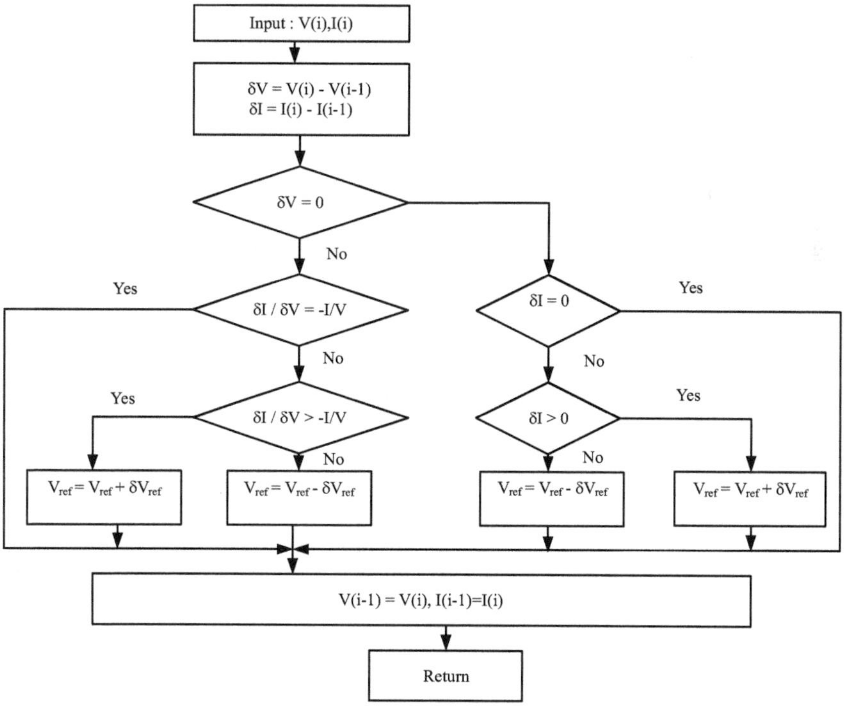

FIGURE 8.2 Incremental conductance–based control technique flowchart.

3.2 Exponential-based variable-step-size least mean square based control technique

3.2.1 Existing least mean square algorithm

Fig. 8.3.A illustrates that the linear-weighted sum of input x(n) and filter coefficients w(n) makes output y(n) and gives error signal e(n)—i.e., the difference between d(n) and y(n).

Here error can be calculated as

$$e(n) = d(n) - y(n) = d(n) - w^T(n)x(n) \tag{8.4}$$

$$\text{Where, } y(n) = \sum_{i=1}^{N} w_i(n)x_i(n) = w(n)^T x(n) \tag{8.5}$$

According to the filtering principle of the adaptive filter, the weight coefficients $w_{optimal}(n)$ of the filter in best state can be obtained through minimum mean square, $E[e^2(n)]_{min}$, which is a quadratic function of the filter coefficients w(n) shown above.

The gradient formula of the LMS algorithm is expressed as

$$\widehat{\nabla}(n) = \nabla\left[e^2(n)\right] = \left[\frac{\delta e^2(n)}{\delta w_0} \frac{\delta e^2(n)}{\delta w_1} \frac{\delta e^2(n)}{\delta w_N}\right]^T \tag{8.6}$$

By using Eqs. (8.4) and (8.6), the iterative formula of LMS filter weight vector is obtained as

$$w(n+1) = w(n) - \mu\widehat{\nabla}(n) = w(n) + 2\mu e(n)x(n) \tag{8.7}$$

where μ is stated as the iterative step size in Eq. (8.7).

It can be seen that the mean square value $E[e^2(n)]_{min}$ reduces with increases in value n. When n reaches infinity, the mean square value $E[e^2(n)]_{min}$ is smallest, and the performance of the filter becomes optimum.

But the problem with the existing algorithm is that μ is fixed. Hence, this technique can converge only if step size satisfies suitable conditions. Also, when μ is small, its convergence will be small, and similarly when μ is large, it

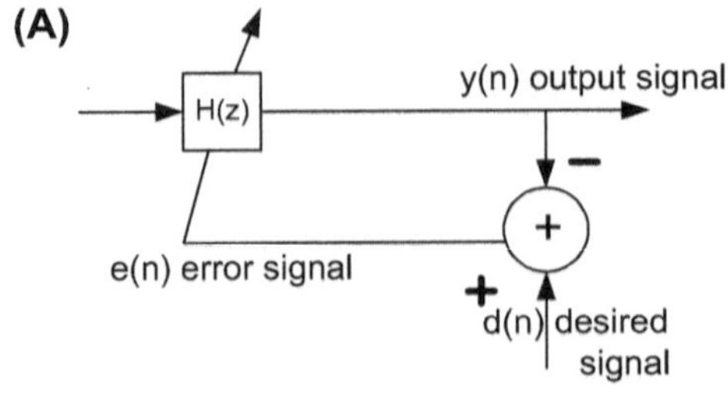

FIGURE 8.3A Illustrative diagram of adaptive filter.

converges fast but may deviate from actual convergence. Therefore, speed and accuracy at the same time, and with stability, are difficult until μ is no longer variable.

3.2.2 Exponential-based variable-step-size least mean square adaptive algorithm

This algorithm overcomes the problem of the existing LMS algorithm and provides a balance between system stability and convergence speed, with step size depending on the error signal of the current moment.

According to this technique, in the initial stage when the system starts to converge, the chosen μ will be large for fast convergence; e(n) is also relatively large and ensures tracking of the time-varying system. After convergence becomes very small, a smaller step size is required. In addition, if a sudden change in the system occurs, the large error signal may control the step size increase—i.e., retracing is required and hence can be easily done by adding one regulator called α to the algorithm, which makes it flexible for easy adjustment of the step-size function.

The new algorithm goes under an effortless process on e(n) to stop random oscillations of the step size caused by peak samples of e(n). The step size is given as

$$\mu = \alpha\left\{1 - \frac{1}{\exp[e(n)]^x}\right\}, x = \frac{1}{2}, \frac{1}{4}, \frac{1}{6} \tag{8.8}$$

The function μ is time dependent and varying, where x determines the smoothing degree of e(n).

It can be seen from Fig. 8.3B that variable step size increases with an increase in e(n), but the increasing rate is not the same for different values—i.e., $x = \frac{1}{2}, \frac{1}{4}, \frac{1}{6}$.

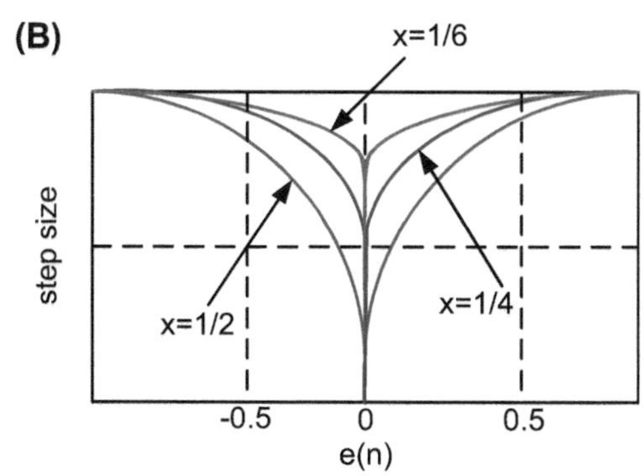

FIGURE 8.3B Wave forms of an improved variable step function.

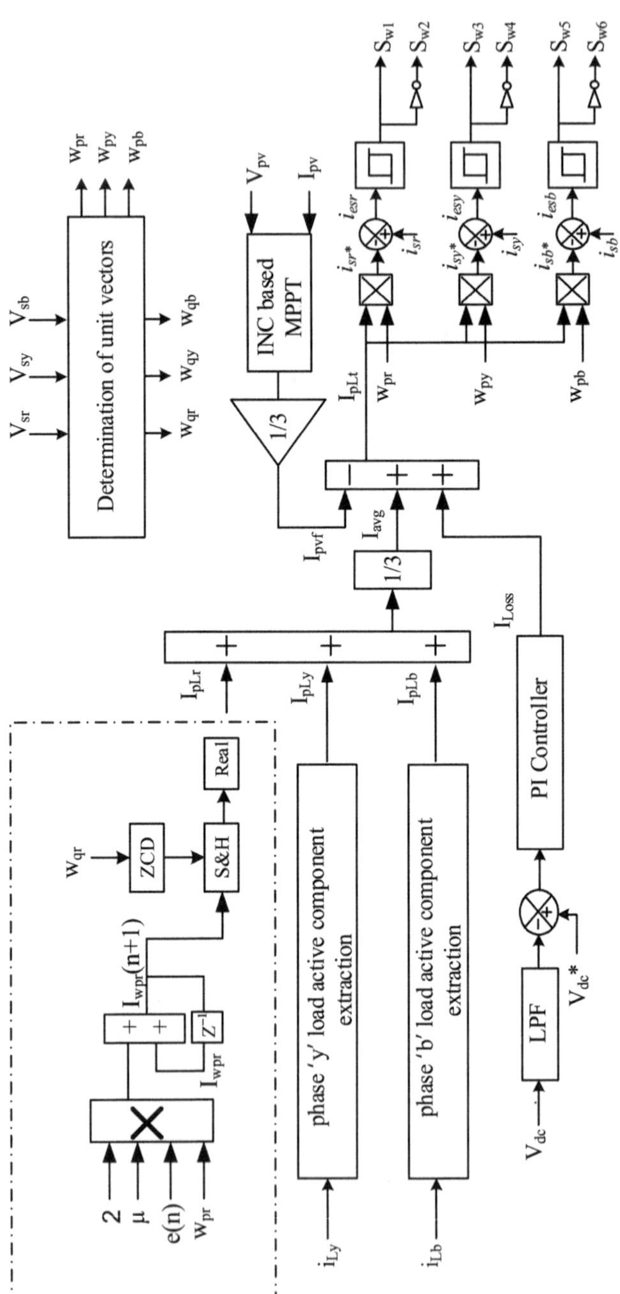

FIGURE 8.4 Exponential-based variable-step-size least mean square control technique. The EVSSLMS control technique is discussed in this section, and a flowchart of the technique is shown as Fig. 8.4.

3.3 Exponential-based variable-step-size least mean square control technique

3.3.1 Extraction of unit vector using unit template

Amplitude of PCC voltage (V_t) is given as

$$V_t = \left\{2\left(v_{sr}^2 + v_{sy}^2 + v_{sb}^2\right)/3\right\}^{1/2} \tag{8.9}$$

For the in-phase unit vector calculation, v_{ry}, v_{yb} (line voltages) are taken and transformed into v_{sr}, v_{sy}, and v_{sb} (phase voltages), which are expressed as [16]

$$\begin{bmatrix} v_{sr} \\ v_{sy} \\ v_{sb} \end{bmatrix} = \frac{1}{3}\begin{bmatrix} 2 & 1 \\ -1 & 1 \\ -1 & -2 \end{bmatrix}\begin{bmatrix} v_{ry} \\ v_{yb} \end{bmatrix} \tag{8.10}$$

These phase voltages are used for the in-phase unit vector calculation, which is given as

$$w_{pr} = \frac{v_{sr}}{V_t}, w_{py} = \frac{v_{sy}}{V_t}, w_{pb} = \frac{v_{sb}}{V_t} \tag{8.11}$$

The quadrature voltage unit vector estimation is

$$w_{qr} = \frac{(-w_{py} + w_{pb})}{\sqrt{3}}, \tag{8.12}$$

$$w_{qy} = \frac{(3w_{pr} + w_{py} - w_{pb})}{2\sqrt{3}}, \tag{8.13}$$

$$w_{qb} = \frac{(-3w_{pr} + w_{py} - w_{pb})}{2\sqrt{3}} \tag{8.14}$$

3.3.2 Extraction of weight of fundamental active component of load current

The proposed control algorithm extracts the weight of fundamental active load current components of phase "a," which can be computed as

$$I_{wpr}(n+1) = w(n) - \mu\hat{\nabla}(n) = I_{wpr}(n) + 2\mu e(n) w_{pr}(n) \tag{8.15}$$

where e(n) is the error in the adaptive component, μ is the step size and also the e(n) error is given by

$$e(n) = i_{Lr}(n) - w_{pr}(n) I_{wpr}(n) \tag{8.16}$$

where $I_{wpr}(n)$ is the weight of the active load current component of the nth sample, $i_{Lr}(n)$ is the load current at the nth sample, and $w_{pr}(n)$ is the in-phase unit template at the nth sample.

Also, step size is given by the given below term consisting of e(n):

$$\mu = \alpha\left\{1 - \frac{1}{\exp[e(n)]^{x}}\right\} \tag{8.17}$$

where regulator α is added to the algorithm that makes the function flexible and hence easy for adjusting step size.

The calculated harmonic less fundamental component that is the same in-phase with i_L has a definite Φ (phase angle). The triggering pulse for the S&H (sample and hold) logic circuit is provided by zero crossing detector. The magnitude of i_{pL} is the output of S&H logic. An absolute block in the circuit is used for conversion of the active load component into the actual signal. Similarly, I_{pLy}, I_{pLb} can be calculated.

3.3.3 Fundamental weight extraction of load current

The three-phase equivalent-weight component can be estimated as

$$I_{pLavg} = \frac{1}{3}[I_{pLr} + I_{pLy} + I_{pLb}] \tag{8.18}$$

where I_{pLavg} = average component of fundamental active load current

Unbalanced nonlinear load compensation is very important. Hence, the I_{loss} (m) component should be calculated. Any unbalance in the load directly influences DC link voltage and must be within prescribed limits. PI controller output is taken as the loss component for the DC bus voltage mentioned previously and as given below:

$$I_{loss}(m) = I_{loss}(m-1) + k_p\{V_{dce}(m) - V_{dce}(m-1)\} + k_i V_{dce}(m) \tag{8.19}$$

In the above Eq. V_{dce} is given as,

$$V_{dce} = V_{dc} * - V_{dc} \tag{8.20}$$

After adding I_{loss} with I_{pLavg} and I_{pvf}, I_{pLt} can be written as,

$$I_{pLt} = I_{pLavg} + I_{loss} - I_{pvf} \tag{8.21}$$

I_{loss} = DC Loss component
I_{pvf} = Feed forward current component
I_{pLt} = net active current component

I_{pvf} is added in I_{pLt} to provide instant response w.r.t. changing surrounding conditions, is calculated as,

$$I_{pvf} = \frac{2}{3}\left[\frac{P_{pv}}{V_{pv}}\right] \tag{8.22}$$

Extraction of reference grid current are evaluated as,

$$i_{sr}^{*} = I_{pLt} * w_{pr},$$
$$i_{sy}^{*} = I_{pLt} * w_{py}, \tag{8.23}$$
$$i_{sb}^{*} = I_{pLt} * w_{pb},$$

The current errors are estimated as,

$$i_{esr} = i_{sr}^{*} - i_{sr},$$
$$i_{esy} = i_{sy}^{*} - i_{sy}, \tag{8.24}$$
$$i_{esb} = i_{sb}^{*} - i_{sb},$$

Switching sequences for a three-leg VSC can be derived using these error signals.

3.3.4 Estimation of switching signals using a hysteretic pulse width modulation controller with a three-phase voltage-source converter

The Proposed control technique is useful for providing switching signals to the VSC. A hysteretic current controller (HSS) with an indirect current-control loop that senses grid current parameters. These sensed signals are subtracted from the reference grid current parameters and hence the difference of these signals is given to the HSS and passed through it with a limiting value ($\delta = 0.01$) to generate pulses for the VSC.

3.4 Design parameters

A diagram illustrating the three-phase VSC is shown in Fig. 8.1. It can be seen that six IGBT-based switches are used to provide necessary compensation using the VSC. The design parameter is explained below for the calculation and selection of DC bus voltage, AC inductors, etc. (8.20).

3.4.1 DC bus capacitor voltage selection

$$V_{DC} = \frac{2\sqrt{2}V_{Line-Line}}{\sqrt{3}m} \tag{8.25}$$

Where m = modulation index

$V_{Line\text{-}Line}$ = the line–line PCC voltage.

For m assumed to be 1 and a value for V_{LL} of 415 V, one can get the magnitude of V_{DC} as 677.6 V. Hence, it is approximated as 700 V.

3.4.2 DC bus capacitor selection

$$C_{DC} = \frac{6Valtk_t}{V_{DC}^2 - V_{DC1}^2} \tag{8.26}$$

Where I = capacitor current,
V_{DCpp} = ripple in capacitor voltage,
a = overloading factor (1.2), and
t = time (30 ms)

Select V_{DC} = 700 V and V_{DC1} = 677.69 V and further assume the variation of energy during dynamics is 10%. Thus the capacitance value is 4000 μF.

3.4.3 AC inductor selection

$$L_r = \frac{\sqrt{3}mV_{DC}}{12af_sI_{cr,pp}(\%)} \tag{8.27}$$

Here fs = 1800 Hz, m = 1 and V_{DC} = 700 V while considering a 15% ripple in the current and a = 1.2, so the AC inductor value L_r can be calculated as 4 mH.

3.4.4 Ripple filter selection

$$R_f C_f = \frac{T_s}{10} \tag{8.28}$$

Here R_f and C_f = resistance and capacitance of ripple filter, respectively, and T_s = switching time period. For the particular switching frequency = 1800 Hz, ripple filter parameters are supposed to be taken as R_f = 10 Ω and C_f = 5.5 μF. The calculated ripple filter value is 10 Ω, 5.5 μF.

3.4.5 Selection of voltage and current ratings

Under dynamic condition, V_{sw} (voltage rating) can be estimated as

$$V_{sw} = V_{DC} + V_d \tag{8.29}$$

Under dynamic condition, a 10% overshoot is likely to happen in the DC link capacitor voltage, which is why V_d is taken. V_{sw} is calculated as 770 V. Hence, IGBTs are suitable for these voltage ratings if they include an appropriate safety factor. Hence, IGBTs rated at 1200 V will be needed for selection of the VSC.

Under dynamic condition, I_{sw} (current rating) can be given as

$$I_{sw} = 1.25\left(I_{cr,pp} + I_{peak}\right) \tag{8.30}$$

Here $I_{peak} = 67\sqrt{2}$ A and $I_{cc,pp}$ is 15% of I_{peak}

Eqs. 8.24 and 8.25 are used for calculating the ratings of switches. The ampere rating of the switch is calculated as 136 A. Hence, IGBTs can work at the current rating with an appropriate safety factor. Therefore, IGBTs at the current rating of 300A are selected for the VSC. IGBT as a switch is required for the fulfillment of the design parameter for the VSC.

4. Results

The response of a grid-integrated solar PV array using the EVSSLMS control technique is analyzed by performing a simulation in MATLAB for steady-state and dynamic conditions. A nonlinear load with values R = 20 Ω and L = 200 mH is selected. The grid is modeled with a series and load inductance of 3 mH. Here, 1Soltech 1STH-215-P is used as a PV module in MATLAB-Simulink.

4.1 MATLAB-simulink results for normal condition

Figs. 8.5–8.7 shows the behavior of a single-stage solar PV-3P3W grid-connected system at the steady-state condition for an interval of 0.2–0.3 s. Fig. 8.5 shows that the grid current parameters are sustained, balanced, and sinusoidal, whereas load current is distorted and sinusoidal. It also shows that VSC currents are the same in magnitude and opposite in phase and can maintain balanced and sinusoidal grid currents while keeping DC link capacitor voltage unaltered at unchanging atmospheric solar conditions. Fig. 8.6 depicts the behavior of solar PV power, solar PV current, and voltage. The negative power of the grid means that power is provided to the grid, whereas load requires positive power as shown in Fig. 8.6. Fig. 8.7 shows the internal signal of the proposed controller, and it can be clearly seen that the proposed controller extracts the sinusoidal fundamental load current component from the load current. Fig. 8.7 depicts internal parameters such as active load component (i_{pLr}), equivalent weight load component (I_{pLavg}), feed-forward current component (I_{pvf}), loss component (I_{loss}), and net active load component (I_{pLt}), which remain at fixed values under a steady-state condition because only system disturbance can affect these parameters.

4.2 MATLAB-simulink results for unbalanced nonlinear load

Figs. 8.8–8.10 shows the response of a single-stage solar PV-3P3W grid-connected system at a load unbalanced condition. The load unbalancing is done in the interval of 0.2–0.3 s by disconnection of a load of phase "r" for the

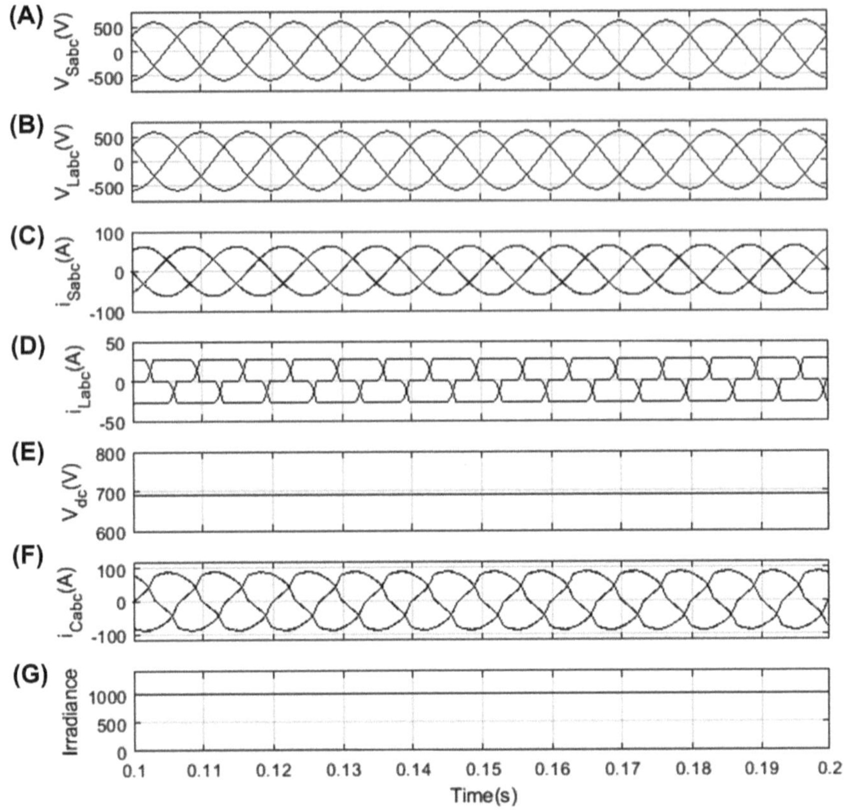

FIGURE 8.5 MATLAB-Simulink results for normal condition (A) grid voltage V_{Sryb} (B) load voltage V_{Lryb} (C) grid current i_{Sryb} (D) load current i_{Lryb} (E) DC voltage V_{dc} (F) filter current i_{Cryb} (G) irradiance.

interval. Fig. 8.8 shows that during this interval, grid current (i_{Sryb}) is maintained sinusoidally even after load current (i_{Lr}) is distorted while making the DC link capacitor voltage almost constant. This shows that sudden disconnection does not affect system stability. The VSC current regulates to make grid current sinusoidal. Fig. 8.8 shows that grid currents have increased due to the decrease in load power requirement and hence more solar PV active power is injected into the grid during load reversal. Fig. 8.9 depicts that under the disconnection of load, i_{Sryb} has increased because the active power injected into the grid has raised as power required by the load has decreased. Moreover, after the transition period, all signals are settled to their steady-state values. Fig. 8.10

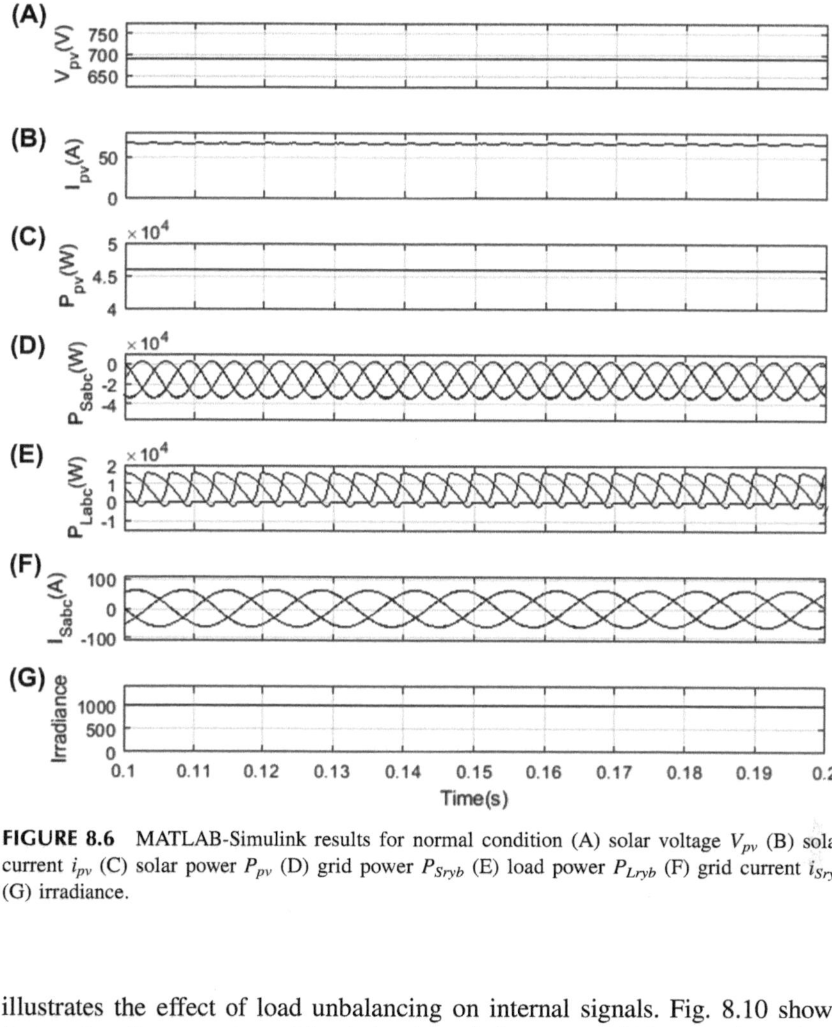

FIGURE 8.6 MATLAB-Simulink results for normal condition (A) solar voltage V_{pv} (B) solar current i_{pv} (C) solar power P_{pv} (D) grid power P_{Sryb} (E) load power P_{Lryb} (F) grid current i_{Sryb} (G) irradiance.

illustrates the effect of load unbalancing on internal signals. Fig. 8.10 shows that under disconnection of load, i_{fLr} is ended to zero within a cycle as load current (i_{Lr}) becomes zero which shows that the system has fast dynamic behavior.

The I_{pLr} component has also settled to zero position under load perturbation, whereas the I_{loss} component of the system has increased as shown in Fig. 8.10. But it does not affect the feed-forward composition component because it is derived from the PV voltage and the current, which is only affected by variation in solar irradiance as depicted in Fig. 8.10. Finally, after the connection of the load of phase "r," all values are settled to steady-state value.

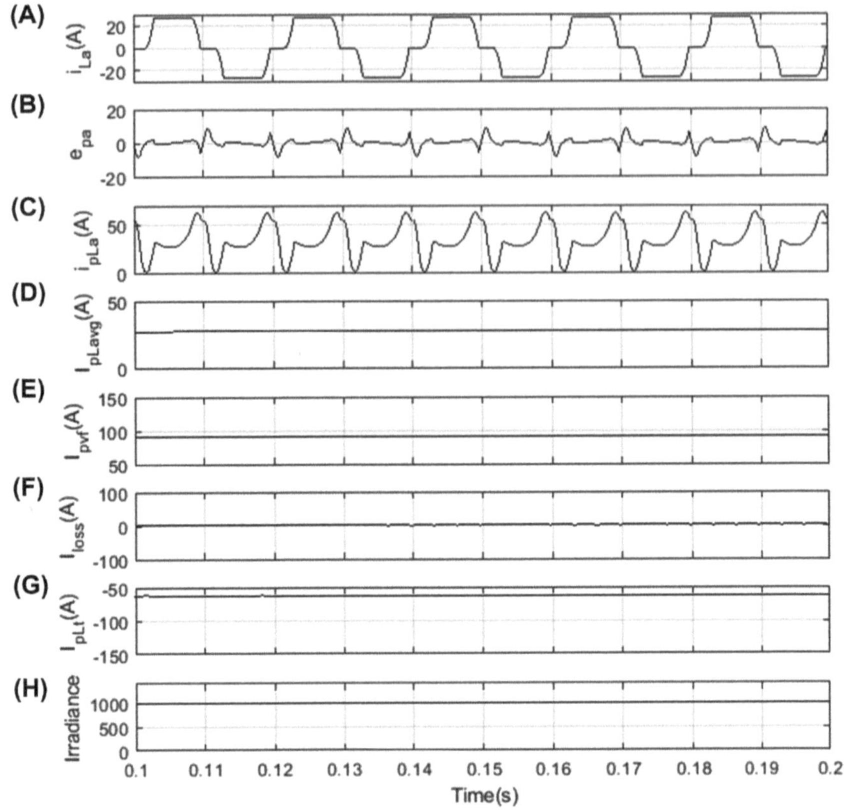

FIGURE 8.7 MATLAB-Simulink results for normal condition (A) load current i_{Lr} (B) fundamental load component i_{fLr} (C) active load component i_{pLr} (D) equivalent weight load component I_{pLavg} (E) feed-forward current component I_{pvf} (F) loss component I_{loss} (G) net active load component I_{pLt} (H) irradiance.

4.3 MATLAB-simulink results under changing atmospheric conditions

Figs. 8.11–8.13 illustrate the dynamic response of a single-stage solar PV-3P3W system under changing atmospheric conditions for an interval of 0.5–0.6 s. Fig. 8.11 represents decreased grid current signal magnitude due to decreased solar PV insolation, with the load current requirement satisfactorily supplied while keeping DC link voltage constant at the desired level. The dynamic behavior of the solar PV system is proportional to variation in solar PV irradiance. Fig. 8.11 depicts VSC current variation due to a decline in

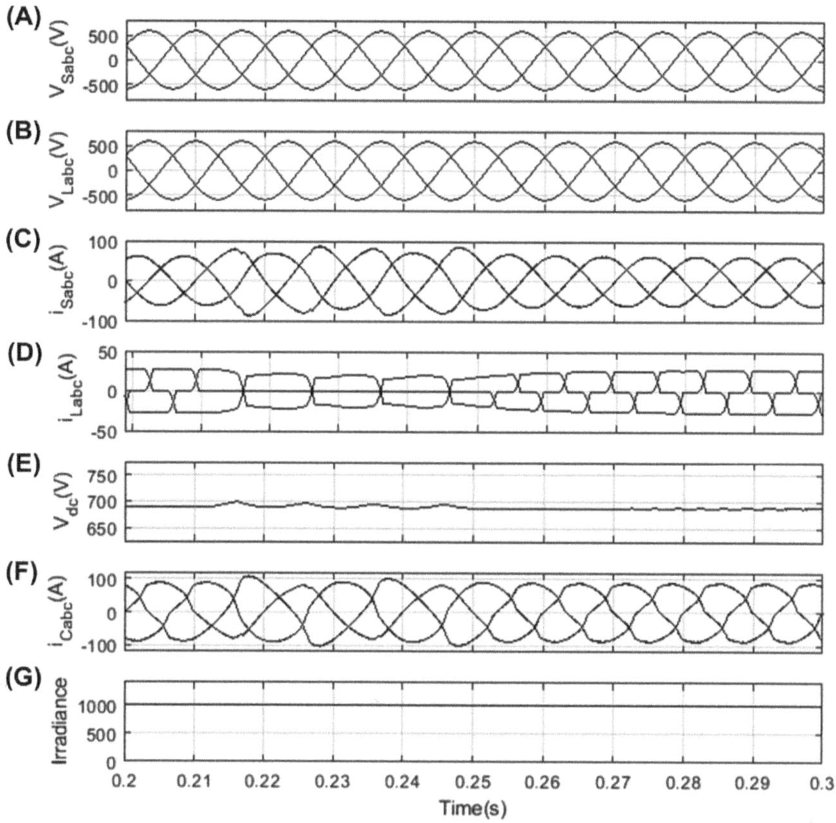

FIGURE 8.8 MATLAB-Simulink results for load unbalanced condition (A) grid voltage V_{Sryb} (B) load voltage V_{Lryb} (C) grid current i_{Sryb} (D) load current i_{Lryb} (E) DC voltage V_{dc} (F) filter current i_{Cryb} (G) irradiance.

irradiance at a magnitude of 1000 to 500 W/m^2, leading to declination of grid currents that results in reduced solar power. This indicates that the INC control technique can work properly such that peak power can be drawn from a solar PV array under changing surrounding conditions. Fig. 8.12 depicts that under a variation in insolation level, solar PV parameter—i.e., voltage and current—remain at their open-circuit voltage (V_{oc}) and short circuit current (I_{sc}). When solar irradiance is decreased to 500 W/m^2, it can be seen that voltage remains near V_{oc}, whereas I_{sc} is reduced to 50% of its maximum value, resulting in decreased power supplied to the grid. Fig. 8.13 shows that under varying

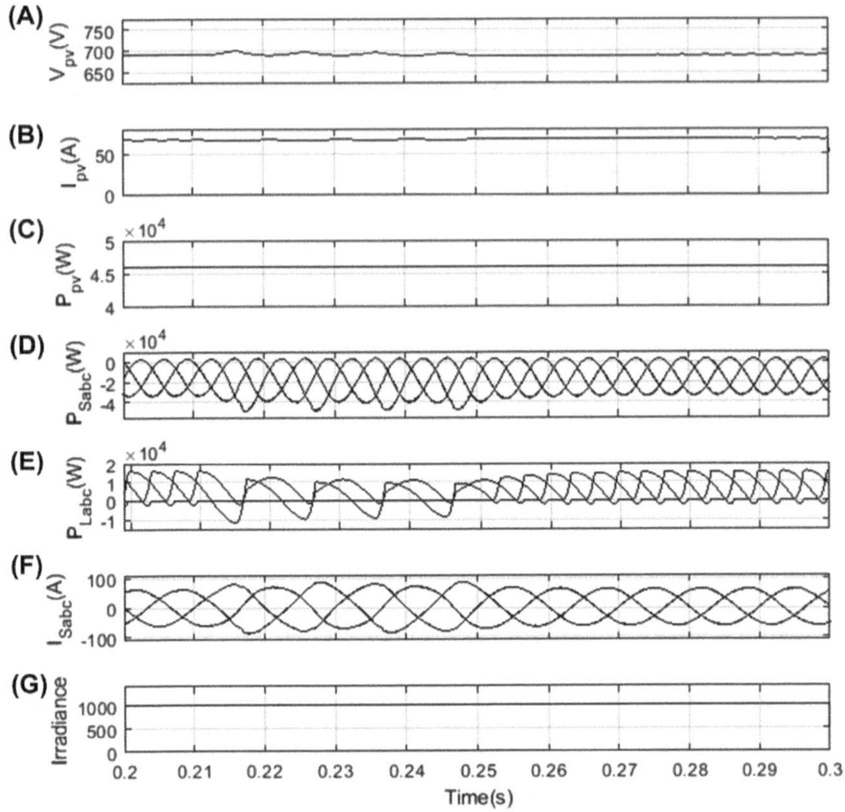

FIGURE 8.9 MATLAB-Simulink results for load unbalance condition (A) solar voltage V_{pv} (B) solar current i_{pv} (C) solar power P_{pv} (D) grid power P_{Sryb} (E) load power P_{Lryb} (F) grid current i_{Sryb} (G) irradiance.

atmospheric conditions due to variable irradiance i_{Lr}, i_{fLr}, i_{pLr} is not affected because it depends on dynamic disturbance of the system, whereas I_{pvf}, I_{loss}, and I_{pLt} directly depend on solar irradiance; hence it changes but finally settles on the corresponding value within a cycle that shows a fast system response.

4.4 Total harmonic distortion analysis

Grid current signals and load current signal THD are depicted in Fig. 8.14. It is found that the THD of the load current signal is 41.00%, and the THD of the grid current is 1.42%. The EVSSLMS algorithm is capable of compensating load current harmonics. The THD of the grid current signal is reduced to below 5%, confirming the limits and procedures outlined in the IEEE-519 standard.

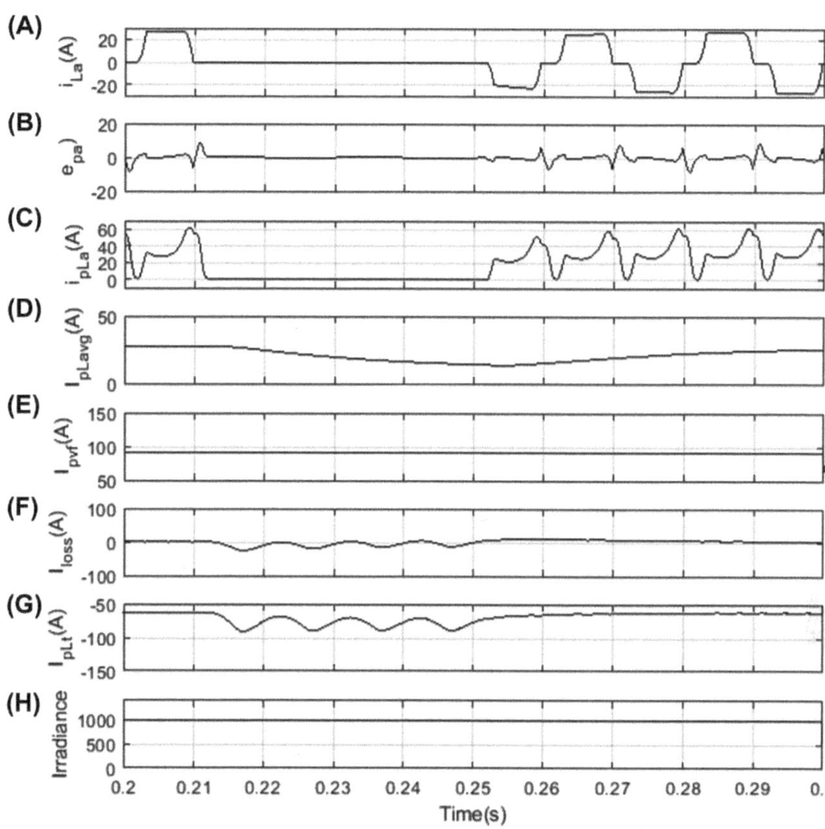

FIGURE 8.10 MATLAB-Simulink results for load unbalance condition (A) load current i_{Lr} (B) fundamental load component i_{fLr} (C) active load component i_{pLr} (D) equivalent-weight load component I_{pLavg} (E) feed-forward current component I_{pvf} (F) loss component I_{loss} (G) net active load component I_{pLt} (H) irradiance.

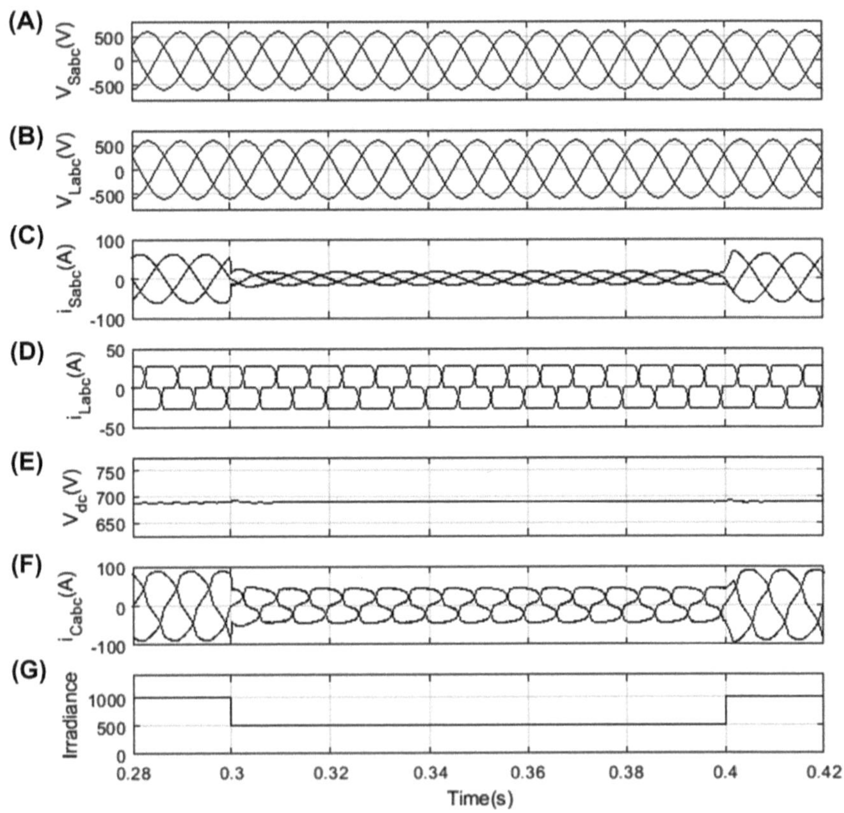

FIGURE 8.11 MATLAB-Simulink results for changing surrounding condition (A) grid voltage V_{Sryb} (B) load voltage V_{Lryb} (C) grid current i_{Sryb} (D) load current i_{Lryb} (E) DC voltage V_{dc} (F) filter current i_{Cryb} (G) irradiance.

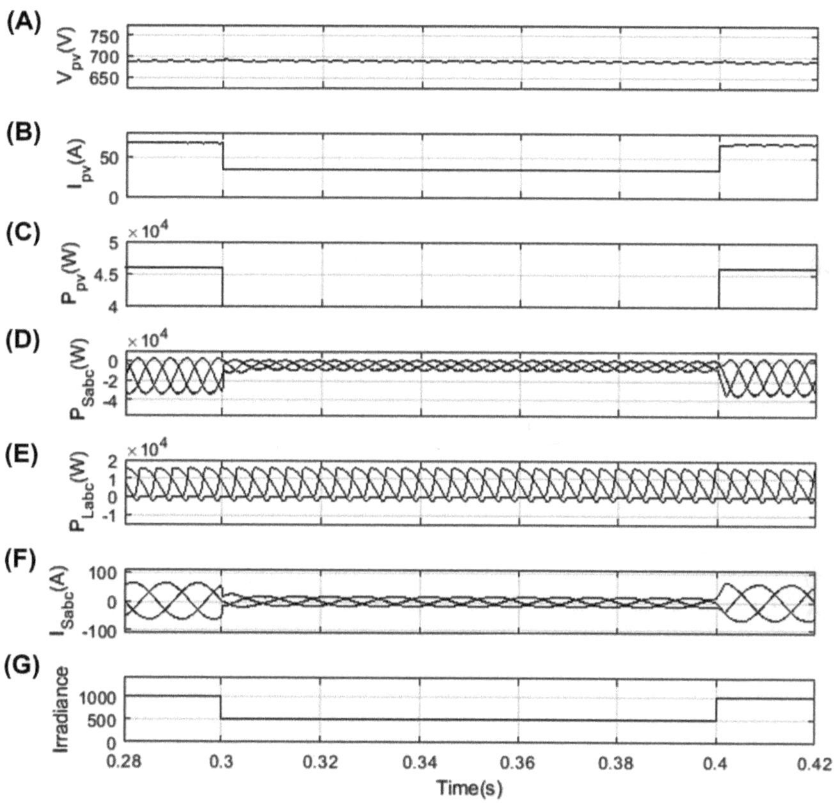

FIGURE 8.12 MATLAB-Simulink results for changing surrounding condition (A) solar voltage V_{pv} (B) solar current i_{pv} (C) solar power P_{pv} (D) grid power P_{Sryb} (E) load power P_{Lryb} (F) grid current i_{Sryb} (G) irradiance.

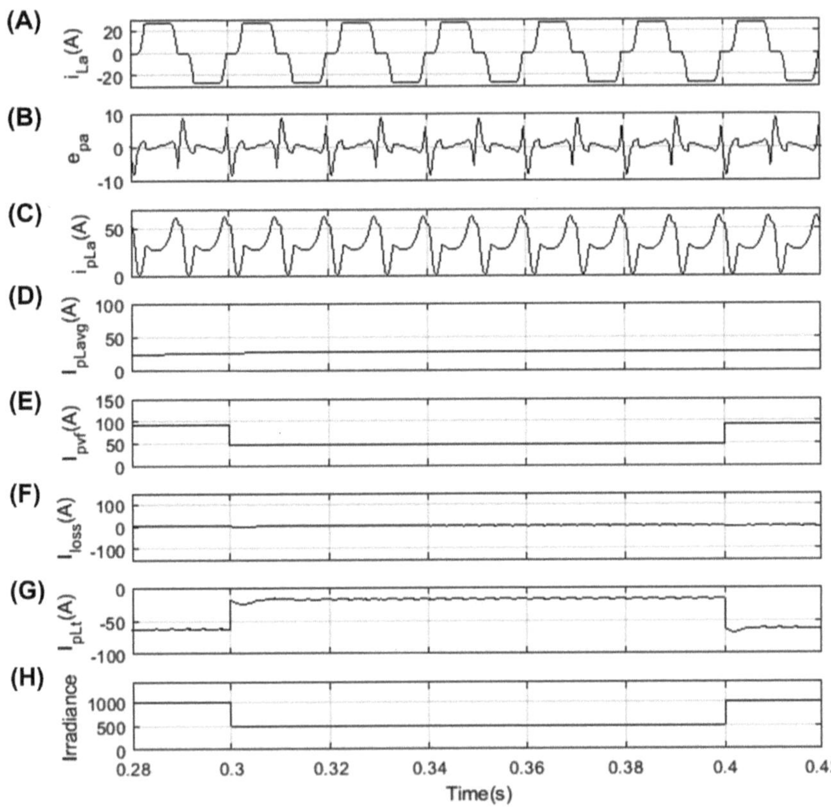

FIGURE 8.13 MATLAB-Simulink results for changing surrounding condition (A) load current i_{Lr} (B) fundamental load component i_{fLr} (C) active load component i_{pLr} (D) equivalent-weight load component I_{pLavg} (E) feed-forward current component I_{pvf} (F) loss component I_{loss} (G) net active load component I_{pLt} (H) irradiance.

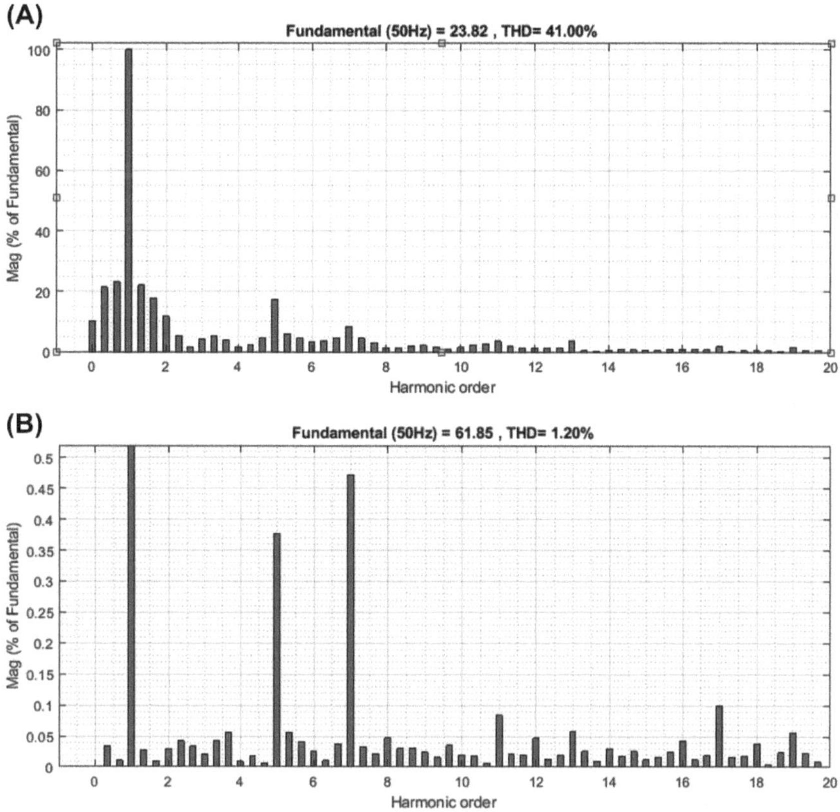

FIGURE 8.14 (A) Source current total harmonic distortion (THD) and (B) load current THD.

5. Conclusion

Extraction of the fundamental load current component is free from harmonics with the EVSSLMS technique, which has a reduced computational burden. The EVSSLMS technique is used for single-stage 3P3W and can reduce THD from 41% to 1.42%, confirming the limits and procedures outlined in the IEEE-519 standard. During solar irradiance variation, the INC technique has been successfully implemented for the extraction of peak power from the solar PV array. During the load unbalance period, the balance condition of the grid current signal is not affected. Hence, the dynamic performance of the system is healthy under load unbalancing and changing environmental conditions.

Appendix

Solar PV array Vmpp = 698.97 V, Impp = 70.56 A (at standard 1000 W/m^2), capacitor at DC bus = 4000 μF, 3P3W grid voltage = 415 V, 3P3W grid frequency = 50 Hz, transmission line inductance = 3 mH, filter value - R = 10Ω, C = 5.5 μF, Kp = −4, Ki = −7.5, Vdc = 700 V, nonlinear loads—20 Ω, 200 mH

References

[1] C. Singh Solanki, Solar Photovoltaics (Fundamentals, Technologies and Applications), PHI Learning, New Delhi, 2018.

[2] A. Kumar, Seema, B. Singh, R. Jain, ROGI with FsLMS based control technique for solar PV system under weak grid, in: International Conference on Environment and Electrical Engineering and 2019 IEEE Industrial and Commercial Power Systems Europe (EEEIC/ I&CPS Europe), IEEE, 2019.

[3] R. Abid, S. Mahjoub, F. Masmoudi, N. Derbel, MPPT control strategies for photovoltaic applications: algorithms and comparative analysis, in: 16th International Multi-Conference on Systems, Signals & Devices (SSD), IEEE, 2019.

[4] T. Bose, M.-Q. Chen, Adaptive bilinear filtering, in: IEEE International Symposium on Circuits and Systems (ISCAS), IEEE, 1993.

[5] S. Rajan, K. Sub Joo, T. Bose, M.-Q. Chen, Conjugate gradient method for adaptive nonlinear filtering, in: 37th Midwest Symposium on Circuits and Systems, IEEE, 1994.

[6] M. Givens, P.M. Clarkson, The application of the median LMS algorithm to ADPCM systems, in: International Conference on Acoustics Speech and Signal Processing, ICASSP, 1991.

[7] M.D. Macleod, Robust normalised LMS filtering, in: International Conference on Acoustics Speech and Signal Processing, IEEE, 2005.

[8] R. Długosz, T. Talaśka, T. Nikolić, G. Nikolić, A parallel adaptive LMS FIR filter realized in CMOS technology, in: 31st International Conference on Microelectronics (MIEL), IEEE, 2019.

[9] Q. Ye, L. Hu, F. Chen, Diffusion LMS strategies for distributed estimation based on sub-regions, in: International Conference on Signal Processing, Communications and Computing (ICSPCC) IEEE, 2017.

[10] C.-C. Tseng, S.-L. Lee, Adaptive Graph Filter Based on LMS Algorithm, in: International Conference on Consumer Electronics-Taiwan (ICCE-TW), IEEE, 2018.

[11] W. Villegas-Ch, S. Luján-Mora, Analysis of data mining techniques applied to LMS for personalized education, in: World Engineering Education Conference (EDUNINE) IEEE, 2017.

[12] C. Turan, M.S. Salman, A block LMS-type algorithm for sparse system identification with analysis, in: International Siberian Conference on Control and Communications (SIBCON) IEEE, 2017.

[13] http://www.ccs.fau.edu/~bressler/EDU/CompNeuro/Resources/Widrow_HBTNN_Perceptrons.htm.

[14] Y. Qian, A new variable stepsize algorithm applied in LMS adaptive signal processing, in: Chinese Control and Decision Conference (CCDC), IEEE Conference Publications, 2016, pp. 4326–4329, https://doi.org/10.1109/CCDC.2016.7531744.

[15] IEEE recommended practices and requirements for harmonic control on electric power system, IEEE Stand. 519 (1992).

[16] IEEE Recommended Practices and Requirements for Harmonic Control in Electrical Power Systems.

Chapter 9

Environmental and technoeconomic aspects of distributed generation

Neeraj Gupta[1], Karan Singh Joshal[1], Anuradha Tomar[2,3]
[1]*Department of Electrical Engineering, National Institute of Technology, Srinagar, Jammu and Kashmir, India;* [2]*Electrical Engineering Department, Eindhoven University of Technology, Eindhoven, The Netherlands;* [3]*Electrical Engineering Department, JSS Academy of Technical Education, Noida, India*

1. Introduction

Distributed generation (DG) is the term used for the different technologies to generate electricity near the point where it will be used, such as wind generation, solar panels, and combined heat and power (CHP). The DG is connected to the electric utility's lower-voltage distribution system, which helps in the delivery of clean, reliable power to customers and in turn reduce electricity losses along transmission and distribution lines. It may be part of microgrid, such as for industry, a college campus or military base, or a single structure such as a home or business [1].

Fig. 9.1 shows the physical structure of a traditional power system, which consists of large power stations to generate electricity and that reaches the distribution system through the transmission utility—e.g., the Power Grid Corporation of India [1]. The distribution system satisfies the load conditions of domestic and commercial costumers. However, in the present scenario, with the growth of distributed system technologies, large power stations are connected to the transmission system, and small generating stations called DGs are connected to distribution networks. Fig. 9.2 shows the physical structure of a contemporary power system with DG. Some characteristics of DG are listed below [2]:

- Distributed or embedded generation: If generation is connected to the distribution network, it is called distributed or embedded generation.
- The generated power can be used on-site or exported to the distribution system.

Advances in Smart Grid Power System. https://doi.org/10.1016/B978-0-12-824337-4.00009-6

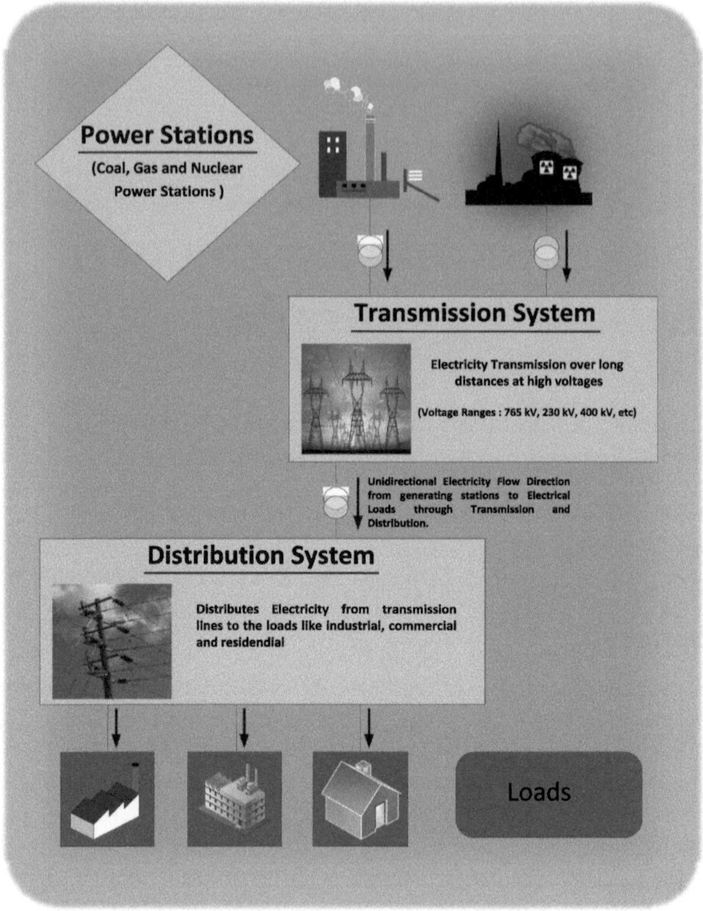

FIGURE 9.1 Structure of a basic power system [1].

- There is bidirectional power flow—i.e., from the generating utility to customers and from DG customers to the distribution system.
- As DGs consist of small-scale wind or solar plants, there is increased unpredictability.
- The nature of the power system is no longer vertical in nature, as power also flows horizontally.

In the industrial and commercial sectors, DG resources include the following:

- wind
- solar photovoltaic (PV) system
- combined heat and power
- biomass combustion or co-firing

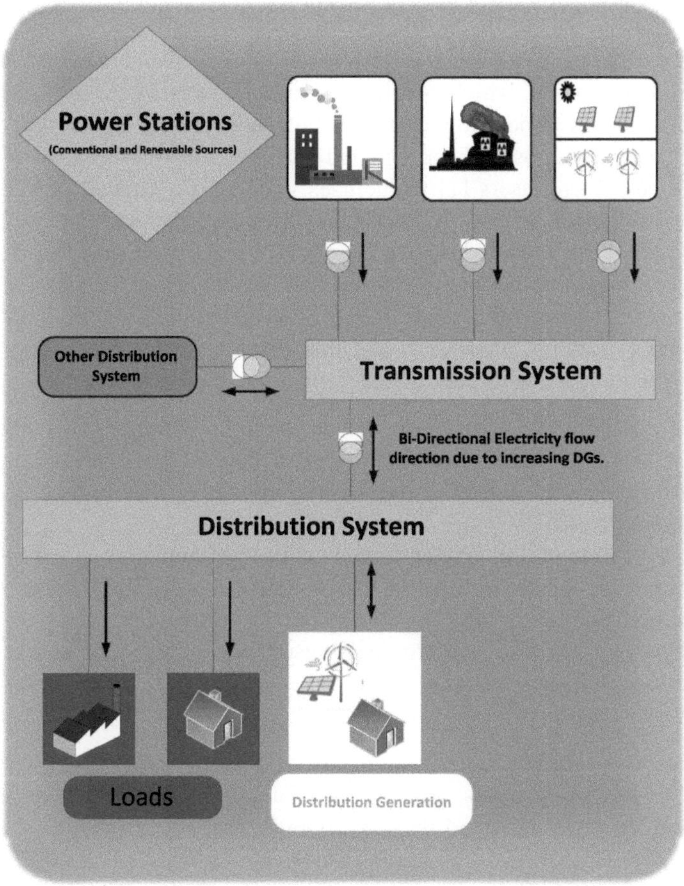

FIGURE 9.2 Structure of a contemporary power system with distributed generation [1].

- hydropower
- fuel cells fired by natural gas or biomass
- municipal solid waste incineration
- reciprocating combustion engines, including backup generators, that may be fueled by oil

Common DG systems in the residential sector include the following:

- small wind turbines
- solar PV system
- gasoline or diesel fueled emergency backup generators
- fuel cells fired by natural gas

2. Impact of centralized generation on the environment

The generation of large-scale electricity at a centralized facility is called centralized generation, which is generally located away from load centers/end users. Centralized generation is generally connected to high-voltage transmission lines, and power is distributed through the transmission utility to various kinds of end users [1]. It includes wind farms, nuclear power plants, hydroelectric power plants, and fossil-fuel-fired power plants.

Centralized generation leads to localized concerns (those that affect the area directly surrounding a power plant) and regional environmental concerns based on the source of the plant. Along with generation are problems associated with extracting, producing, and transporting certain fuels. The various emissions associated with centralized generation are listed below:

1. Air emissions: This includes carbon dioxide emissions (as shown in Fig. 9.3), sulfur dioxide, nitrogen oxides, mercury, and particulate matter from burning fuel, which varies by the amount of fuel burned and other plant characteristics.
2. Water contamination: Water may contain contaminants when used for steam production or cooling that can be returned at warmer temperatures to water bodies.
3. Solid waste: Burning certain fuels results in solid waste that contains hazardous substances like radioactive waste and heavy metals like mercury.

In addition to these concerns, large spaces are required for the operation of these types of power plants, and the total energy content of power plants is wasted during generation and delivery to end users.

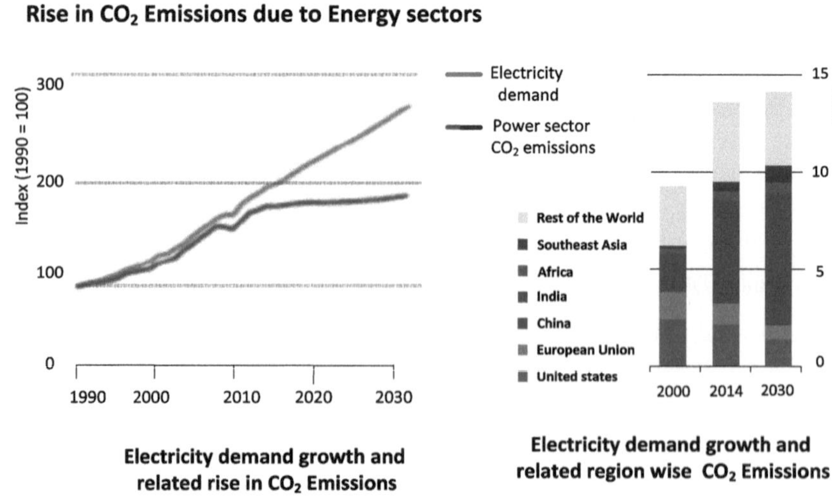

FIGURE 9.3 Electricity demand growth and related region-wise CO_2 emissions. *IEA Briefing for COP21 (October 2015).*

3. What is driving distributed generation?

The amount of DG is increasing in conventional power systems. Several factors are driving this increase [1]:

1. Environmental concerns: With increased environmental pollution due to the emission of greenhouse gases from conventional power-generating plants, the need is felt globally for clean power. So an energy mix of conventional and renewable generation is an optimal solution for satisfying energy needs.
2. Technological innovation: With technological innovations in generating power from clean sources of energy, connections from these sources to the grid are required. Since these sources are not numerous and are near load centers, they are used in distribution systems as DG.
3. Supply security: Many countries depend on imports of fuel, coal, and oil to fuel their various power plants and meet their energy needs. These fossil fuels will be consumed faster than they are being created, so we are eventually moving toward scarce resources, which will result in increased energy prices. To ensure energy supply security in the future, alternative resources are encouraged.
4. Increased energy mix: DG technologies are mostly a mix of renewable energy resources—wind, solar, etc.—and thus lead to a lower mix of carbon energy resources in power systems.
5. Reduced need for transmission and distribution infrastructure: DG is used in a distribution system near load centers, so there is reduced need for transmission and distribution infrastructure, as in case of conventional power plants, which are generally remotely located.
6. Reduction in transmission losses: DG is connected to a distribution system that caters to local load demand; as a result, transmission losses are reduced because less power is required from conventional power plants.
7. Commercial benefits: There are many commercial benefits for customers who adopt DG:
   - Self-consumption: Electricity is generated at the point of consumption, resulting in reduced electricity bills.
   - Selling electricity: The surplus power generated can be sold to the distribution system utility, resulting in revenue generation.
   - Climate change levy exemption certificates: In some countries, climate change levy exemption certificates are issued by authorities to those who generate renewable energy, which can be sold to supplier to reduce or avoid paying climate change levy taxes [1].
   - Ancillary services: Generators with capacity of more than 3 MW and that can provide a reactive power requirement of ± 15 MVAr can provide ancillary services to generate additional revenue.

These advantages lead to the implementation of targets for distributed energy generation as shown in Fig. 9.4. However, certain barriers need to be addressed when adopting DG, as shown in Fig. 9.5.

4. Environmental issues related to distributed generation

DG, though beneficial in reducing some negative effects of conventional generation, is not free from negatives, which can be summarized as follows:

1. Water is required for steam generation and cooling in some DG technologies, such as biomass combustion and combined heat and power.
2. Combustion-based DG systems are less efficient than centralized power plants.
3. DG technologies involving combustion produce the same impacts as large power plants that are smaller in scale but may be closer to populated areas.
4. DG systems located closer to end users might be unpleasant to the eye or cause land-use concerns.

Details of the impacts of each DG technology are given in the following section.

4.1 Environmental impacts of solar power plants

With growing concerns about energy security and to meet growing demand, governments are pressing their utilities to provide increased power from solar PV systems. Fig. 9.6 shows the growth in the global PV market. Many governments are enforcing renewable portfolio standards—i.e., generators must supply a percentage of their power from renewable sources. Consequently, solar PV systems are expanding to a wider range of locations and ecosystems. Energy from solar power plants is the future in an energy-deficient world, which leads to energy security but also has environmental impacts—e.g., greenhouse gas emissions, emissions of hazardous materials, impacts on wildlife, intensity of land use, and albedo effects [3].

4.2 Characteristics of the installation and operation of solar power plants

A variety of locations and environments such as forests, deserts, and other tropical locations are well suited for the development of solar power plants. Solar power plants may have environmental impacts depending on where they are located. Classified under biomes such as forests, desert shrublands, farmlands, grasslands, and true deserts, significant location characteristics for the installation of a solar power plant are the main area of discussion in this section. This discussion is mainly undertaken by keeping certain factors in

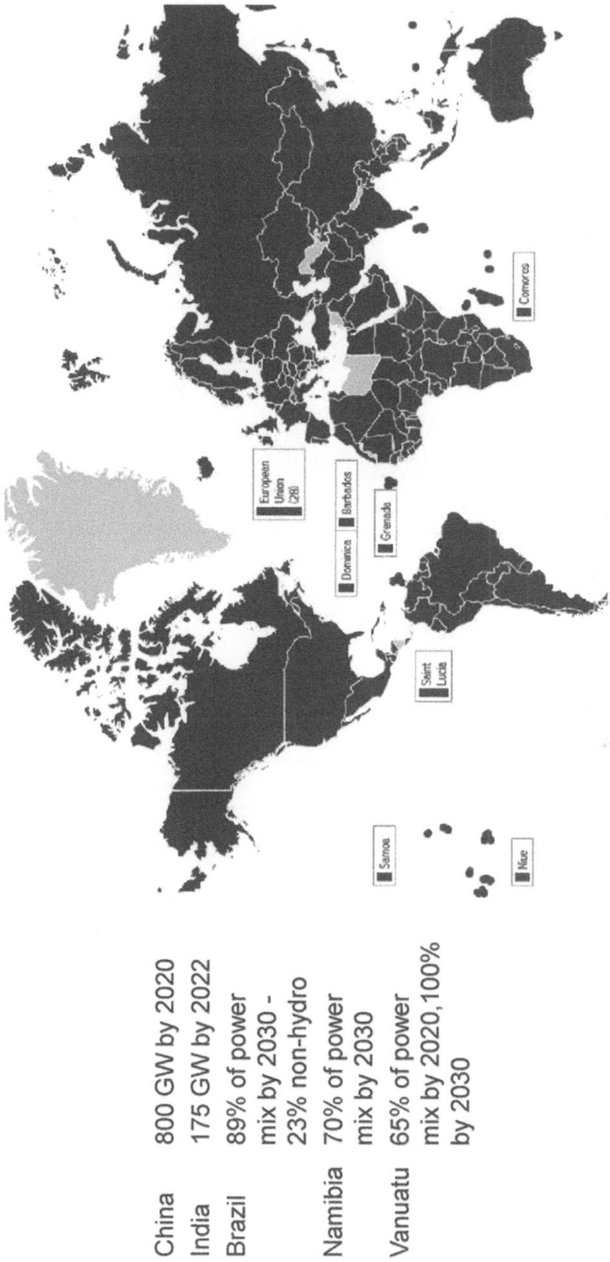

FIGURE 9.4 Renewable energy targets worldwide. *Map from World Resources Institute, CAIT Climate Data, 2015.*

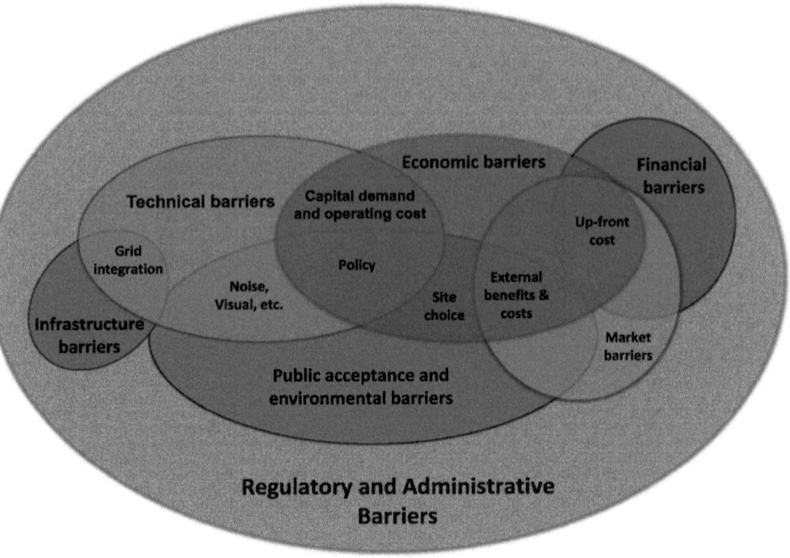

FIGURE 9.5 Barriers in connecting distributed generation [1].

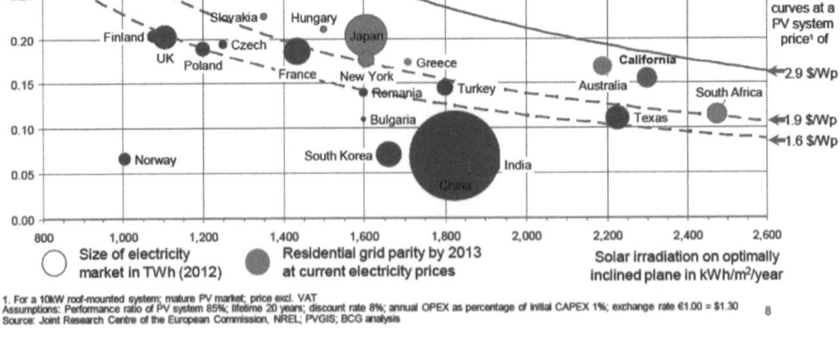

FIGURE 9.6 Growth of the global photovoltaic market. *Duke Energy, CCAP Energy Dialogue, California, 2015.*

focus. These factors are the main environmental parameters, which include biodiversity, solar insolation, and biomass density.

Biodiversity can be associated with sunshine and precipitation [4]. It is basically measured as density of species in species/ha. Continued stages of freezing or drought are not present in forests, and required precipitation is at least 50 cm/yr [5]. Solar insolation for forests is commonly reduced by 25%−50% due to cloud cover. Forest age, tree species, and local climate are the reasons for the wide-ranging biomass density of forests, from 100 to 500 Mg C/ha [6]. As estimated by species density, tropical rainforests have more biodiversity than any other biome. The benefits and services provided by forests include the generation of timber, food, pulp, and fuel. Other ecological services include air purification, wildlife habitat maintenance, and flood prevention. A lower biomass density, ranging from 10 to 30 Mg C/ha, exists in desert shrublands. Compared with forests, desert shrubland has approximately 25% less biodiversity and receives precipitation between 5 and 30 cm/yr [3]. It has a lower grazing capacity, and flood risk mitigation is also low. Its natural benefits include, as with forests, the expected generation of wood. Grasslands do not have a dense population of trees because of frequent freezing or drought. Except for the production of wood and pulp, grasslands provide similar natural benefits to those of forests but with a high grazing capacity. The received precipitation of grasslands is generally between 30 and 100 cm/yr. It has biodiversity that is approximately similar to that of desert shrublands, and its biomass density varies from 10 to 50 Mg C/ha [3]. Depending on location, farmlands constitute the full range of variation in cloud coverage. They are exclusively artificial, or manufactured. They have a similar biomass density and lower biodiversity compared with those of grasslands and desert shrublands. But compared with true deserts, the biodiversity of farmlands is higher. True deserts have various advantages that are required for better installation and operation of solar power. These advantages include extremely minimal cloud cover, the fewest wildlife, and the lowest human population. With precipitation of less than 3 cm/yr, it is categorized as a region with extremely low rainfall. Fig. 9.7 shows measures for various biomass (forests, grasslands, desert shrublands, true deserts, and farmlands) of important environmental parameters (cloud cover, biomass density, and biodiversity), compared with those of tropical rainforests, that are quite significant for solar power plant installation and operation. In Fig. 9.7, the tropical rainforest is chosen for comparison because it has the highest biodiversity, cloud cover, and biomass density. The lower values of these parameters lessen the environmental impact for installations of large-scale solar power. Because solar panels are situated in an open operating environment, dust accumulates on the solar panel surfaces and obstructs sunlight. If not cleaned for a month, power output decreases by about 30%. Therefore, it is necessary to wash the panels regularly, and nearly 5,001,000 gallons of water per MWp of panels per year are required [3]. The installation of solar panels together with ground support (approximately 1 m

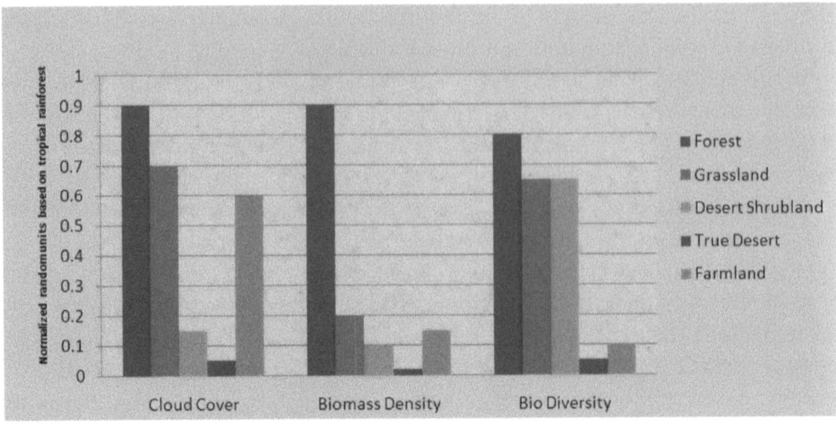

FIGURE 9.7 Measures of important environmental parameters.

above ground) typically requires 58 acres per MWp. This area also includes space for other equipment like batteries, inverters, charge controllers, and collector boxes. Operation of solar power plants requires mowing (less than 1 m) of surrounding vegetation in order to achieve full sunlight.

4.3 Environmental impact metrics

Together, the physical units of measurement and methods of data collection and analysis constitute a metric. Successful construction of life cycle analysis (LCA) for best comparing power generation technology requires the most consistent and objective metrics. Some comprehensive metrics for environmental impacts include emissions of greenhouse gases (kg CO_2-eq/yr), noise measurements above the audible threshold for noise impacts, albedo effects, heavy metals, or priority pollutants. To add qualitative methods to the quantitative ones, the US Forest Service has constructed and manages metrics for impacts associated with visual resources [7]. There are several other impacts for which no well-defined metrics exist. These climate impacts are related to human health and well-being and affect food security and disease-related problems. In Section 4.8, we discuss these impacts on geohydrologic resources resulting from large-scale power. Other underdeveloped metrics are those with wildlife- and habitat-related impacts on the environment. In Section 4.4, we discuss another important environmental impact—i.e., land use intensity and related metrics used for describing land effects and the diversity of land utilization.

4.4 Land use

To compute land use intensity, we require land area transformation (per unit of electrical energy produced km^2 TW/h) and occupation (per unit of electrical energy produced km^2 yr TW/h) metrics. The transformation metric is used to capture one-time changes in the physical character of land during the installation phase only. On the other hand, the occupation metric measures the impacts of land use intensity over a known period and is defined as the product of land area and the length of time the land is kept for use. Therefore, unlike the transformation metric, the occupation metric measures the impact of both installation and operating periods. These transformation and occupation metrics are well suited for comparing the life cycles of solar and coal power if land use intensity is considered. Calculating these metrics for comparison requires additional information. This includes information about the area used for generating equipment (solar panels, turbines, etc.) and for gathering and transporting fuel. It also requires knowledge about the life span of the power plant, land transformation recovery time, and an understanding of the land and energy requirements for building the components. PV power plants have an operational life span of about 30 years, and no land for fuel is required for their operation. On the other hand, coal power plants are more dependent on the fuel requirement that comes from mining coal. Therefore, a coal power plant requires land not only for its operations but also for gathering and transporting fuel. A coal power plant occupies about 15% more land than that of a solar power plant when both are 30 years old [3]. A solar power plant's land use intensity decreases relative to that of a coal power plant as the age of the plant increases.

4.5 Human health and well-being

Compared with traditional electricity generation, solar energy in electricity production has more beneficial impacts. Traditional power generation processes emit higher amounts of NOx, SO_2, and numerous other pollutants compared with solar energy power plants, which alone bring great trouble for human health [8]. The emission of roughly 65% of greenhouse gases worldwide comes from combustion of the fossil fuels required for traditional power plants [9]. This includes emission of CO_2, which has significant risk for human health and well-being and results in severe climatic conditions, rises in sea level, food security issues, and socioeconomic change [9]. Direct mercury (Hg) emission from solar power equipment is 1/1000 to 1/50 of the emission that comes from traditional electricity generation. Combustion of fossil fuels is responsible for approximately 65% of Hg deposits in lakes and reservoirs [10]. Therefore, it can be observed that increased use of solar energy over fossil fuel-based plants is much more beneficial for human health and well-being. Another impact of solar power on human well-being is increased recreational opportunities. This requires some important steps to improve esthetics as well. The US Forest Service has worked on, prioritized, and expanded the use of visual resources during land development [7]. More power generated from solar leads to lower Hg deposition in lakes and rivers, which promotes an increased scope for fishing and other recreational activities.

4.6 Wildlife and habitat

Wide-ranging, significant improvement efforts are required to prevent disturbances to wildlife and habitats caused by the increased construction and operation of solar power plants. These large-scale power plants are comparatively new technologies that create an urgent need to understand the science behind its ecological impacts. Land occupation by solar power plants is the foremost impact threatening wildlife and habitats. Impacts coming from solar power plants basically affect movement by animals [11], their preying strategies, their hiding spots, and food availability. McCrary et al. [12] presented analytical measures of impacts on wildlife related to the fatality of bats, birds, and insects due to a solar power tower. Several reports have considered and stated that solar power plants have a significant impact (a legal term used under the US Endangered Species Act) that can threaten the lives of endangered animal species and special-status plants that live near it. The proper sunlight for solar panels requires continuous mowing of vegetation and herbicide use for vegetation-free land [13]. Solar panels also cause alteration of the microclimate because of the shadows they cast. This altogether affects vegetation, which is still an open, unconsidered area for study. If we differentiate the production of electrical power coming from solar and coal power plants, it is noticeable that the recovery of soil and the ecosystem occurs more quickly in the case of solar power plants. Coal power production requires coal strip-mining, which leads to serious land deterioration that requires about 50–100 years for the soil to regenerate [14]. On the other hand, power produced by solar plants requires less removed soil from the land, and therefore recovery is faster than it is with coal power plants. It can also be seen that land used for operating solar power plants is less hazardous (due to less deposition of Hg, NO_x, and sulfates) than it is for power plants based on fossil fuels. Considering these burdens on wildlife from solar energy generation, large-scale solar power projects have come up with the lifetime funding required to take the significant beneficial actions necessary for the protection of wildlife [15]. Some of these beneficial actions include extensive monitoring of the status of the ecosystem and wildlife, restoration of habitats, elimination of recreational off-highway vehicles, and building of appropriate surroundings for endangered species. More efforts are being considered over time in order to make these solar energy projects beneficial rather than a burden on wildlife and habitats.

4.7 Climate and greenhouse gas emissions

Installation and operation of a solar power plant has comparative advantages and disadvantages associated with its impact on the climate as well as greenhouse gas emissions. Lower emission of CO_2 is a major inspiration for its operational deployment, but at the same time, its installation results in the removal of trees and brush together with mowing of vegetation. These removed trees, branches, and woody debris result either in emission of CO_2 into the atmosphere (due to burning and mulching) or delayed emission via carbon sequestration when it is used for durable wood products [9]. Removal of forested areas for solar power plants results in a lower carbon sequestration rate. The growth of trees and soil horizons is responsible for the carbon sequestration that takes place for the 75 years that follow. This stage of carbon sequestration is disturbed by continuous trimming of vegetation, leading to CO_2 emission into the atmosphere. The life cycle of solar power shows that the emission of important greenhouse gases such as CO_2, CH_4, and NO_2 is much lower than that of of fossil fuel-based power plants [16]. A concerning fact about PV panels is that they reflect less sunlight and absorb more solar energy. A large part of this energy is further converted to heat, which may have an effect on global or local climates as well as microclimates. Extensive installation of PV panels leads to albedo change that affects the global climate. But the benefits of lower greenhouse gas emission are still quite favorable compared with the climatic effects of energy generation from fossil fuels.

4.8 Geohydrologic resources

Geohydrology is about gaining knowledge about groundwater together with its physical and chemical interactions with the physical surroundings. Topsoil erosion, reduced filtration of air and rainwater pollutants, higher chances of flooding, thickness in local streams, and reduced groundwater are the potential impacts on geohydrologic resources [17]. Soil erosion in vineyards severely depletes soil quality, and similar soil erosion can happen between PV panels that contain access roads. Solar power plants reduce native vegetation, which leads to negative impacts on soil infiltration rates, runoff rates, and evapotranspiration. Forested land used by solar power plants have increased chances of flooding and reduces protection of stream banks. Therefore, proper study and monitoring must be required to ensure the conservation of geohydrologic resources.

4.9 Environmental impacts of wind power plants

Similar to solar power generation, growing energy demand and the scarcity of fossil fuels lead to the installation of an increasing number of wind power plants with capacities ranging from MW to kW. The energy from wind is clean and intermittent in nature and satisfies a large percentage of total energy demand in many countries. However, it has many negative environmental impacts.

4.10 Greenhouse gas emissions in the case of wind power plants

Constructing foundation wind turbines requires the manufacture of concrete and steel, which is the main cause of greenhouse gas emissions from wind power. The emission of greenhouse gases is estimated using different methods for LCA of wind farms. These methods include two hybrid LCAs (integrated hybrid LCA and output-based) and a process-based LCA [18]. The variable estimation of emitted greenhouse gases is dependent on these different LCA methods as well as wind farm size and location. It has been observed that compared with the process-based LCA, the two hybrid-based LCAs show higher figures for greenhouse gas emissions. This because the process-based LCA does not consider the inputs required to set up the boundaries of the system related to higher upstream production processes.

4.11 Bird fatality

Due to a lack of understanding of the concern for bird fatality, early wind power plants were situated at places where birds were in abundance [19]. The major cause of bird fatalities from wind power plants is collision with the turbines and other structures of the plants. The collision rate depends on various factors including wind farm location, weather conditions, the season, bird activities (e.g., migrating and moving around feeding areas), wind strength and direction, and the spatial distribution and other characteristics of the turbines [20]. Raptors are more vulnerable to death than other species are, which occurs mostly during spring due to postbreeding migration [20]. Other species reported include the black kite, or *Milvus migrans* (0.004 collisions per turbine per year), short-toed eagle, or *Circaetus gallicus* (0.008 collisions per turbine per year), kestrel, or *Falco tinnunculus* (0.14 collisions per turbine per year), and Eurasian griffon, or *Gyps fulvus* (0.12 collisions per turbine per year) [25]. Agricultural landscapes and dense forests are locations where scavengers usually remove the remains of dead birds before researchers can discover them, which creates some uncertainty for the correct estimation of bird fatalities [21]. From the early establishment of wind power till now, due to improved understanding and technologies, a declining collision rate of birds with the turbines has been observed, indicating continuous improvement in lowering bird fatalities.

4.12 Bat fatality

The understanding and estimation of bat fatalities have been taken more lightly than those of birds, since most methods for monitoring were developed for bird fatalities, and therefore the reports regarding bat fatalities were very few till 2001 [22]. After that, monitoring studies showed that utility-scale wind energy facilities (e.g., utility-scale turbines exceeding 100 kW) are the main reason for large-scale bat fatalities, especially around forest ridgetops and agricultural regions [23]. Bats are killed by the rapid decreases in air pressure around wind turbines, which ruptures the lungs of bats. Both bats' attraction to turbines and turbines standing in the way of flying bats are leading causes of bat deaths. Some possible causes for bats getting close to turbines and thus increasing their risk of death include (1) landscape modification that attracts aerial insects to the turbines, further attracting bats for foraging; (2) increased heights and sizes of turbines; (3) visually mistaking turbines as trees where bats look for food, shelter, and social opportunities; (4) attraction to ultrasonic sounds produced by turbines; (5) active populations of bats along ridgetops (because of cool and foggy weather in valleys) increase the possibilities for bats to come into contact with turbines. All of these causes are still under consideration, and therefore more studies and research are required in order to understand the altered behavior of bats around turbines.

4.13 Land surface impacts

Wind power plants have a visible effect on the land surface temperature around wind turbines. This basically happens because turbines generate turbulence wakes that mix cool air upward and hot air downward during nighttime and vice versa for daytime. Hence, turbines increase the surface temperature during nighttime, whereas for daytime they have no significant cooling effect because of the sun's heat. Observed data coming from satellites shows an increase in temperature to about 0.72°C per decade (especially at nighttime) near wind power plant locations compared with other locations [24]. This puts further evidential impact on local climate and weather conditions like cloud formations, local precipitation, and atmospheric moisture content. More studies on these impacts on the land surface coming from wind power plant operations are ongoing, as it is a relatively fresh subject.

4.14 Noise pollution

Mechanical and aerodynamic noise together constitute the total generation of noise coming from wind turbines. Mechanical noise generated from mechanical and electrical parts of turbines is not considered significant compared with the aerodynamic noise that comes from the interaction of air with blades. This is because of developed mechanical designs like appropriate insulation and vibration damping that eliminate much of the mechanical noise coming from towers. Using semiempirical models [25] and following international standards are the two main methods for measuring wind turbine noise emission. Following standards and guidelines is not suitable for measuring noise emission from wind turbines, as it is exterior to the scope of standards for windy conditions. Based on simplistic and complex computational solvers, quite a few semiempirical models have been produced, but their application consumes a reasonable amount of time [25]. Other problems related to measurement of noise from wind turbines include higher wind speed and background noises (e.g., road traffic, rail tracks). Higher wind-speed (8 m/s and above) noise together with background noise is fully capable of masking turbine noise [26]. As discussed in Section 4.12, bat fatalities may also be connected to the noise of the rotating blades, bats appear to be attracted to it.

5. Technical and other challenges of moving toward distributed generation

The integration of DG technologies consists of connection, development of hardware, accounting and metering, balancing of load, protection of bidirectional power, maintaining stability, etc. Connection means the connection of DG at a point of common coupling, development of hardware means development of an appropriate network to accommodate DG, and metering means to have appropriate meters for bidirectional power. Connecting DG includes several technical challenges, as follows.

5.1 Measurements

Significant locations on the system require measurements as well as appropriate communication between system operators and measured quantities. A lack of capability to modify the generation of active power together with a shortage of communication of measured quantities between generators and consumers leads to reduced reliability of the distributed network that considers system operating functions [27]. System operating function solutions are contributed by large-distribution generators controlled by system operators; however, distribution network companies offer no forthwith operational role. Distribution network companies need to broaden their services beyond simply dealing with faults and outages.

5.2 More complex system

For a system to be considered complex, it must have nonlinear activities from the interplay of a large number of interrelated elements that cannot be extracted solely by simple summation of those elements' activities [28]. System complexities become visible as it becomes harder to predict the state of activities working as a whole. Interaction among various constraints as well as the requirements for optimal utilization of network facilities are the leading reasons for complexities in control problems. Even if we are only connecting a distributed energy resource (DER) into the distribution network, it has a significant impact on transmission systems. It can result in problems like reduced demand and excess supply over demand, which are observed in transmission networks whenever new DG is connected to the distribution system [29]. Instead of standards-based solutions, complexities in power systems require more principle-based solutions in terms of policy-making and regulations.

5.3 Flexibility in reliability needs

More uninterrupted power supply to customers means a more reliable power system. Higher value-engineering quality together with energy market liberalization is making customers more aware of reliability requirements. Competition in energy markets due to liberalization requires an incentive structure for cost-effectiveness, which makes it difficult to also ensure high reliability levels. Developing countries require modern manufacturing techniques to lift their economies, and this requirement needs a reliable supply of power. A highly reliable power supply is a must for some industries like metal, paper, chemical, refining, and telecommunications, and that power comes in the form of electricity supplied by the grid. Therefore, these companies opt for DG units for dealing with high-reliability issues [27].

5.4 Connection issues

Considerable human and management resources are needed in order to contribute a certain amount of generating capacity through compact- to average-scale DERs. A multitude of connection choices must be managed by providing guidance and support to connection applicants, but distribution network operators (DNOs) are not well equipped with resources and are somewhat less qualified for such functions. These connection applicants (since they are smaller in size) basically lack the understanding related to connection charges and also need guidance in deducing the significant industry codes. Therefore, both utility networks and regulators require more human resources with improves skills when considering the development of control of future prices [27].

5.5 Power quality impacts

If the supply of electricity is coming from the grid, operators on the transmission side can deal with frequency deviation in the system. But when DG units come into play, they cause frequency deviations, and dealing with such deviations is out of the hands of transmission grid operators. The regulatory authorities require that more hard work is put into solving this problem, and installation of DG units should be done with proper care and planning. The DG unit helps by contributing voltage support to grids with low-voltage problems. But DG is itself is considered a problem in maintaining power quality in the system, as it has a significant impact on variations in the local voltage level [30,31]. DG connected to the distribution grid brings problematic issues related to harmonic distortions, voltage fluctuations, and voltage flickers. Voltage fluctuation is due to power variation caused by solar or wind energy generator usage in DG, which in turn causes flickers. These voltage flicker levels can become poorer because the squirrel cage induction generator lacks the ability to actively control terminal voltage. Harmonics in the grid are usually produced by the use of nonlinear loads, inverters, and rectifiers. Harmonics can lead to excessive heating, load tripping, equipment failure, and interference with communication lines. Solar generators and fuel in DG lead to production of harmonics within the system. High harmonic currents in the system are due to the use of thyristor and line-commuted inverters, although recent PWM inverters produce very low distortion of harmonic currents compared with that of thyristor and line-commuted inverters. Grounding and design of synchronous distribution generators can also produce harmonics. These produced harmonics can be restrained by having a required grounding arrangement for distribution generators [32].

5.6 Unpredictability of power output

Forecasting power outputs for distributed generators is usually difficult in practice, as it uses renewable resources or heat-driven power generation. This makes it more difficult to maintain a balance in the equilibrium between generation and contracted supply. This means that power injected into the system must be equal to power withdrawn from the grid; otherwise, market participants who have the right to access the transmission grid are penalized. These penalties cannot be included anywhere other than with market participants, who are thus unable to balance their portfolios [27].

5.7 Management and information risk

In power systems, conventional practices and regulatory arrangements require suitable changes in order to manage the best allocation of risk. Risk is the outcome of uncertainties related to planning and operation of a power system. Attaining a lower-priced electricity system is difficult without proper distribution of risk. In order to get support for debt-financed DER projects, the regulatory structure requires sufficient assurance concerning revenue streams to ensure an appropriate splitting of risk. Understanding distribution network constraints, together with their forecasting to support DNOs, facilitates the development of DERs [27]. Increasing participation by DERs is leading to relatively more small-scale projects compared with large-scale utility projects, which is altering the appropriate risk distribution, and policy-makers must understand this change for better risk distribution among large utilities and DER developers.

5.8 Implications of distributed generation for distribution network operators

To control, sustain, and develop the power distribution system requires the accountability of DNOs. The inclusion of DG has made it more difficult for DNOs to perform their role in the electricity distribution network, and DNOs need to fulfill connection requests for DG on their distribution network. In order to accommodate DG, new advanced network designs and technology implementations are required by DNOs, which gives further rise to expenses that must be carried out by DNOs. For more dynamic participation by DNOs for the required growth and development of DG within their distribution network areas, electricity regulators should encourage DNOs by providing financial incentive mechanisms. These incentive mechanisms should provide DNOs with adequate and reasonable investment so that they are encouraged to become intensely involved in supporting more DG connections [33].

6. Economics of distributed generation

6.1 Quantifying the economic benefits of distributed generation

The benefits of having DG in power systems also provide economic benefits for customers and distribution utilities [34]. Depending on the economic benefits provided by a particular DG, regulators accordingly set the level of incentives to encourage development of DG. Economic compensation is a must for those who provide benefits to others. One technical benefit of DG to distribution utilities is reducing electrical losses, which can be considered an economic benefit. Compared with the wholesale price of electricity from spot markets, it is profitable for distribution utilities to purchase electricity from customer-owned DGs during peak-load hours. The large-scale employment of DGs on distribution networks has a significant impact on reducing energy prices in electricity markets [35]. By tactically locating DGs for peak-load hour usage, costly network upgrades can be put off. DG improves the reliability of supply, which is especially significant for customers with high interruption costs. DG's application as an alternative source of power is an efficient economical solution for customers as well as distribution utilities.

6.2 Less choice between different (costly) primary fuels

Diversification of leading energy supplies with fewer choices between primary fuels has occurred because DG's contribution to installed generating capacity is growing. Compared with central generation, the supply of primary fuel is expensive in the case of DG applications. Dependence on gas is growing, as gas-based generation is a major part of the DG that is taking over coal-based generation. Compensation for the adverse effects of dependency on gas can be achieved by increasing fuel efficiency and decreasing overall adoption of fuel for CHP [36].

6.3 Flexibility in price response

Unlike central power generation, features of DG like smaller size, fast construction, and on-site production of energy make it more flexible in response to changing energy prices. Therefore, DG has become a way to get around the price fluctuations of electrical energy in the system. The DG application includes introducing renewable resources, improvements in potential efficiency, and better substitutes for grid investments, which increases market demands for adopting DGs. Basically, various benefits coming from the use of DGs make it flexible in responding to price variations [36].

6.4 Financial cost

The major difficulty in installing DGs on a large scale is the high capital cost of energy/kW produced by various DG technologies. DG applications require relatively expensive technologies. The high capital cost per kW of installed power is one the main adversities related to DG versus central generation. The different technologies of the DG system depend on different relative costs per kW installed. These technologies include combustion turbines with a lower cost per kW compared with that of fuel cell-based DG. Important cost variations because of technological advancements and unpredictable equipment costs are experienced by renewable DG. This unpredictability of renewable DG equipment costs is due to technological progress and economies of scale [36].

6.5 Substitute for grid investments

Depending on the sufficient availability of alternative primary fuels for DG application, DG should contribute to savings in transmission and distribution costs for DG customers. If this availability is not considered, it will result in problems like dependency on gas for DG generation. A large share of transmission and distribution costs must be billed to smaller customers, which makes DG a much better substitute because it can reduce these costs [36]. It saves about 30% of electricity costs if transmission and distribution costs are bypassed by using on-site production [37].

7. Incentives by different governments for distributed generation

7.1 Who is affected when a customer installs distributed generation?

There are financial implications of installing DG for utilities and consumers. Under metering and billing arrangements of DG, DG system owners must satisfy load requirements before exporting, and as a result, utilities lose electricity sales. This is because the utility tariff structure consists of fixed and variable parts, and they ofter recover some of their network cost from the variable part [38].

Large-scale adoption of DG systems requires infrastructure to meet the needs of new technology. Utilities may pass those additional infrastructure costs on to consumers, and thus the consumer's cost of energy increases. This sometimes leads to financial loss for nonadopting DG consumers, as the net gain of the system depends on an entire set of costs and benefits from DG adoption. This is sometimes called cross-subsidization—i.e., shifting costs to nonadopting DG consumers.

So before installing a DG, the compensation mechanism must be one that reinforces positive effects that mitigate negative effects and is profitable for distribution utilities, DG system owners, and non-DG-owning consumers.

7.2 Compensation mechanism

It is important to include convenient compensation mechanisms with regulatory systems and public policies for better operation of DG applications in a grid-connected system. DG providers get positive consideration for power generation transported to the utility grid through this compensation mechanism design. Use of DG is significantly impacted by the compensation mechanism, as it has a strong impact on the DG investment schemes for individual customers. The DG compensation mechanism has three core components.

7.2.1 Metering and billing arrangements

Measurement and billing of electricity consumed as well as generated between DG providers and utilities is done under this component. Net energy metering (NEM); buy all, sell all; and net billing are the three main types of metering & billing arrangements [38]. In NEM, billing is based on the net energy consumption of electricity (as shown in Fig. 9.8). Net energy consumption means electricity consumed by DG owners that comes from the grid minus electricity generated by the DG owners that is exported to the grid. DG owners basically receive credits in terms of kWh for exporting the generated energy that is injected into the grid. These credits are used to compensate electricity consumption for current as well as future billing cycles. For a common billing cycle, if the exported energy (to the grid) is more than the energy consumed, the earned credits can be banked for the next or future billing cycles. These credits are fixed with a predetermined sell rate and predetermined time according to the NEM agreement. The meter used is bidirectional—i.e., it can spin in a backward direction (if the generated energy is more than consumed) and calculates net energy consumption at the end of the billing cycle.

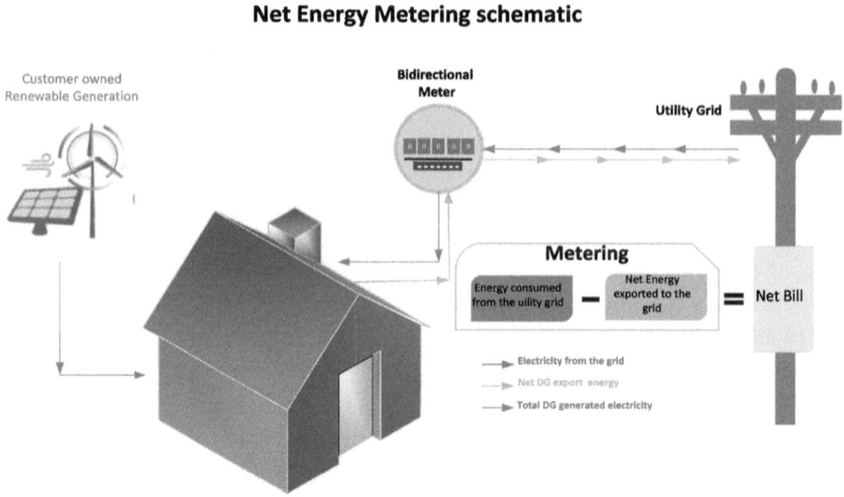

FIGURE 9.8 Net energy metering.

The net billing method has no provision for credits (in kWh) for current and future billing cycles (as shown in Fig. 9.9). In net billing, there are two types of meters, net export and net consumption. If the generation of electrical energy (for exporting to grid) by DG owners is more than the on-site consumed energy, the net electricity *export* meter spins in a forward direction and measures the net energy injected into the grid. A predetermined sell rate is applicable for this exported electricity, which is received by the customer. Similarly, a customer must pay a predetermined retail rate if on-site consumed energy is more than the generated electrical energy required for injection into the grid. In this case, the net electricity *consumption* meter spins in a forward direction to calculate the net energy consumed by customers.

In a buy all, sell all arrangement, DG owners receive a standard sell rate (in cash or through utility bill credits) for all the electricity they produce (independent of consumed energy) to inject into the grid (as shown in Fig. 9.10). Similarly, DG owners have to pay for all the energy they consume (independent of generated energy) at a required retail rate. Unlike NEM and net billing arrangements, buy all, sell all customers do not consume their own generated electricity; rather, all generated energy is exported to the grid. Here two types of meters are used, DG production and preexisting consumption [38].

7.2.2 Sell-rate design

The DG owners generate their own electricity and export it to the grid. For the export of electricity, the DG owners get some compensation that is defined by the sell-rate design [38]. The type of metering & billing arrangement selected determines the sell-rate that is applied for distinct quantities. Sell-rates can be

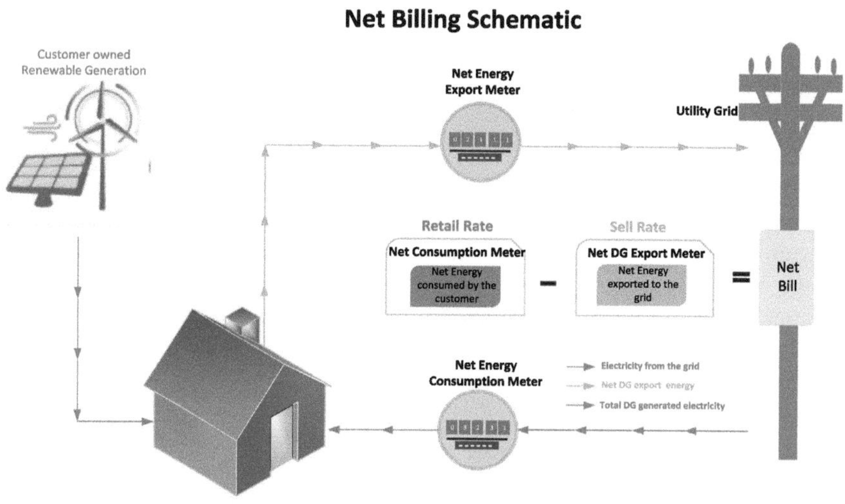

FIGURE 9.9 Net billing method.

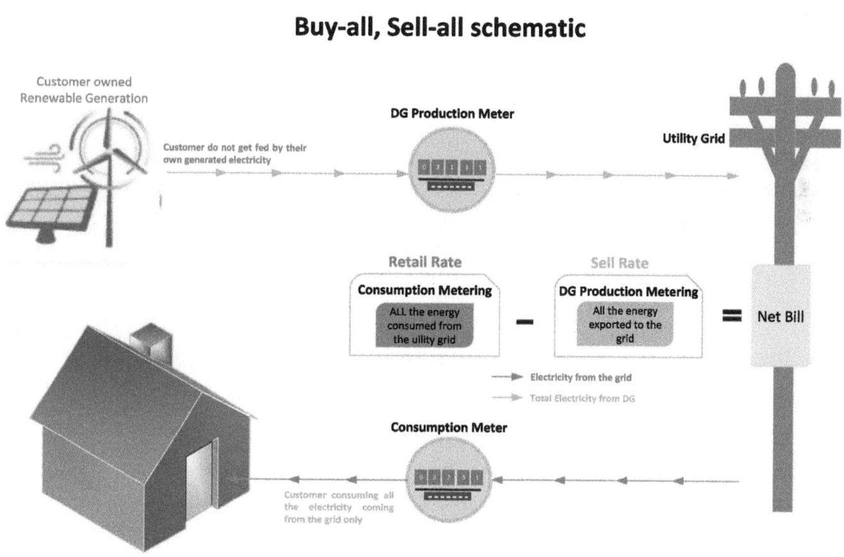

FIGURE 9.10 Buy all, sell all method.

both static and dynamic. A static sell-rate remains fixed over a predetermined period based on the contract. A dynamic sell-rates keeps changing depending on time and location.

7.2.3 Retail-rate design

A DG system owner consumes electrical energy from the grid. DG owners also have to pay for consumed electrical energy from the grid. The payment for consumed energy is defined by the retail-rate design [38]. This retail-rate design structure also depends on the type of metering and billing scheme. Practically, DG owners in NEM and net billing schemes consume less energy from the grid because they first consume their own produced energy. This is not the same case for buy all, sell all customers.

7.3 Selling electricity—feed-in tariffs

Incentives required to financially support DGs and small-scale renewable generation are provided through feed-in tariffs (FITs). Hydropower, solar PV, wind energy, anaerobic digestion units, and domestic CHP units are encouraged by FITs provided they are up to 5 MW. Generation tariffs, export tariffs, and import reductions are ways for the aforementioned generation technologies (under FITs) to gain financial advantage. The fixed price for the each unit of electricity generated by the generation projects under FITs is decided through a generation tariff, and similarly, each unit price of exported electricity to the grid is decided through the export tariff. Import reduction is the benefit from using self-generated electricity for utilization that further reduces import from the grid. The generation and export tariffs are published [39] by the Office of Gas and Electricity Markets, which is a government regulator of gas and electricity markets in the UK. An example of benefits from FIT after installation of DG is shown in Fig. 9.11. The DG technology shown in Fig. 9.11 is wind power with a total installed capacity between 0 and 50 kW.

7.4 Incentives in India for connection of distributed generation

In India, to support rooftop solar installations, the Ministry of New and Renewable Energy provides financial incentives to distribution companies of up to INR 3.75 million/MW for 1350 MW of rooftop solar capacity [40,41]. The distribution companies can use this fund for upgrading the IT infrastructure and distribution network, consumer awareness, training employees, setting up consumer help lines, etc. The Indian government is planning to increase the installed capacity of rooftop solar to 1 GW in the next 5 years. However, despite the large incentive, distribution companies are quite reluctant about large-scale market adoption due to the additional implementation burdens of modernizing the grid and potential loss of revenue.

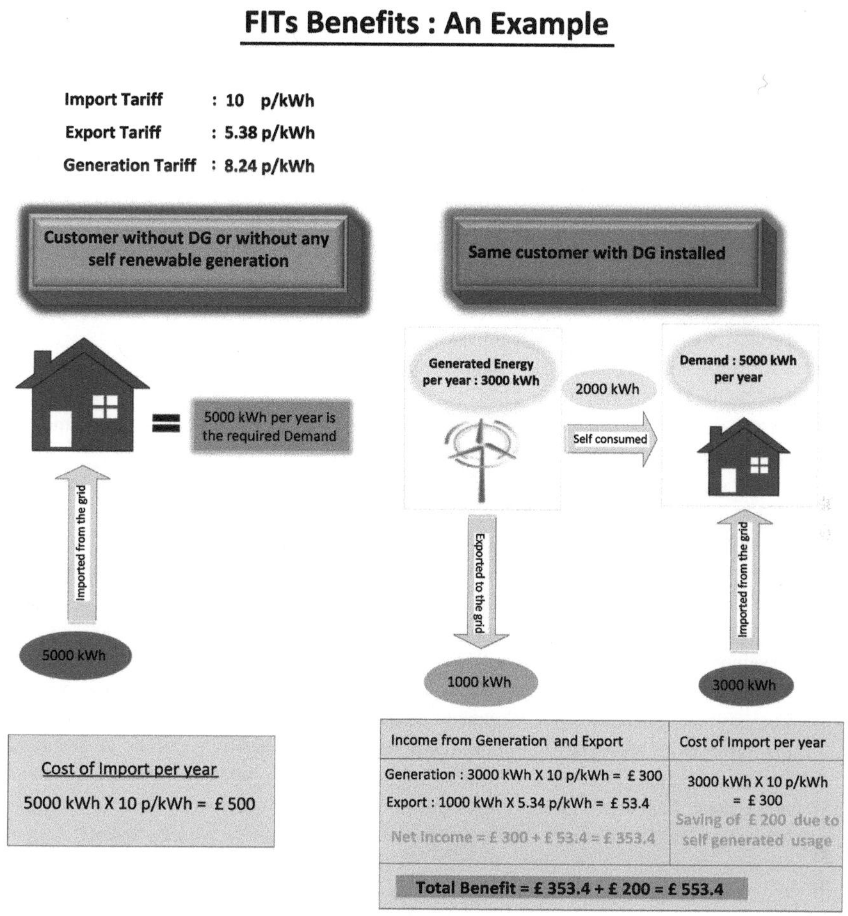

FIGURE 9.11 Feed-in tariffs: an example.

8. Conclusion

In this chapter, the various environmental and technical aspects of DG are explained in detail. The tariff structure of connecting a DG from consumer and utility perspectives were discussed along with the economics involved in connecting to the present power system.

References

[1] D.G.C. Guide, Distributed generation connection guidea guide for connecting generation that fallsunder g59 to the distribution network (typically by developers, industry, commercial or farms) april, Tech. Rep. (2016).

[2] https://www.epa.gov/energy/distributed-generation-electricity-and-its-environmental-impacts.
[3] D. Turney, V. Fthenakis, Environmental impacts from the installation and operation of large-scale solar power plants, Renew. Sustain. Energy Rev. 15 (6) (2011) 3261–3270.
[4] B.A. Hawkins, R. Field, H.V. Cornell, D.J. Currie, J.-F. Guégan, D.M. Kaufman, J.T. Kerr, G.G. Mittelbach, T. Oberdorff, E.M. O'Brien, et al., Energy, water, and broad-scale geographic patterns of species richness, Ecology 84 (12) (2003) 3105–3117.
[5] S.E. Jørgensen, Ecosystem Ecology, Academic press, 2009.
[6] K.S. Pregitzer, E.S. Euskirchen, Carbon cycling and storage in world forests: biome patterns related to forest age, Global Change Biol. 10 (12) (2004) 2052–2077.
[7] US-FS, State of Knowledge: Ecosystem Services from Forests, Tech. rep., U.S. Forest Service, Washington, DC, 2007, p. 2007.
[8] V.M. Fthenakis, H.C. Kim, E. Alsema, Emissions from photovoltaic life cycles, Environ. Sci. Technol. 42 (2008) 2168–2174.
[9] IPCC, Fourth assessment report of the international panel on climate change: mitigation of climate change, Tech. Rep. (2007).
[10] E.G. Pacyna, J.M. Pacyna, F. Steenhuisen, S. Wilson, Global anthropogenic mercury emission inventory for 2000, Atmos. Environ. 40 (22) (2006) 4048–4063.
[11] US-DOE, Environmental Assessment for Bp Solar Array Project, 2009.
[12] M. MD, M. PAF, W. WD, Wildlife interactions at solar one: final report. rosemead, ca: Southern California edison, Tech. Rep. (1984).
[13] US-BLM, US-DOE, Draft Programmatic Environmental Impact Statement for Solar Energy Development in Six Southwestern States, 2010.
[14] J.A. Burger, Achieving Forestry Goals on Mined Land: Mine Site Conditions, Species Composition, and Stand Development of Economically Viable Hard-Woods Forests. Market-Based Approaches to Mined Land Reclamation and Reforestation: A Technical Interactive Forum, Tech. rep., U.S. Office of Surface Mining, Mid-continent Region, 2002.
[15] A. Fesnock, Renewable Energy Development on California blm Lands: Can We Leverage Mitigation into Recovery for the Desert Tortoise? Tech. rep., The Wildlife Society, Annual Meeting. Snowbird, Utah, USA, 2010.
[16] US-EPA, Us-epa, Global Anthropogenic Non-co2 Greenhouse Gas Emissions: 19902020, Tech. rep., U.S. Environmental Protection Agency, Washington, DC:, 2006.
[17] E.A.T. Karl, Efficient atmospheric cleansing of oxidized organic trace gases by vegetation, Science 2010 (2010). Tech. Rep.
[18] T.O. Wiedmann, S. Suh, K. Feng, M. Lenzen, A. Acquaye, K. Scott, J.R. Barrett, Application of hybrid life cycle approaches to emerging energy technologies—the case of wind power in the UK, Environ. Sci. Technol. 45 (13) (2011) 5900–5907.
[19] A. W. E. Association, et al., Avian Interactions with Wind Energy Facilities: A Summary, Prepared by Colson & Associates for AWEA, Washington, DC.
[20] European Wind Energy Association (EWEA), Wind Energy-the Facts, Wind facts Project, Tech. Rep, 2009.
[21] C.K. Willis, R.M. Barclay, J.G. Boyles, R.M. Brigham, V. Brack Jr., D.L. Waldien, J. Reichard, Bats Are Not Birds and Other Problems with Sovacools, Analysis of Animal Fatalities Due to Electricity Generation, 2009.
[22] T.H. Kunz, E.B. Arnett, W.P. Erickson, A.R. Hoar, G.D. Johnson, R.P. Larkin, M.D. Strickland, R.W. Thresher, M.D. Tuttle, Ecological impacts of wind energy development on bats: questions, research needs, and hypotheses, Front. Ecol. Environ. 5 (6) (2007) 315–324.

[23] G. Johnson, A review of bat mortality at wind-energy developments in the United States, Bat Res. News 46 (2) (2005) 45–49.

[24] L. Zhou, Y. Tian, S.B. Roy, C. Thorncroft, L.F. Bosart, Y. Hu, Impacts of wind farms on land surface temperature, Nat. Clim. Change 2 (7) (2012) 539.

[25] J. Kaldellis, K. Garakis, M. Kapsali, Noise impact assessment on the basis of onsite acoustic noise immission measurements for a representative wind farm, Renew. Energy 41 (2012) 306–314.

[26] E. Pedersen, F. Van den Berg, R. Bakker, J. Bouma, Can road traffic mask sound from wind turbines? response to wind turbine sound at different levels of road traffic sound, Energy Pol. 38 (5) (2010) 2520–2527.

[27] K. Bell, S. Gill, Delivering a highly distributed electricity system: technical, regulatory and policy challenges, Energy Pol. 113 (2018) 765–777.

[28] http://www.informatics.indiana.edu/rocha/publications/complex/csm.html.

[29] http://www2. nationalgrid.com/UK/Industry-information/Future-of-Energy/Electricity-Ten-Year- Statement/.

[30] T. Ackermann, G. Andersson, L. Söder, Distributed generation: a definition, Elec. Power Syst. Res. 57 (3) (2001) 195–204.

[31] D. G. CIRED, reportPreliminary Report of Cired (International Conference on Electricity Distribution), Working Group WG04, Brussels, Belgium.

[32] I. El-Samahy, E. El-Saadany, The effect of dg on power quality in a deregulated environment, in: IEEE Power Engineering Society General Meeting, IEEE, 2005, pp. 2969–2976, 2005.

[33] Distributed generation incentive - innovation funding incentive - registered power zones - regulatory instructions and guidance - version 2.

[34] P.A. Daly, J. Morrison, Understanding the potential benefits of distributed generation on power delivery systems, in: Rural Electric Power Conference. Papers Presented at the 45th Annual Conference (Cat. No. 01CH37214), IEEE, 2001, 2001, pp. A2–1.

[35] H.A. Gil, G. Joos, Models for quantifying the economic benefits of distributed generation, IEEE Trans. Power Syst. 23 (2) (2008) 327–335.

[36] K. Purchala, R. Belmans, L. Exarchakos, A. Hawkes, Distributed Generation and the Grid Integration Issues, KULeuven, Imperial College London, 2006, p. 9.

[37] P. Dondi, D. Bayoumi, C. Haederli, D. Julian, M. Suter, Network integration of distributed power generation, J. Power Sources 106 (1–2) (2002) 1–9.

[38] A.Y. Aznar, O.R. Zinaman, Grid-connected Distributed Generation: Compensation Mechanism Basics, Tech. Rep., National Renewable Energy Laboratory (NREL), Golden, CO (United States) OSTI, 2017.

[39] www.ofgem.gov.uk/environmental-programmes/feed-tariff-fit-scheme/tariff-tables, (Tech. rep).

[40] https://bridgetoindia.com/mnre-announces-financial-incentives-discoms-support-rooftop-solar/.

[41] http://www.indiaenergyportal.org.

Chapter 10

Forecasting of renewable generation for applications in smart grid power systems

Debesh Shankar Tripathy[1], B. Rajanarayan Prusty[2]

[1]*National Institute of Science and Technology, Berhampur, Odisha, India;* [2]*School of Electrical Engineering, Vellore Institute of Technology, Vellore, India*

1. Introduction

1.1 The growing energy industry

The energy industry has a vital role in determining the progress of civilization. Energy is required for the sustainable development of society as well as the economy. In the modern era, energy, or more specifically electrical energy, is essential for a huge number of applications across various domains. Due to this ever-increasing demand for energy by numerous sectors for their multiple activities, the world faces severe challenges in procuring the required amount of energy. Fossil fuel sources have been leading the energy industries for a long time. But these conventional sources are exhaustible and primarily responsible for environmental pollution and degradation. In addition, conventional energy sources are insufficient to fulfill the rapidly growing demand for electricity worldwide due to their gradual depletion. For a while now, renewable energy sources such as solar photovoltaic (PV), wind power, hydroelectric power, tidal energy, biomass energy, and a variety of others are becoming popular due to their long-term availability and eco-friendliness. The reason for this is the growing demand for clean and efficient energy sources that are inexhaustible for a long time. Though these sources cannot yet completely replace the existing power infrastructure, their integration with existing power systems can effectively satisfy domestic, industrial, and commercial load demands.

In the smart grid era, these integrations are becoming easier with efficient monitoring and control operations. But these bring with them a new set of challenges. Higher penetration of renewable energy technologies requires tackling of issues related to power system stability in the short, medium, and

Advances in Smart Grid Power System. https://doi.org/10.1016/B978-0-12-824337-4.00010-2

long term. The new challenges demand the rapid evolution of energy markets. That being said, technological development is a top priority for utilizing these renewable sources for distributed generation alongside smart grid power systems. The flow of data in smart grids is one of the critical concepts driving technological growth today [1]. New and innovative technologies capable of acquiring, storing, analyzing, and transferring real-time information more efficiently help in adapting to uncertain and volatile scenarios.

With the advent of smart grids, the concept of distributed generation is gaining popularity. It is an approach using small-scale technologies for power generation near end users. It uses compact and modular design technologies to provide secure, reliable, and cost-effective electricity. It reduces dependence on existing power grids, thus reducing load demand to some extent. Compared with their larger centralized counterparts, local generation usually produces less environmentally damaging or disrupting energy. Due to its compactness and decentralized nature, distributed generation is becoming a popular choice in the present energy crisis. One of the main types of distributed generation is renewable energy. The dropping costs of renewable technologies are responsible for their accelerating growth over the past few years. Wind and solar energy harnessing technologies have shown the most prominent growth in the renewable energy market. Though other renewable sources have their importance, these two are the significant contributors to renewable energy today.

1.2 New technological developments

New and efficient technologies have been researched and developed over the last few years. Many contributions have been made in the mainstream technologies of PV arrays, wind energy generators, hydropower turbines, etc. Also, many new technologies are emerging with broader and deeper research [2]. Some renewable energy technologies are discussed below.

1.2.1 Developing photovoltaics

A considerable amount of solar energy reaches the Earth every hour (about 430 quintillion joules). This amount is estimated to be equivalent to global annual energy consumption. The engineering of PV cells demands the use of efficient, reliable, cost-effective, and scalable materials that are market-ready. The discovery of such materials by the use of machine learning with high throughput is in the experimentation stage and helps in accelerating the analysis and discovery of new and efficient materials for PV cells.

1.2.2 Concentrated solar power

Concentrated solar power uses the thermal power of sunlight to generate electricity. A thermal fluid (molten salt or oil) is heated and used to generate electricity by running a steam turbine. A recent project involves the development of a solar thermal aerogel receiver being built at the Bates Research and Engineering Center, Middleton, Massachusetts. It uses Fresnel reflectors to concentrate solar thermal power on the thermal fluid using transparent aerogel (that can trap heat at too-high temperatures), which eliminates the use of expensive vacuum tubes.

1.2.3 Wind energy extraction

Typically, energy from high-velocity winds is extracted with the help of a wind turbine-generator setup. Effective design and orientation of different parts of the wind energy generator—blades, aerofoil, etc.—help in deriving maximum efficiency. Indirectly, concentrated wind energy can also be harnessed in terms of wave energy.

1.2.4 Harnessing wave energy

Despite consistent availability and a high energy density, respectively, in comparison with PV and wind energy, harvesting wave energy requires substantial evolution. Researchers at the Massachusetts Institute of Technology have been working on efficient wave energy converters with optimal geometry for using wave hydrodynamics to capture wave energy and convert it into electricity.

1.2.5 Geothermal energy

Generally, geothermal sources are present in the Earth's crust, from which energy is extracted in the form of superheated steam. Enhanced geothermal systems help to skip the localized nature of geothermal energy reservoirs by engineering/enhancing them, known as engineered geothermal energy, for increased flexibility, productivity, and life span.

1.3 Visions for the future

Technologies for distributed generation are improving over time. The demand for clean energy is fueling research and development of these technologies for efficiently providing the necessary power for different applications. In coming years, many new ideas resulting in state-of-the-art technologies are expected. Distributed generation using renewable energy technologies is good, but much improvement is still required for its efficient integration into smart grids. Technical, economic, environmental, and social issues need to be addressed for further development and application [2]. Matters related to resources, policies, devices, infrastructure, etc. also need attention. The upcoming technologies need to revolutionize existing energy markets, driving them toward more environmentally friendly yet economic technologies that nullify the usage of fossil fuels entirely.

2. Renewable generation forecasting

The penetration of distributed generation, in terms of renewable energy sources, into smart grids brings many opportunities and challenges. Addressing them requires proper planning and decision-making; calculated-risk decisions are needed for a balanced future. A glimpse into the future can work wonders, so forecasting of the future is imperative. As the main focus of this chapter is on renewable generation, the forecasting of these energy sources is discussed in subsequent sections. Forecasting of intermittent renewable energy generation such as PV and wind needs the use of proper techniques to understand the patterns and make worthy estimates of the future. Prior knowledge about future production helps to manage available resources and technologies efficiently and cost-effectively.

2.1 Point forecasting

Typically, a forecast is thought of as a single number representing the best future estimate of the desired variable at a particular instant. In statistics, this is termed as point estimation or point forecasting. This is also known as deterministic forecasting because an exact future value is predicted. A vast literature exists on point forecasting techniques due to their comparatively lower complexity and easy interpretation. However, these predictions are prone to forecast errors resulting from a lack of knowledge about the future. Uncertainties in prediction are not considered in point forecasting models; this makes these forecasts less reliable. Point forecasting models can be mainly classified into the following types [3]:

(a) time-series models
(b) regression models
(c) exponential smoothing models

Though point forecasting models are mainly classified into the above types, other models are also available [3]. Time-series and regression models are focused on in this chapter due to their wide usage. Exponential smoothing models have their advantages and are easy to understand [4].

2.2 Probabilistic forecasting

Point predictions have the disadvantage of giving an incomplete picture of potential variability of the target variable. In other words, uncertainty in modeling is not considered in point forecasting techniques. Consequently, it is a good practice to associate point forecasts with an estimate of forecast error/ uncertainty. In this context, probabilistic forecasts provide comprehensive information about potential future outcomes. They do so by incorporating information about the possible variations of future forecasts and providing an interval or density forecast. An interval forecast, known as a prediction interval (PI), is a range of future values of the observations. A density forecast or more appropriately a probability density forecast, as the name suggests, provides a probability distribution of the target variable in terms of a probability density function (PDF) or cumulative distribution function (CDF).

2.2.1 Quantile forecasts

Let Y_t be a random variable at time t, τ be the nominal level or quantile level, $t+l$ be the lead time, and $Q_{t+l,\ \tau}$ be the τth quantile for $Y_{t+l,\tau}$. If the parameters of a model, M, are β, then the quantile forecast $\widehat{Q}_{t+l,\ \tau}$ at time t for $t+\ l$ by M, given the information set $\mathscr{I}$, is an estimate of the τth quantile for $Y_{t+l,\ \tau}$—i.e., $P\left[Y_{t+l,\ \tau} \le \widehat{Q}_{t+l,\ \tau} \,\middle|\, M, \mathscr{I}, \beta\right] = \tau$. The quantile forecast tells the forecaster that there is probability τ that the value of the random variable will be less than $\widehat{Q}_{t+l,\ \tau}$ at time $t+l$ [5].

2.2.1.1 Interval forecasts

PIs, inferred from quantile forecasts, are threshold levels associated with a probability. They can be directly used for many operational problems and give an intuitive idea about future forecast uncertainties [5]. PIs with various nominal coverage rates are obtained from the set of predicted quantiles and provide the upper and lower bounds for where the actual value is expected to lie, with a probability equal to the nominal coverage rate [6]. These are equivalently referred to as interval forecasts. The PI is generally centered on the median for an interval forecast. A central PI with $(1-\alpha)100\%$ nominal coverage at time t for lead time $t+l$ is given as

$$\widehat{\text{PI}}_{t+l,\ (1-\alpha)100\%} = \left[\widehat{Q}_{t+l,\ \tau=\frac{\alpha}{2}},\ \widehat{Q}_{t+l,\ \tau=1-\frac{\alpha}{2}}\right] \tag{10.1}$$

In Eq. (10.1), the terms in square brackets represent the lower- and upper-bound quantiles generated at $\frac{\alpha}{2}$ and $1-\frac{\alpha}{2}$ respectively.

2.2.2 Density forecasts

The just-discussed quantile and interval forecasts describe a part of the whole information, whereas density forecasts provide complete information by giving a density for each point [5]. Thus, density forecasts represent the entire conditional PDF or CDF of future realizations. Though parametric density estimation can be useful in many cases by assuming a suitable parametric distribution, the nonparametric density estimation method is chosen when the probability distribution cannot be defined with the help of any standard distributions. Nonparametric density estimation is based on kernel functions for approximating the density using some suitable bandwidth.

2.3 Scenarios

Despite the factual information provided by probabilistic forecasts about desired stochastic process attributes, they detail only the marginal densities for each time/space independently. In other words, dependence structures and levels are not considered. Information about temporal and spatial dependencies may be crucial for numerous problems. Scenarios can be seen as sample trajectories from multivariate probabilistic forecasts [5] that model the interdependence structure. In the case of renewable energy generation, scenarios represent the forecast marginal densities for each temporal/spatial horizon and the interdependence structure through time and space.

2.4 Forecast combination

Forecast accuracy can be improved by combining multiple individual forecasts. Many forecasting combination methods have been proposed in the past that help combine the forecasts of individual forecasters/experts to gain a significant improvement in performance. Though simple combination techniques perform reasonably well for point forecast combination, they fail to give probabilistic forecasts [7]. This is because different laws govern the combination of probabilities; simple averaging of multiple individual forecasts is not appropriate in this case [8]. Other methods, such as linear regression with ordinary least squares (OLS) or a least absolute deviation (LAD) loss function, offer an advanced approach for forecast combination [9,10]. Still, they cannot be applied to combine or produce probabilistic forecasts. Quantile regression averaging (QRA) [11−13] is a recent approach to combine multiple point forecasts, thereby providing a probabilistic forecast of the intended variable. It does so with the pinball loss (PL) function applied in a quantile regression (QR) setting. A forecast combination using QRA helps to generate probabilistic forecasts of the data.

2.5 Forecast evaluation

Before utilizing forecasts for any real-life applications, they must be evaluated and analyzed based on the type of forecast. Two essential things to keep in mind are the *quality* of the forecasts to accurately present future realizations by proper modeling of the process, and the *value* of the forecast derived by using them for decision-making [14]. The quality of forecasts is assessed quantitatively by validating them over a period whose data has not been used for building (identifying and learning/training) the model.

2.5.1 Evaluating point forecasts

Point forecasts are widely assessed by calculating the forecast error via the use of different error measures. The most popular error measures are root mean square error (RMSE) and mean absolute error (MAE), which are discussed below. Apart from these, a variety of other error measures are available: mean square error, mean bias error, mean absolute percentage error, etc.

2.5.1.1 Root mean square error

The RMSE between the ith observation, Y_i, and the corresponding forecast $\widehat{Y}_i$ for n forecast instants is given as

$$\text{RMSE} = \sqrt{\frac{1}{n}\sum_{i=1}^{n}\left(Y_i - \widehat{Y}_i\right)^2}. \tag{10.2}$$

2.5.1.2 Mean absolute error

MAE is given as

$$\text{MAE} = \frac{1}{n}\sum_{i=1}^{n}\left|Y_i - \widehat{Y}_i\right|. \tag{10.3}$$

In Eq. (10.3), the terms have the same meanings as in Eq. (10.2).

Note: RMSE and MAE are used to assess point forecasts but also can be extended to the probabilistic framework by replacing the point forecast with a quantile forecast to get the error measure for the corresponding quantile. Then the errors across different quantiles can be averaged to get a single value.

2.5.2 *Evaluating probabilistic forecasts*

Probabilistic forecasts, such as quantile, interval, and density forecasts, have different measures for evaluating their quality. Quantile forecasts can be assessed by the quantile score (QS) [12], which uses the PL function to differentially weigh the quantiles. Measures to evaluate the interval forecast include the Winkler score (WS) [12], prediction interval coverage probability (PICP) [15], and prediction interval normalized average width (PINAW) [15], which assess coverage and interval widths. Continuous ranked probability score (CRPS) [15] is the most widely used measure to determine the reliability and sharpness of density forecasts.

2.5.2.1 Quantile score

The QS uses the PL as a measure for the error in quantile forecasts. The PL function is defined as

$$\text{PL}\left(Y_i, \widehat{Y}_{i,\tau}, \tau\right) = \begin{cases} (1-\tau)\left(\widehat{Y}_{i,\tau} - Y_i\right) & \text{if } Y_i < \widehat{Y}_{i,\tau} \\ \tau\left(Y_i - \widehat{Y}_{i,\tau}\right) & \text{if } Y_i \geq \widehat{Y}_{i,\tau} \end{cases}. \tag{10.4}$$

In Eq. (10.4), Y_i is the ith observation, and $\widehat{Y}_{i,\tau}$ is the τth quantile forecast of the ith observation. The QS is the average PL across the n_q predicted quantiles across the forecast horizon n and is given as

$$\text{QS} = \frac{1}{nn_q} \sum_{i=1}^{n} \sum_{j=1}^{n_q} \text{PL}\left(Y_i, \widehat{Y}_{i,\tau}, \tau\right). \tag{10.5}$$

In Eq. (10.5), j is an index variable that represents the number of the quantile forecast. A lower score indicates a better prediction of the quantiles. The QS helps to assess both the reliability and sharpness of the quantile forecasts.

2.5.2.2 Winkler score

The WS is used to assess the accuracy of the coverage as well as the interval width for the PIs. For a PI centered at the median with $(1-\alpha)100\%$ nominal coverage, this score is defined as the mean across the forecast horizon of

$$\text{WS} = \begin{cases} \Delta_i & \text{if } L_i \le Y_i \le U_i \\ \Delta_i + 2(L_i - Y_i)/\alpha & \text{if } Y_i < L_i \\ \Delta_i + 2(Y_i - U_i)/\alpha & \text{if } Y_i > U_i \end{cases}. \tag{10.6}$$

In Eq. (10.6), L_i is the lower-bound quantile, U_i is the upper-bound quantile, and $\Delta_i = U_i - L_i$ is the interval width for the ith observation. It penalizes observations for lying outside the PI and rewards narrower interval widths. Hence, a lower WS is an indication of a better PI.

2.5.2.3 Prediction interval coverage probability

The PICP helps to ensure whether the observed probability distribution is bounded within the PI. The PICP index is calculated by counting the fraction of the number of observations lying within the forecast PI. It is mathematically represented as

$$\text{PICP} = \frac{1}{n} \sum_{i=1}^{n} \epsilon_i, \tag{10.7}$$

where ϵ_i is defined as

$$\epsilon_i = \begin{cases} 1 & \text{if } Y_i \in [L_i, U_i] \\ 0 & \text{if } Y_i \notin [L_i, U_i] \end{cases}. \tag{10.8}$$

In Eqs. (10.7) and (10.8), the terms have their usual meaning as defined earlier. A higher PICP is desirable, as it indicates that more observations lie within the constructed PI. PICP quantitatively expresses reliability. A valid PICP should always be greater than the nominal confidence level.

2.5.2.4 Prediction interval normalized average width

The sole use of PICP can sometimes be misleading due to concentrating only on coverage of the PIs. The widths of PIs also play a significant role in assessing their informativeness. The quantitative expression of PI width is provided by the PINAW, which is given below:

$$\text{PINAW} = \frac{1}{nR}\sum_{i=1}^{n}\Delta_i. \tag{10.9}$$

In Eq. (10.9), Δ_i is the interval width as in Eq. (10.6), and R is the normalizing factor given by the difference between the maximum and minimum forecast values. Hence, combined use of PICP and PINAW is required for proper evaluation of PIs.

2.5.2.5 Continuous ranked probability score

The CRPS robustly measures the reliability and sharpness of the probabilistic forecast. It is analogous to the MAE for point forecasts because it reduces to the absolute error for the point forecasts, hence allowing comparison of probabilistic and point forecasts. Since it is a measure of the absolute error of the forecast distribution, a lower CRPS is preferable. It considers the entire distribution of the forecasts contrary to QS, which uses the PL to weigh the different quantiles asymmetrically. It can be expressed as below:

$$\left.\begin{aligned}\text{CRPS}(F, y) &= \int_{-\infty}^{\infty} \left(F(u) - \mathbb{I}\{y \le u\}\right)^2 du \\ &= \mathbb{E}_F|Y - y| - \frac{1}{2}\mathbb{E}_F|Y - Y'|\end{aligned}\right\}. \tag{10.10}$$

In Eq. (10.10), y is the observation, $F(u)$ is the CDF of density forecasts, $\mathbb{I}$ is the Heaviside step function, and Y and Y' are two random variables with F as the distribution function.

3. Time-series-based forecasting of renewable generation

3.1 Basic idea

The topic of time-series analysis is a vast domain that includes numerous areas. The most fundamental concepts of stationarity, trend, seasonality, and basic time-series models are briefly discussed for an introductory idea about them. For more details about time-series analysis and its related areas, readers are referred to Ref. [3,14].

3.1.1 Stationarity

Based on observations of the stochastic process, statistical inferences can be drawn by making certain assumptions about the structure for a simplified understanding of the process. Stationarity is the essential assumption relating to statistical properties of a time-series over time. A stationary time-series exhibits similar "statistical behavior" across time. The probability laws governing the process remain constant over time [3,14]. Thus, a stationary time-series has the following attributes:

(a) The expected value of the time-series is time-invariant.
(b) The covariance, or more appropriately the autocovariance function (γ), depends only on the lag l' and is independent of time—i.e., $\gamma(l') = Cov(Y_t, Y_{t-l'})$.

The white-noise time-series is a typical stationary process. It is a sequence of independent and identically distributed (IID) random variables that have a zero mean and constant variance over time. A mean stationary process has zero mean, whereas a standard deviation/variance stationary process has a constant variance throughout the period.

3.1.2 Trend and seasonality

Though the mean function is a time-invariant function for a stationary time-series, this is not the case for practical time-series in general. Usually, these are arbitrary functions of time, but their simpler assumptions can provide relatively useful insights about the trends in a time-series. A trend in a time-series can be defined as the time-dependent mean function of the time-series [14]. In addition to trend, time-series are also characterized by a seasonal component. Seasonality in a time-series can be defined as cyclic behavior or seasonal patterns that appear periodically. Trends can be constant, linearly increasing/decreasing, exponential, etc., whereas seasonalities are either additive or multiplicative.

3.2 Time-series point forecasting models

A time-series model uses historical observations of a random variable (Y) to learn the patterns in data, such as the trend and seasonalities. It can then be utilized to forecast the future values of the random variable that follow a similar pattern.

3.2.1 Models for stationary time-series

3.2.1.1 Autoregressive model

An autoregressive (AR) model is a simple time-series model that says that the value of a random variable at time t depends on p' past lags—i.e., it depends on the observations at $t-1, t-2, \ldots, t-p'$. It is given by

$$Y_t = \phi_0 + \sum_{i=1}^{p'} \phi_i Y_{t-i} + \varepsilon_t. \tag{10.11}$$

In Eq. (10.11), p' is known as the order of the AR model, which is taken as the maximum lag after which the partial autocorrelation function (PACF) shuts off. The parameters of the model are ϕ_i, and ε_t is the random error.

3.2.1.2 Moving-average model

The moving-average (MA) model assumes that the present value of a random variable depends on q' past errors/innovations—i.e., it depends on the errors at $t-1, t-2, \ldots, t-q'$. The model is given by the following equation:

$$Y_t = \phi'_0 + \sum_{j=1}^{q'} \phi'_j \varepsilon_{t-j}. \tag{10.12}$$

The order of the MA model, q', is decided from the autocorrelation function (ACF) plot and is taken as the maximum lag (of the random variable) beyond which the ACF shuts off. The model parameters are ϕ'_j.

3.2.1.3 Autoregressive moving-average model

Combination of the AR(p') and MA(q') models given by Eqs. (10.11) and (10.12), respectively, results in the autoregressive moving-average, or ARMA(p', q') model. When neither the ACF nor the PACF shuts off—i.e., when both show an exponential decay or a damped sinusoid pattern—an ARMA model is suitable. The model is given as

$$Y_t = \phi_0 + \sum_{i=1}^{p'} \phi_i Y_{t-i} + \varepsilon_t + \sum_{j=1}^{q'} \phi'_j \varepsilon_{t-j}. \tag{10.13}$$

Deciding on the order from ACF and PACF is relatively difficult. So the selection of appropriate order is generally made by additional sample functions such as inverse, generalized-sample partial, and extended-sample ACFs.

3.2.2 Models for nonstationary time-series

This section discusses models for time-series that are either nonstationary in mean but stationary in variance or vice versa.

3.2.2.1 Autoregressive integrated-moving-average model

The previously discussed time-series models are applicable only for stationary time-series with zero mean and a constant variance. An autoregressive integrated-moving-average (ARIMA) model can be used for nonstationary time-series having a constant variance. An ARIMA(p', d', q') model with p' as the order of AR, d' as the order of integration (differencing), and q' as the order of MA is a simple alteration of Eq. (10.13) and given as

$$Y_t = \phi_0 + \sum_{i=1}^{p'} \phi_i \mathscr{D}^{d'} Y_{t-i} + \varepsilon_t + \sum_{j=1}^{q'} \phi'_j \varepsilon_{t-j}. \tag{10.14}$$

In (10.14), $\mathscr{D}$ represents the difference operator, and the other terms have their usual meanings.

3.2.2.2 Autoregressive conditional heteroscedastic model

When the variance of a time-series is time-dependent, or the time-series shows volatility clustering, the autoregressive conditional heteroscedastic (ARCH) model comes in handy. It models the time-dependent variance σ_t^2 of the residuals ε_t given by $\varepsilon_t = \sigma_t Z_t$, where σ_t is the time-dependent standard deviation, and Z_t is the stochastic part representing the IID, zero mean, and unit variance strong white-noise process—i.e., $Z_t \sim \text{IID } N(0, 1)$. The ARCH$(q)$ model with parameters α_j' is given as follows:

$$\sigma_t^2 = \alpha_0' + \sum_{j=1}^{q} \alpha_j' \varepsilon_{t-j}^2. \tag{10.15}$$

The order of the model is decided similarly to the method in the AR(p') model given by Eq. (10.11).

3.2.2.3 Generalized autoregressive conditional heteroscedastic model

The generalized autoregressive conditional heteroscedastic (GARCH) model is a generalization of the ARCH model where additional variance terms are modeled along with the residuals. Since the variance is time-dependent, its use as the GARCH terms is self-explanatory. A GARCH$(p,\ q)$ model with p and q as the respective orders of GARCH and ARCH terms is given by

$$\sigma_t^2 = \alpha_0 + \sum_{i=1}^{p} \alpha_i \sigma_{t-i}^2 + \sum_{j=1}^{q} \alpha_j' \varepsilon_{t-j}^2. \tag{10.16}$$

Eq. (10.16) is a generalized version of Eq. (10.15) where α_i and α_j' are the model parameters.

3.2.3 *Models with exogenous inputs*

Exogenous inputs are external information independent of the model, but they affect model output. The inclusion of exogenous variables can result in reduced forecasting errors due to the use of additional information. The models in Eqs. (10.11), (10.13), and (10.14) can be modified to accommodate exogenous variables resulting in ARX, ARMAX, and ARIMAX models, respectively. An ARMAX $(p',\ q', b')$ model where b' is the number of exogenous inputs E_t with parameters η_e is given as

$$Y_t = \phi_0 + \sum_{i=1}^{p'} \phi_i Y_{t-i} + \varepsilon_t + \sum_{j=1}^{q'} \phi_j' \varepsilon_{t-j} + \sum_{e=1}^{b'} \eta_e E_{t-e}. \tag{10.17}$$

In Eq. (10.17), the terms have their usual meanings. The details about these models can be found in Ref. [16,17].

4. Regression-based forecasting of renewable generation

4.1 Regression-based point forecasting

Regression analysis is a statistical technique for modeling and investigating the relationship(s) between a response/target/dependent variable, or predictand or regressand, and one or more explanatory/independent variables, or predictors or regressors. This study helps to develop a model for forecasting future responses given specified values for the predictor variable(s) [3]. Though many regression-based forecasting models are available, the following methods are frequently used in the literature.

4.1.1 Multiple linear regression

Let Y denote the dependent variable linearly related to $\wp$ independent variables $X_1, X_2, \ldots, X_\wp$ via the parameters $\omega_1, \omega_2, \ldots, \omega_\wp$. Then,

$$Y = X_1\omega_1 + X_2\omega_2 + \ldots + X_\wp\omega_\wp + \varepsilon. \tag{10.18}$$

Eq. (10.18) is the general form of a multiple linear regression (MLR) model [18]. The parameters $\omega_1, \omega_2, \ldots, \omega_\wp$ are the regression coefficients associated with $X_1, X_2, \ldots, X_\wp$, respectively, and ε is the random error component that reflects the difference (due to the joint effect of variables not included in the model, random factors that cannot be accounted for by the model, etc.) between the observed and fitted linear relationship. The jth regression coefficient, ω_j, represents the expected change in Y per unit change in j^{th} independent variable, X_j. Eq. (10.18) can also be represented as $Y = X\omega + \varepsilon$, where Y is an $n\times 1$ vector of n observations of the dependent variable, X is an $n \times \wp$ matrix of n observations of the independent variables, ω is a $\wp \times 1$ vector of regression coefficients, and ε is an $n \times 1$ vector of random errors. The parameters can be estimated by the OLS algorithm, which minimizes the sum of squared errors for each point.

4.1.2 Nonlinear regression

Contrary to MLR, nonlinear regression (NLR) is nonlinear in parameters—i.e., the regression equation defining the model is a nonlinear function. An NLR model assumes the form $Y = \mathscr{f}(X, \omega) + \varepsilon$, where Y is the response/dependent variable, X is a vector of $\wp$ predictors/independent variables, ω is a vector of $\wp$ parameters, $\mathscr{f}(\cdot)$ is the NLR function, and ε is the error term. Unlike MLR, whose parameters are obtained by OLS, the procedure for estimating NLR parameters is quite complicated. It requires the use of nonlinear least squares [19] or some other nonlinear approximation techniques [20] to obtain the model's parameters.

4.1.2.1 Varying coefficient regression

As the name suggests, a varying coefficient regression model has coefficients/parameters that are not constant across time. In other words, the coefficients vary depending on the time instants, which helps model the dynamic relationship across time. For instance, the effect of regressors X on regressand Y can vary depending on the season [21]. A single model for all the seasons is not much use in that case. So a varying coefficient regression model can come in handy in modeling the regression relationship's seasonal effect. The model can be written as $Y_t = \omega_0(t) + \omega(t)X_t + \varepsilon_t$, where t indicates the time instant, $\omega_0(t)$ and $\omega(t)$ are the time-dependent coefficients of the model, and ε_t is the error term.

4.1.3 Support vector regression

Like support vector machines, which classify points by finding a hyperplane to separate them into different classes in the multidimensional space, support vector regression (SVR) fits a hyperplane to the data. A function (hyperplane) is found in such a way that it satisfies (passes through) the maximum number of points within a decision boundary. Decision boundaries are present on either side of the hyperplane at a distance (say, ξ) from it such that the hyperplane for SVR should satisfy $-\xi < Y - \omega_0 - \omega X < +\xi$ when $Y = \omega_0 + \omega X$ is the equation (similar to a linear regression) of the hyperplane. The main objective is to find a decision boundary such that data points nearest to the hyperplane (known as support vectors) should lie within it. Only points within the decision boundaries that minimize the error are chosen to get the best fit model. An important hyperparameter of SVR is the kernel or kernel function, which transforms data from a lower to higher dimension. Though this increases computational cost, it is useful when finding a hyperplane in the lower dimension is not possible. For a more detailed understanding of SVR involving the different equations and constrains to be considered, readers are referred to Ref. [22,23].

4.1.4 Other models for regression

Apart from the aforementioned regression models, some other models capable of regression and classification are also available. Two such models are discussed below.

4.1.4.1 Random forests

Before understanding the random forests (RF) model [24], it is important to know about the concept of decision trees. A decision tree can be seen as a flowchart-like structure with the topmost node called the root node, each internal/decision node denoting a test on an attribute, each branch representing a test outcome, and each terminal node/leaf holding a label. It can be used to describe decisions (in a treelike model) and decision-making visually and explicitly. It is like a sequence of yes/no questions asked about data that leads to a continuous value (in regression) or a predicted class (in classification). Regression trees have numeric values, in contrast to classification trees that have true/false or some other discrete categories in the leaves. The simple structure of a decision tree is shown in Fig. 10.1.

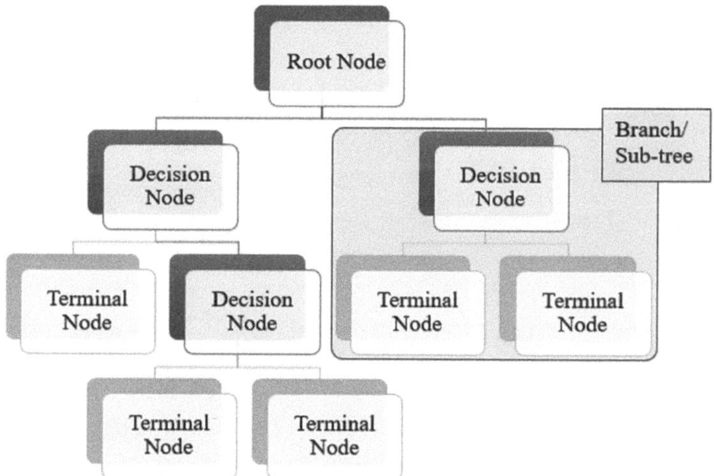

FIGURE 10.1 Basic structure of a decision tree.

The tree predictors (models) are weak learners by themselves. If the model is very flexible, it may lead to overfitting, because the model essentially memorizes both the actual relationship and the noise present in the training data, thus fitting it closely. This leads to a high variance due to the considerable variation of learned parameters with training data. On the other hand, inflexible models make assumptions about the training data, thus introducing a bias. The balance between a flexible model, which memorizes the training data, and an inflexible model that does not learn the training data, is known as the bias–variance trade-off. Unlimited flexibility occurring as a result of precisely one leaf per observation, which overfits the data, can be handled by limiting the depth. This reduces variance but at the cost of increasing bias. Hence, the solution is to use several decision trees as an ensemble model known as the RF.

The simple averaging of the prediction of all the decision trees is known as a forest. Fig. 10.2 gives an intuition of the RF. It is a combination of tree predictors where each tree depends on the values of a randomly and independently sampled vector having the same distribution for all trees in the forest. It uses two key concepts:

- random sampling of data in growing trees while training the model
- random feature selection while splitting the nodes

The RF is trained by considering all available features at every node in the case of regression. The learning of each tree in an RF occurs via randomly sampled data from the training sample. These samples are drawn by bootstrapping—i.e., a few samples may be used more than once in a single tree. Averaging the individual predictions of the trees gives the prediction of

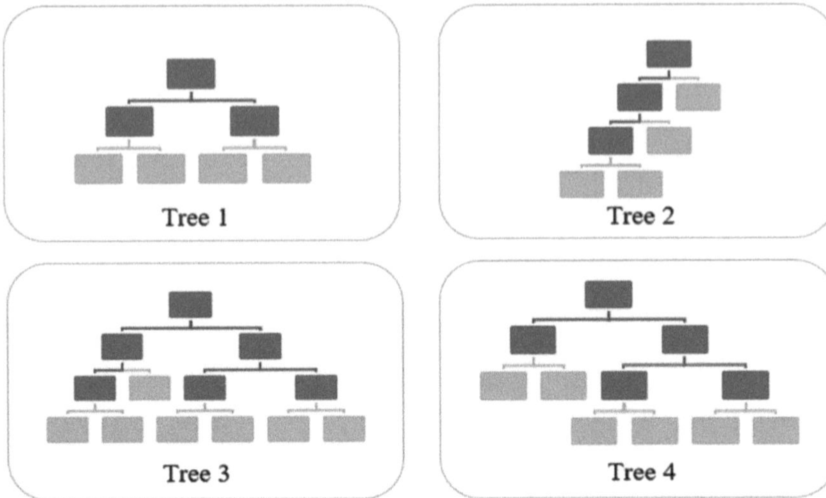

FIGURE 10.2 Random forest—a collection of decision trees.

the forest. The process of training the individual trees on different bootstrapped subsets of the given data, followed by averaging the predictions, is known as bootstrap aggregating or "bagging." Though each tree's variance might be high with respect to the randomly sampled training dataset, the overall variance of the entire forest is low but without an increase in bias. Thus, bootstrapping in RFs helps avoid the bias–variance trade-off.

θ is the random parameter vector to determine the growth of trees, which is represented by $T(\theta)$. Let $\mathfrak{B}$ be the space in which X lives—i.e., $X : \Omega \mapsto \mathfrak{B} \subseteq R^{\wp}$—where $\wp \in \mathbb{N}_+$ is the predictor dimensionality. Every leaf $\ell = 1, 2, \ldots, \mathcal{L}$ of a tree corresponds to a rectangular subspace, $R_\ell \subseteq \mathfrak{B}$. For every $x \in \mathfrak{B}$ dropped down tree $T(\theta)$, there is only one leaf $\ell(x, \theta)$ such that $x \in R_\ell$.

For a new data point $X = x$, the prediction of single tree $T(\theta)$ is the average of all observations in leaf $\ell(x, \theta)$. For observation X_i belonging to leaf $\ell(x, \theta)$, the weight vector $w_i(x, \theta)$ is given by a positive constant, while it is zero when not a part of that leaf. Thus, the weights summing up to one are given as

$$w_i(x, \theta) = \frac{1_{\{X_i \in R_{\ell(x, \theta)}\}}}{\#\{j: X_j \in R_{\ell(x, \theta)}\}}. \tag{10.19}$$

The prediction of a single tree, given covariate $X = x$, is then the weighted average of the original observations Y_i, $i = 1, 2, \ldots, n, \ldots$ and given by

$$\hat{\mu}(x) = \sum_{i=1}^{n} w_i(x, \theta) Y_i.$$

Using RF, the conditional mean $E(Y|X = x)$ is approximated by the averaged prediction of r single trees, each constructed with IID vector

$\theta_m, m = 1, 2, \ldots, r$. The average of $w_i(\theta)$ over this collection of trees is given by

$$w_i(x) = \frac{1}{r}\sum_{m=1}^{r} w_i(x, \theta_m). \tag{10.20}$$

Hence, the prediction of RF is

$$\widehat{\mu}(x) = \sum_{i=1}^{n} w_i(x) Y_i.$$

The approximation of the conditional mean of Y, given $X = x$, is thus given by a weighted sum over all observations. The weights vary with the covariate $X = x$ and tend to be large for those $i \in \{1, \ldots, n\}$ where the conditional distribution of Y, given $X = X_i$, is similar to the conditional distribution of Y, given $X = x$.

4.1.4.2 *k*-nearest neighbors

Its easy implementation and ability to handle nonlinearities well makes the *k*-nearest neighbors (*k*NN) algorithm pretty impressive. The *k*NN algorithm is based on "feature similarity," meaning similar things occur in proximity. Hence, new data points are predicted by considering similar observations of features. Consequently, the new output is predicted by its close resemblance with points in the training set. This proximity is mathematically expressed by suitable distance measures, some of which are given below.

(1) Euclidean distance gives the straight-line distance between two points. Let X^{New} denote the new data point and X^{Old} denote the existing data point for $\wp$ dimensional features. The Euclidean distance is given by the square root of the sum of the squared differences between the two points for all features. Mathematically, it is written as

$$D_{\text{Eucl}} = \sqrt{\sum_{i=1}^{\wp} \left(X_i^{\text{New}} - X_i^{\text{Old}}\right)^2}. \tag{10.21}$$

(2) Manhattan distance is point-to-point travel time, mostly used for binary features. It is the sum of the absolute differences between two points across all features. It can be written as follows:

$$D_{\text{Manh}} = \sum_{i=1}^{\wp} \left|X_i^{\text{New}} - X_i^{\text{Old}}\right| \tag{10.22}$$

The terms have their usual meanings in Eq. (10.22).

(3) Minkowski distance is a generalized version of the two distance measures just discussed and is given by

$$D_{\text{Mink}} = \left\{ \sum_{i=1}^{\wp} \left| X_i^{\text{New}} - X_i^{\text{Old}} \right|^{\mathscr{P}} \right\}^{1/\mathscr{P}}. \tag{10.23}$$

For $\mathscr{P} = 1$, Eq. (10.23) is the Manhattan distance as in Eq. (10.22), and for $\mathscr{P} = 2$, it is the Euclidean distance in Eq. (10.21). These three distances are used for continuous variables—i.e., in regression problems.

(4) Hamming distance is used for categorical variables that are discrete and used for classification problems. It is given as follows:

$$D_{\text{Hamm}} = \sum_{i=1}^{\wp} D_i. \tag{10.24}$$

In Eq. (10.24), $D_i = \begin{cases} 0, & \text{if } X_i^{\text{New}} = X_i^{\text{Old}} \\ 1, & \text{if } X_i^{\text{New}} \neq X_i^{\text{Old}} \end{cases}.$

The Euclidean distance is most widely used for kNN; hence, D_{Eucl} in Eq. (10.21) is chosen as the appropriate measure to express the proximity of feature data points.

4.2 Well-established regression-based probabilistic forecasting methods

4.2.1 Quantile regression

The quantile level is the probability associated with a quantile. It is often denoted by τ. The corresponding conditional quantile of random variable Y given another random variable X is often written as $Q_\tau(Y|X)$. The τth quantile of random variable Y is the value y such that the probability ($P[\cdot]$) that Y will be less than y ($P[Y \leq y = Q_\tau(Y|X)|X]$) equals $\tau \in (0, 1)$. The entire conditional distribution of the response can be described by fitting regression models for a grid of values of $\tau \in (0, 1)$. The optimal choice of the grid depends on the amount of data; the more data, the more detail that can be captured in the conditional distribution [25]. The percentile is the quantile expressed in percent—e.g., a 0.75 quantile is the 75th percentile.

The sum of asymmetrically weighted absolute residuals is the quantile objective function, the minimization of which provides the regression quantiles. A linear model is considered to explain the quantile objective function, $y = X\beta + \varepsilon$, where y is an $n \times 1$ vector of outputs/responses, X is an $n \times \wp$ matrix of inputs/predictors, β is a $\wp \times 1$ vector of unknown coefficients/parameters, ε is an $n \times 1$ vector of random errors, n is the number observations, and $\wp$ is the number of predictors.

The QR problem can be formulated as the following weighted loss function (with respect to β):

$$\underset{\beta \in \mathbb{R}^{\wp}}{\text{Min}} \left[\sum_{iy_i < X_i\beta} (1-\tau)|y_i - X_i\beta| + \sum_{iy_i \geq X_i\beta} \tau|y_i - X_i\beta| \right], \tag{10.25}$$

where the vector $\widehat{\beta}_\tau$ (the solution) gives the τth quantile. $\widehat{y} = X\widehat{\beta}_\tau$ is the estimated conditional quantile. In the objective function in Eq. (10.25), all observations lying below and above the estimated hyperplane $X\widehat{\beta}_\tau$ are weighted by $(1 - \tau)$ and τ respectively. Hence, the objective function written as a sum of piecewise linear functions (check functions) is given as follows:

$$\underset{\beta \in \mathbb{R}^{\wp}}{\text{Min}} \sum_{i=1}^{n} \rho_\tau(y_i - X_i\beta). \tag{10.26}$$

In Eq. (10.26), the check function is $\rho_\tau(u) = u(\tau - I(u < 0))$, and $I(\cdot)$ is an indicator function equal to 1 when the condition $(u < 0)$ is true, and otherwise 0. The check function is nondifferentiable at the origin as seen from Fig. 10.3. Thus, the absence of a general closed-form solution requires the use of computational methods to optimize Eq. (10.25).

If $F(\cdot)$ denotes the CDF of Y, the τth quantile is given as $F^{-1}(\tau) = \mathit{inf}\{y : F(y) \geq \tau\}$, where *inf* stands for the inverse function. QR extends the idea of quantiles to regression analysis, where the objective is

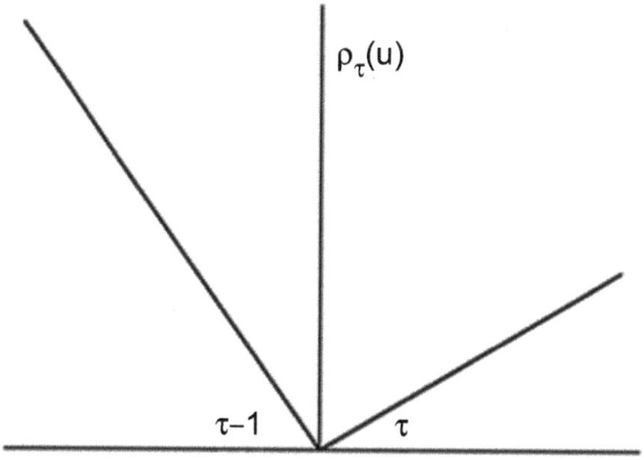

FIGURE 10.3 Quantile regression check function (ρ).

to estimate conditional quantile functions [26]. The set of quantiles gives a better picture of the underlying regression relationships than provided by the conditional mean. Regression quantiles have some desirable properties like equivariance to monotone transformation and reparameterization of design, robustness against outliers of the response, comparable efficiency (to OLS) for Gaussian, and higher efficiency for a wide range of non-Gaussian error distributions. Despite these merits, QR requires higher computational power (compared with OLS) in minimizing the objective function.

The QR model for the τth quantile can thus be written as

$$y_i = X_i\beta_\tau + \varepsilon_i. \tag{10.27}$$

In (10.27), y_i is the response, X_i are the predictors/covariates, β_τ are the unknown coefficients/parameters associated with the τth quantile of the response, and ε_i are random errors.

4.2.2 Quantile regression forests

The quantile regression forests (QRF) [27,28] is an evolved version of the RF with the potential to predict the entire conditional distribution via forecasting different quantiles rather than predicting just the expected mean. It is applied to regression problems and uses the concepts of quantile and RF, hence the name. Generally, the growth of trees in a forest involves considering multiple parameters. These are collectively known as the hyperparameters of an RF model. In general, the selection of the number of trees for a QRF model is random. The algorithm for the QRF model is as follows:

(1) Like RF, an ensemble of r trees $T(\theta_m)$, $m = 1, 2, \ldots, r$ is grown first using n independent observations (Y_i, X_i), $i = 1, 2, \ldots, n$. For each node in each tree, instead of keeping only the mean of the observations, the values of all observations are kept to assess the conditional distribution functions.

(2) For a given value of X (say, x), x is dropped down all the trees and the following weights are computed:

  (a) The weight $w_i(x, \theta_m)$ of all the observations i for every tree is calculated as in Eq. (10.19).

  (b) The weight $w_i(x)$ for each observation i as an averaged prediction of r trees as in Eq. (10.20).

(3) The distribution function for all real values y of the response is estimated from the weights computed before as

$$\widehat{F}(y|X = x) = \sum_{i=1}^{n} w_i(x) 1_{\{Y_i \le y\}}.$$

(4) The τth conditional quantiles are estimated as $\widehat{Q}_{\tau}(Y|X) = inf\left\{y : \widehat{F}(y|X = x) \geq \tau\right\}$.

4.2.3 Quantile k-nearest neighbors

The versatile nature of *k*NN, resulting in its usage in classification as well as regression problems, also finds application as an extension of QR. The use of *k*NN in the probabilistic framework for QR formulates the quantile k-nearest neighbors (Q*k*NN) model [29].

The Q*k*NN model is the same as the QR model in Eq. (10.27). However, instead of considering all data points in the training sample, k data points closest to the data point to be predicted are used for training the model. This significantly improves forecasting performance by eliminating less important data points and thus reducing computation time.

4.2.3.1 Selection of *k*

Proper selection of the hyperparameter k is imperative because it controls the smoothness of the forecast regression curve. A low k value causes noisy estimates that might overfit the data. In contrast, a high value gives a smoother measure, which may not characterize the actual patterns in the data (note: $k = 1$ corresponds to interpolation). The most popular procedure for the selection of an appropriate value of k is by minimization of cross-validation error [30]. The idea is that cross-validation error is calculated for a grid of values of k, and the value of k corresponding to the minimum error is said to be optimal. The cross-validation error for a given k is calculated as follows:

$$\text{CV}(k) = \sum_{i=1}^{n} \left(Y_i - \widehat{Y}_{i,\tau}\right)^2. \tag{10.28}$$

In Eq. (10.28), Y_i is the ith observation, $\widehat{Y}_{i,\tau}$ is the τth quantile forecast of the ith observation, and n is the number of data points excluding the holdout point. For each different τ, a different minimum CV is obtained, leading to different optimal values of k. A different k for different quantiles can create comparison issues, hence a single k is chosen by considering the median regression. The median ($\tau = 0.5$) seems an appropriate choice for the quantile because it plays a balancing role for all other quantiles. Also, since the median is calculated by the absolute loss, it makes sense to use it in Eq. (10.28), as it is closer to the squared error loss of CV than any other quantile loss.

4.2.3.2 Algorithm

(1) At first, the distances from each observation to every other observation of the feature matrix are calculated by Eq. (10.21) and stored in an $n \times n$ matrix where n is the number of observations.

(2) The next step is selecting an appropriate value of k. The optimal k is obtained as discussed previously.

(3) Once an optimal k is found, the kNN for a new (to-be-forecast) data point are used for training a QR model, and the different quantiles for that data point are forecast as usual.

The same process is repeated for all data points in the forecast horizon. It is important to note that for each data point, the kNN are selected from the training sample, consisting of all data points prior to the to-be-forecast point.

4.2.4 Quantile regression averaging

A simple averaging of the point forecasts from different models yields the best point estimate of the data. Another averaging process is the use of linear regression to regress the individual point forecasts on the given data to get a weighted average. It uses OLS to estimate the parameters (weights) of the model. A variant to this is replacing OLS with LAD to get robust point forecasts. Using PL in the QR setting, known as QRA, helps predict the different quantiles of the data instead of a point prediction.

When several point forecasters are available, the selection of one or more models to combine predictions by averaging methods is a problem [12]. Though simple averaging gives a decent prediction, the weighted average over different quantiles helps provide an efficient probabilistic forecast of the data. On this note, the averaging of point forecasts by QR is very useful. In QRA, observed data and individual point forecasts are treated as the dependent variable and independent variables, respectively, for use in a QR model. Then the weights to be given to each method are computed by estimating the parameters of the QR problem. This approach yields PI forecast without using the PIs of the individual models.

The QR problem for the τth quantile can be written as $\widehat{y}_\tau = X\beta_\tau$, whose parameters are estimated by minimizing the PL function for the τth quantile given by Eq. (10.25).

The number and type of individual models can be arbitrarily chosen, or a set of sister forecasts can be considered [12]. The forecasts of these individual point forecasters serve as the dependent variables for QRA. The weights of the individual forecasts are computed in a standard QR setting by minimizing the PL function.

5. Flowcharts of well-established regression-based probabilistic forecasting methods

This section includes flowcharts of all the quantile-based approaches used in simulations for probabilistic renewable generation forecasting. Each flowchart represents the complete step-by-step approach for forecasting using a

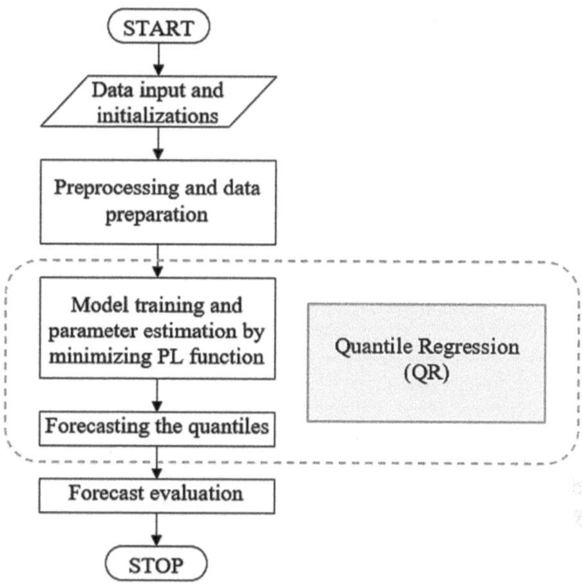

FIGURE 10.4 Steps in forecasting by quantile regression.

particular method in a simplified way. Figs. 10.4–10.7 are for the QR, QRF, Q*k*NN, and QRA models, respectively.

6. Results and discussion

6.1 Data description

Simulations are done for PV data, but the models are applicable to other renewable generation such as wind power probabilistic forecasting. The historical PV generation data for the analyses are collected from rooftop PV installations of three nearby locations in the USA, namely Monkton, Parkesburg, and Lincoln [31], denoted PV_1, PV_2, and PV_3, respectively. The collected PV generation data correspond to the years 2012 to 2016. The data are collected daily corresponding to noon, excluding leap days. The technical and locational details of the three PV arrays are provided in Table 10.1. The effect of daylight time-shifting (popularly known as "daylight saving time") is removed from the data. In a few cases of historical observations, missing values are noticed, and these are less than 2% of the entire database. Such data are updated with recent previous values [32]. Since historical observations are often vulnerable to outliers [33], detection of such anomalies and their appropriate correction are accomplished using the B-spline smoothing approach [34].

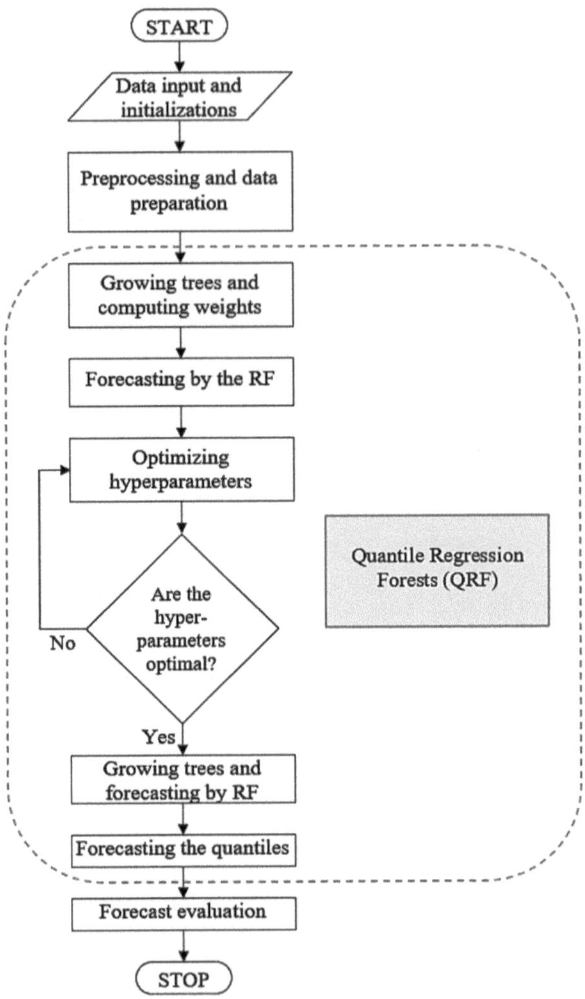

FIGURE 10.5 Steps in forecasting by quantile regression forests.

The QRA model uses ARCH, MLR, and RF as the individual point forecasting models (regressors). The individual point forecasting models are trained using the data from 2012 to 2014, and forecasts are made for 2015 and 2016. For the four probabilistic forecasting models, 2015 is the training period, and out-of-sample forecasts are made for 2016. The regressors for the QR, QRF, and Q*k*NN models are the theoretically formulated physically relevant regressors [35]. The same regressors are also used for the individual point forecasters of MLR and RF. The ARCH model uses the lagged versions of the residuals as the regressors. The regressand for all quantile-based probabilistic forecasting models and individual point forecasting models is from observed

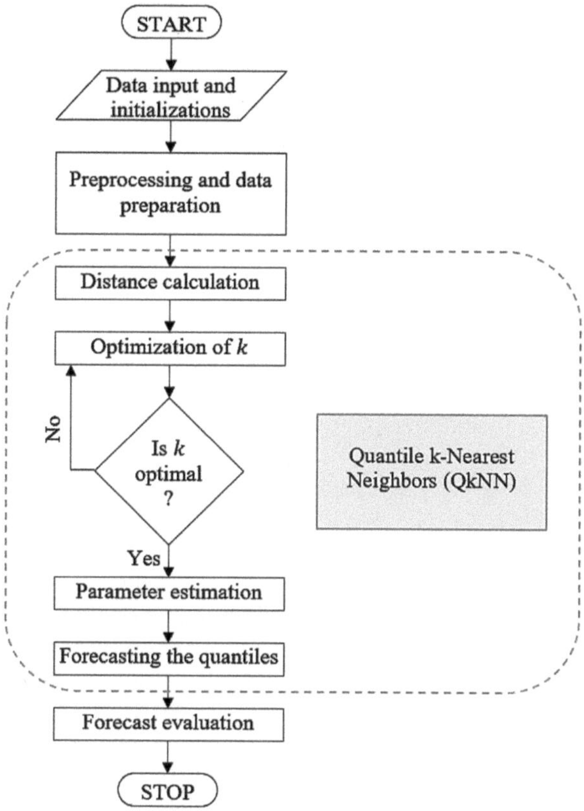

FIGURE 10.6 Steps in forecasting by quantile k-nearest neighbors.

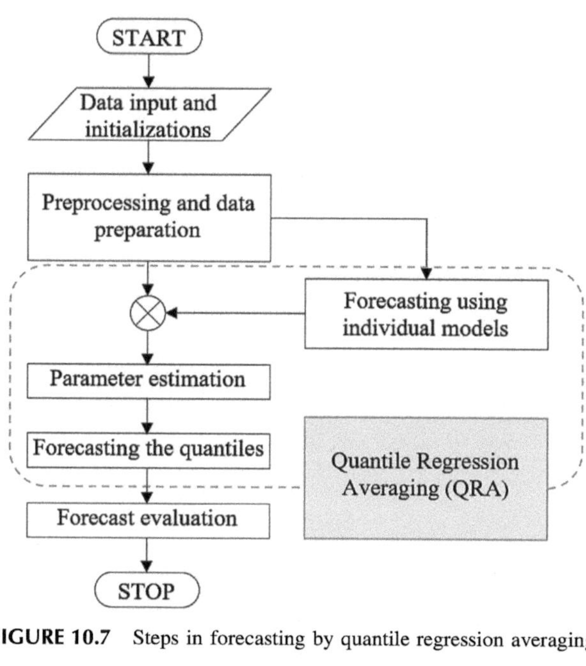

FIGURE 10.7 Steps in forecasting by quantile regression averaging.

TABLE 10.1 Details of photovoltaic arrays.

PV array	Technical details			Locational details	
	Capacity	Orientation	Tilt angle	Place	Coordinate
1	10 kW	South	30 degrees	Monkton	39.58 degrees N, 76.61 degrees W
2	7.65 kW	South	25 degrees	Parkesburg	39.96 degrees N, 75.92 degrees W
3	8.88 kW	Southwest	30 degrees	Lincoln	38.87 degrees N, 75.42 degrees W

PV generation data. For the Q*k*NN model, the optimal value of *k* is selected as discussed earlier. Each model is trained using the training period, and day-ahead out-of-sample forecasts are made for the testing period—i.e., for each forecast day, all the days before that day are used for training. The selection of daily data corresponding to noon is because peak generation occurs around that time each day [36].

6.2 Crucial parameters of forecasting models and their effects

The different parameters of the QR, Q*k*NN, QRF, and QRA probabilistic forecasting models are given in Table 10.2. These parameters are termed crucial parameters because they mainly affect the forecast accuracy and computation time of the respective models. The effects of these crucial parameters on model performance and their appropriate selection procedures are discussed in the following paragraphs.

TABLE 10.2 Crucial parameter(s) of different well-established probabilistic forecasting models.

Model	Crucial parameter(s)
QR	Type and number of regressors
Q*k*NN	Number of nearest neighbors (*k*)
	Dimensionality of predictors
QRF	Number of trees
	Minimum leaf size
	Depth of a tree
	Number of features
QRA	Type and number of individual point forecasters

6.2.1 Effect of change in crucial parameters

For the QR model, the choice of regressors plays a crucial role in forecasting performance. Sensible regressors capable of characterizing the attributing features of the data are essential. The number of regressors also affects model performance. A greater number of regressors can help to describe the data accurately but at the same time increases computational complexity. Time to train the model increases with increases in the dimensionality of the predictors. But since the QR model suffers from the "curse of dimensionality," it fails when too many regressors are used. The Q*k*NN model is similar to the QR model in terms of choice of the type and number of predictors. In addition, the selection of k is very crucial. A lower value reduces computation time due to the use of a smaller number of neighbors but gives degraded model performance due to the presence of noise in the data. It may also lead to overfitting to the training dataset, thus providing worse out-of-sample forecasts. On the other hand, a higher k helps smoothen out the effect of noise at the cost of a bit of a reduction in speed. But it fails in capturing the actual patterns. In spite of this, the Q*k*NN is faster than the QR for the same number and type of regressors.

The QRF model has many parameters to account for, but only some important ones are discussed here. The number of trees to be grown in a forest affects forecasting accuracy, because a greater number of trees can improve forecasting performance but also takes a lot of time to train, and vice versa. The minimum leaf size decides the number of observations to be kept in each leaf. The lower the value, the higher the depth of the tree, which reduces bias but at the cost of increasing the variance and vice versa. The features or predictors affect the performance of the model in learning the data, and in turn forecasting from the learned model. QRF is suitable for a high-dimensional dataset, but this results in increased computation time. A QRF model is better suited to learning the actual characteristics of the data than are the previous two models. But the appropriate selection of features is an issue that requires the use of different feature-selection procedures for improving the performance of the forecasting model and keeping an optimal computation time by eliminating the use of unnecessary features for model training. The critical parameters of the QRA model are the type and number of individual point forecasters. The number of selected unique models affects the computation time as usual. More importantly, the variety of models to be used is mainly responsible for forecast accuracy. Since the performance of the QRA model solely depends on the forecasting performance of the individual models, proper selection of these is imperative.

6.2.2 Appropriate selection of crucial parameters

Appropriate choice of the crucial parameters has a substantial effect on forecasting accuracy and speed. Optimal selection of these parameters is therefore required to build a balanced forecasting model with the necessary degree of accuracy and speed. So there is a trade-off between speed and accuracy in the selection of these parameters.

The choice of regressors plays a significant role in forecasting results. In other words, the quality of a forecast depends on a sensible choice of regressors. In these simulations, theoretically formulated physically relevant regressors [34] are chosen that define the intricate production patterns of renewable energy generation. Due to the limited availability of regressors, the number of regressors is not optimized. Still, it can be done by trying different numbers of regressors and calculating the forecasting error, and then choosing the case with the least error.

Optimization of k has been clearly described in Section 4.2.3.1, where CV error is minimized to get the optimal value of the parameter.

The crucial parameters of the QRF model, as mentioned in Table 10.2, are also known as the hyperparameters of the model. As indicated earlier, due to the limited availability of regressors, the number of regressors/features is not optimized. But number of trees and minimum leaf size are optimized by searching from a grid of the values of these two parameters and minimizing the out-of-bag (OOB) quantile error, which is the error in estimating the OOB observations for a given tree. A large number of trees will provide better prediction but at the cost of increasing computation time. Usually, the maximum number of trees is selected depending on available computing power. But there is a need to tune the process of selection of this hyperparameter for optimum model performance. The minimum leaf size is tuned along with the number of trees to tackle the bias–variance trade-off. Hence, it helps to keep optimal variance by neither increasing the bias nor overfitting to the data.

Here, the choice of the individual point forecasting models is made by intuitively selecting different models that are self-able to characterize the different aspects of the given data. The ARCH model helps in describing the heteroscedastic nature of renewable energy generation. The MLR model smoothens out the noisy estimates of the ARCH model due to faulty shocks (outliers) in the data. The RF model prevents overfitting and compensates for the ARCH and MLR models.

6.3 Significance of performance metrics and comparison of forecasting models

Table 10.3 shows some metrics for assessing the forecasting performance of different models for the three PV arrays. RMSE is calculated by considering the median forecast as the point forecast of the models. The MAE is computed for the quantiles $\tau = 0.01, 0.02, \ldots, 0.99$, and the average is shown. The QS is calculated for the same quantiles, and the WS is calculated for the 50% and 90% PIs. The RMSE helps assess the forecasting error for the median forecast, while MAE assesses it for all quantiles. QS measures the accuracy of the forecasted quantiles of the different models, while WS indicates the accuracy

TABLE 10.3 Metrics for the forecasting models at noon for the three photovoltaic arrays.

PV array	Model	RMSE	MAE	QS	WS (50%)	WS (90%)
PV_1	QRA	2.546	2.089	1.101	5.073	8.380
	QRF	3.195	2.497	1.475	6.210	8.627
	Q*k*NN ($k=340$)	3.244	2.576	1.442	6.030	7.857
	QR	3.281	2.623	1.477	5.991	7.769
PV_2	QRA	1.820	1.503	0.788	3.553	5.851
	QRF	2.527	2.036	1.145	4.824	6.001
	Q*k*NN ($k=300$)	2.496	2.022	1.076	4.562	5.801
	QR	2.531	2.062	1.127	4.627	5.853
PV_3	QRA	2.025	1.645	0.818	4.009	6.759
	QRF	2.621	2.021	0.888	5.174	8.379
	Q*k*NN ($k=260$)	2.616	2.041	0.952	4.936	6.672
	QR	2.725	2.179	1.052	5.155	7.258

of the PIs constructed from these quantile forecasts. Since lower errors and scores are appreciated, their lower values indicate better performance in forecasting and vice versa.

As seen from Table 10.3, RMSEs for all PV arrays are lowest for the QRA model and highest for the QR model. The QRF and Q*k*NN models have moderate RMSEs. This indicates that the median forecasts of QRA are the best, whereas those for QR are the worst. The MAE gives a similar inference but is computed for all quantiles rather than the median. The MAE gives an intuitive measure of the error in quantile forecasts. So the QRA seems to have superior performance over the other forecasting models in estimating quantiles. A better evaluation of quantiles is provided by the QS. Due to individually weighing each quantile, the inference from QS is more accurate than that of the MAE. It can also be observed that the magnitudes of QS are lower than those for MAE in all cases. This is because of the separate weights given to each quantile rather than simple averaging of the errors for each quantile. Again, QRA shows the best performance. WS (50%) and WS (90%) assess the 50% and 90% PIs, respectively. It can be observed that WS (50%) is lower than WS (90%). This is because higher PIs have greater coverage and wider interval widths, resulting in higher WS. As before, QRA gives the best 50% PIs, but some discrepancy is observed for 90% PIs. This is due to overestimation/underestimation of the extreme quantiles by the QRA model. However, the Q*k*NN model constructs the best 90% PI for PV_2 and PV_3, while QR does so for PV_1. In many cases, the QS and WS values for the QRF model are the worst. This could be because QRF is good for high-dimensional datasets, whereas only three regressors are used here.

A box-plot is a five-number summary to represent the distribution of data compactly. It shows the median (0.5 quantile), first and third quartiles (0.25 and 0.75 quantiles), and minimum and maximum values. The IQR is the difference between the first and third quartiles. It gives the spread of the data, or the variance. Each box represents the IQR with the median as the horizontal line in the middle. The vertical dotted lines join the minimum and maximum values to the first and third quartiles, respectively. The outliers (points far from the distribution relative to the IQR) are represented by red dots beyond the extrema.

Fig. 10.8 shows the box-plots for the different models at the 0.95 quantile (95th percentile) for the three PV arrays. The horizontal line across each plot indicates the true 0.95 quantile. From each plot, it is observed that the variance for the QRA is highest, indicated by a higher spread due to a greater IQR. This is because it uses ARCH as one of its point forecasters, whose forecasts are spiky due to the modeling of variance of the data. The QRF, QR, and Q*k*NN models have a low variance (due to the use of theoretical regressors), indicating a narrower distribution. For PV_1 and PV_3, it is seen that the median for QRA is closest to the true quantile, indicating a forecast closer to the true value. However, for PV_2, the Q*k*NN model has a median relatively closer to

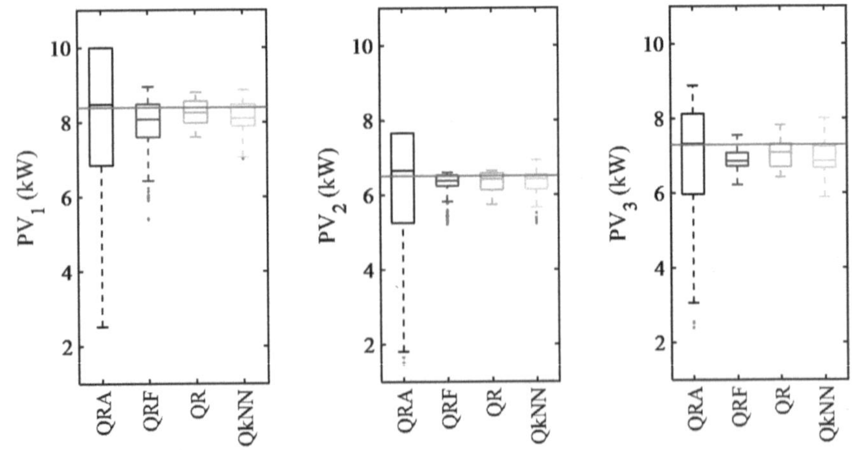

FIGURE 10.8 Box-plots for the forecasting models at the 0.95 quantile for the three photovoltaic arrays.

the true value. The QRF model has the lowest variance for PV_2 and PV_3, indicating the narrowest distributions for the forecast 0.95 quantile. The QR model has a moderate variance compared with that of other models. The box-plots also help to see the skewness in the data, which is known as the shift of the median toward either of the quartiles, from the center. In most cases, the forecast data are skewed toward higher values, which indicates higher generation during the considered time of noon.

6.4 Merits, demerits, and applications of different forecasting models

The QR model is the most basic version of the different quantile-based probabilistic forecasting models. Its simple yet powerful nature provides critical insights about the data by forecasting the conditional distribution. In addition, it is more robust to outliers than mean regression is. It helps to understand the partial effect of each regressor across the different quantiles of the output. The Q*k*NN is an improvement over QR by selecting only the most relevant data from a set. In other words, to forecast a point, the *k* number of data points closest (based on similarity) to it are searched in the dataset to forecast that individual point. Speed is increased by the handling of a smaller amount of data for each forecast. QRF is a forest-based quantile forecasting model. It can handle high-dimensional datasets—i.e., it works well with a greater number of predictor variables/features. It also has the inherent capability to prevent overfitting and handle the bias−variance trade-off. It can estimate all quantiles simultaneously rather than making individual models for each. The tuning of hyperparameters improves forecasting accuracy. A forecast combination approach such as the QRA helps encompass the ability of different individual point forecasters and provides accurate forecasts for the different quantiles, resulting in probabilistic forecasts of the intended variable. A proper choice among these individual models adds to its efficacy. The QRA has the benefits of its constituting point forecasters in addition to the power to make quantile forecasts via QR.

Apart from these advantages, the aforementioned models have some drawbacks, too, that must be addressed. Firstly, QR makes separate models for individual quantiles, which makes the algorithm computationally intensive. Also, independent estimation of the quantiles may result in crossing of different quantiles, though this can be avoided by adding appropriate constraints during parameter estimation. The QR model also suffers from the problem of the "curse of dimensionality," which leads to the failure of the model with high-dimensional data. The Q*k*NN model, being based on QR, also suffers from this issue, but use of the *k*NN algorithm helps in compensating speed issues. The proper choice of *k* is essential for building a balanced forecasting model. A very low value will cause overfitting and noisy estimates, whereas a high value will result in inefficiency in characterizing the actual patterns and also increase computational time. As discussed earlier, the QRF model works well with high-dimensional datasets, so it may not give accurate forecasts when very few features are available. Also, a random selection of model hyperparameters may result in deviation from optimal forecasts. The QRA model may give erroneous forecasts if the chosen individual models are inaccurate. Also, due to a two-stage forecasting process (the first stage being the forecast from individual models and the next step being forecasting by averaging), computational requirements are high. Using multiple variables that characterize similar effects may cause unnecessary delays by slowing down the model training process.

From the above discussions of the different quantile-based forecasting models, their merits and demerits are quite clear. These forecasting models are applicable for the forecasting of any renewable generation time-series data, even though the simulations are done for PV generation. Apart from renewable generation, these models find applications in other domains that involve probabilistic forecasting by regression methods. QR is used when the effect of regressors on different quantiles is simply required to be understood. The Q*k*NN is required when speed yet simplicity is important. The QRF model is applicable when a large number of features are available. The QRA model is useful for applications requiring greater accuracy with high computing capacity. It is also useful when point forecasts are already available and probabilistic forecasts are required.

7. Conclusion

The growth, development, future, and forecasting of renewable energy generation are discussed in this chapter. The mainstream forecasting models are given importance. Some significant developments are highlighted, but most of the chapter deals with probabilistic forecasting of renewable energy generation. The widely used techniques are elaborated, and the quantile-based probabilistic forecasting models are clearly explained along with simulation results. The application of these models is explained by considering the case of PV generation data. Still, the same models also apply to the probabilistic forecasting of other types of renewable energy generation.

References

[1] A. Keyhani, Design of Smart Power Grid Renewable Energy Systems, John Wiley & Sons, Inc., Hoboken, New Jersey, 2016 (Chapter 1).

[2] A. Hussain, S.M. Arif, M. Aslam, Emerging renewable and sustainable energy technologies: state of the art, Renew. Sustain. Energy Rev. 71 (2017) 12–28.

[3] D.C. Montgomery, C.L. Jennings, M. Kulahci, Introduction to Time Series Analysis and Forecasting, John Wiley & Sons, Inc., Hoboken, 2008 (Chapter 1).

[4] V. Prema, K.U. Rao, Development of statistical time series models for solar power prediction, Renew. Energy 83 (2015) 100–109.

[5] J.M. Morales, A.J. Conejo, H. Madsen, P. Pinson, M. Zugno, Integrating Renewables in Electricity Markets: Operational Problems, Springer Science & Business Media, New York, 2014 (Chapter 2).

[6] P. Lauret, M. David, H. Pedro, Probabilistic solar forecasting using quantile regression models, Energies 10 (10) (2017) 1591.

[7] Y. Wang, N. Zhang, Y. Tan, T. Hong, D.S. Kirschen, C. Kang, Combining probabilistic load forecasts, IEEE Trans. Smart Grid 10 (4) (2018) 3664–3674.

[8] K. Holden, D.A. Peel, An empirical investigation of combinations of economic forecasts, J. Forecast. 5 (4) (1986) 229–242.

[9] R. Ramanathan, C.W.J. Granger, Improved methods of combining forecasts, J. Forecast. 3 (2) (1984) 197–204.

[10] J. Nowotarski, E. Raviv, S. Trück, R. Weron, An empirical comparison of alternative schemes for combining electricity spot price forecasts, Energy Econ. 46 (2014) 395–412.

[11] J. Nowotarski, R. Weron, Merging quantile regression with forecast averaging to obtain more accurate interval forecasts of Nord Pool spot prices, in: Proc. 2014 IEEE 11th International Conference on the European Energy Market, 2014.

[12] B. Liu, J. Nowotarski, T. Hong, R. Weron, Probabilistic load forecasting via quantile regression averaging on sister forecasts, IEEE Trans. Smart Grid 8 (2) (2015) 730–737.

[13] J. Nowotarski, R. Weron, Computing electricity spot price prediction intervals using quantile regression and forecast averaging, Comput. Stat. 30 (3) (2015) 791–803.

[14] J.D. Cryer, K.S. Chan, Time Series Analysis: With Applications in R, second ed., Springer Science & Business Media, New York, 2008 (Chapter 2).

[15] D.W. Van der Meer, J. Widén, J. Munkhammar, Review on probabilistic forecasting of photovoltaic power production and electricity consumption, Renew. Sustain. Energy Rev. 81 (2018) 1484–1512.

[16] C. Wan, J. Zhao, Y. Song, Z. Xu, J. Lin, Z. Hu, Photovoltaic and solar power forecasting for smart grid energy management, CSEE J. Power. Energy Syst. 1 (4) (2015) 38–46.

[17] R.H. Inman, H.T. Pedro, C.F. Coimbra, Solar forecasting methods for renewable energy integration, Prog. Energy Combust. Sci. 39 (6) (2013) 535–576.

[18] B.R. Prusty, D. Jena, A spatiotemporal probabilistic model-based temperature-augmented probabilistic load flow considering PV generations, Int. Trans. Elect. Energy Syst. 29 (5) (2019) e2819.

[19] S. Tomioka, S. Nisiyama, T. Enoto, Nonlinear least square regression by adaptive domain method with multiple genetic algorithms, IEEE Trans. Evol. Comput. 11 (1) (2007) 1–16.

[20] E. Tsykin, Multiple nonlinear regressions derived with choice of free parameters, Appl. Math. Model. 8 (4) (1984) 288–292.

[21] H. Li, X. Deng, D.Y. Kim, E.P. Smith, Modeling maximum daily temperature using a varying coefficient regression model, Water Resour. Res. 50 (4) (2014) 3073–3087.

[22] M.W. Ahmad, M. Mourshed, Y. Rezgui, Tree-based ensemble methods for predicting PV power generation and their comparison with support vector regression, Energy 164 (2018) 465–474.

[23] U.K. Das, K.S. Tey, M. Seyedmahmoudian, I. Idris, M. Yamani, S. Mekhilef, B. Horan, A. Stojcevski, SVR-based model to forecast PV power generation under different weather conditions, Energies 10 (7) (2017) 876.

[24] L. Breiman, Random forests, Mach. Learn. 45 (1) (2001) 5–32.

[25] M.A. Rahman, Quantile regression using metaheuristic algorithms, Int. J. Comput. Econ. Econom. 3 (3/4) (2013) 205–233.

[26] R. Koenker, K.F. Hallock, Quantile regression, J. Econ. Perspect. 15 (4) (2001) 143–156.

[27] N. Meinshausen, Quantile regression forests, J. Mach. Learn. Res. 7 (2006) 983–999.

[28] D.S. Tripathy, B.R. Prusty, D. Jena, M.K. Sahu, Multi-time instant probabilistic PV generation forecasting using quantile regression forests, In Proc. 2020 IEEE 9th Power India International Conference, 2020.

[29] A. Bracale, G. Carpinelli, P. De Falco, Developing and comparing different strategies for combining probabilistic photovoltaic power forecasts in an ensemble method, Energies 12 (6) (2019) 1011.

[30] X. Ma, X. He, X. Shi, A variant of k nearest neighbor quantile regression, J. Appl. Stat. 43 (3) (2016) 526–537.

[31] The Hourly PV Generation Data, 2016. Available from: www.pvoutput.org.

[32] B.R. Prusty, D. Jena, Uncertainty Modeling Steps for Probabilistic Steady-State Analysis, Applications of Computing, Automation and Wireless Systems in Electrical Engineering, Springer, Singapore, 2019.

[33] K.G. Ranjan, B.R. Prusty, D. Jena, Comparison of two data cleaning methods as applied to volatile time-series, in: Proc. 2019 International Conference on Power Electronics Applications and Technology in Present Energy Scenario, 2019.

[34] J. Chen, W. Li, A. Lau, J. Cao, K. Wang, Automated load curve data cleansing in power systems, IEEE Trans. Smart Grid 1 (2) (2010) 213–221.

[35] B.R. Prusty, D. Jena, Preprocessing of multi-time instant PV generation data, IEEE Trans. Power Syst. 33 (3) (2018) 3189–3191.

[36] B.R. Prusty, D. Jena, A sensitivity matrix-based temperature-augmented probabilistic load flow study, IEEE Trans. Ind. Appl. 53 (3) (2017) 2506–2516.

Chapter 11

Power quality issues, modeling, and control techniques

Alka Singh[1], Manoj Badoni[2]
[1]*Electrical Engineering Department, Delhi Technological University, Delhi, India;* [2]*Electrical and Instrumentation Engineering Department, Thapar Institute of Engineering and Technology, Patiala, Punjab, India*

1. Introduction to power quality

Ideal AC supply is expected to be an ideal sine wave of fundamental frequency (50/60 Hz). In addition, the voltage should be of rated value with minimum deviations. However, recent trends show numerous power problems encountered in voltage, current, or frequency that have detrimental effects on the operation and working of the power system [1]. These effects can possibly lead to damage, misoperation, and malfunctioning of consumer equipment. The problems may occur due to natural causes or human interventions. These problems may be classified as humankind and natural; humankind causes include switching operations or the starting of large motor loads, sudden load, or generation changes. Natural causes include bad weather, flashover, faults, and pollution. The power quality (PQ) problems in power systems are transients, voltage sag/swell, voltage interruption, undervoltage, overvoltage, harmonics, low power factor, notching, DC offset, and voltage flicker [2]. The main sources of PQ problems are lightning strikes, switching operations of capacitor banks or large motor loads, tap changing on transformers, nonlinear loads, etc.

Numerous standards have been defined and established by the Institute of Electrical and Electronic Engineers (IEEE) and International Electrotechnical Commission (IEC) to maintain and quantify the levels of voltage sag/swell, undervoltage/overvoltage, transients, and voltage flicker. Some of these are IEEE 519-1992, IEEE 1159-1995, IEEE 1100-1999, IEC 6100-2-2, IEC 6100-3-2, and EN 50160 [3].

The present power environment shows increased use of nonlinear loads and the integration of renewable energy sources such as photovoltaic (PV), wind, and others with the power distribution network. This has resulted in serious PQ deterioration. Advances in power electronics have benefited mankind; however, PQ has become poor. Power electronic devices such as uninterruptable

Advances in Smart Grid Power System. https://doi.org/10.1016/B978-0-12-824337-4.00011-4

power supply, adjustable-speed drives, computer and battery chargers, fax machines, and photocopiers are examples of nonlinear loads and generate harmonic currents that propagate to other connected loads at the point of common coupling (PCC). Some other PQ issues, viz. high reactive power demand, voltage regulation, load unbalancing, and poor power factors, cause concerns for electric utilities and consumers of electric power [4]. The solutions to such problems need to be identified and employed in the most cost-effective manner.

Solutions in the form of passive compensators such as the use of tuned capacitors and inductors are in practice, yet new, fast, and effective solutions in the form of active compensators such as custom power devices and hybrid compensators are necessary. The performance of passive compensators is limited due to resonance problems, large component sizes, and the absence of dynamic compensation [5,6]. Active compensators have been developed to overcome most disadvantages of passive devices. These compensators have shown improved performance in compensating voltage and current harmonics, excess reactive power, sag/swell, and unbalancing [7—10].

A number of power converter—based solutions have been suggested in the recent past. The group of devices designed and developed for alleviation of PQ concerns in distribution systems is referred to as custom power devices [11,12]. These include the dynamic voltage restorer (DVR), distribution static compensator (DSTATCOM), and unified power quality compensator (UPQC) [13—15]. A specific shunt-connected custom power device operated for the alleviation of current-associated PQ concerns is widely recognized as a DSTATCOM [16—18].

The chapter includes design and modeling issues for shunt compensators. The behavior of DSTATCOM is generally based on the type of control technique operated for computation of inphase or active, and quadrature or reactive, components of reference currents. These references are used produce gating pulses for control of the voltage-source converter (VSC) used in DSTATCOM. Modeling, control, and implementation of DSTATCOM using several conventional and state-of-the-art techniques shall be discussed.

The chapter also includes the integration of PV systems to grid as well as the modified control required. The converter used for grid integration can be suitably controlled to also provide improved PQ features. This results in increased utilization of the existing converter for the entire day. The chapter presents suitable results showing power factor correction (PFC), load balancing, and other improved features.

2. Power quality issues

PQ is described as any issue related to electric power exhibited in voltage, current, and frequency leading to impair and maloperation of end user equipment. At present, consumers demand not only reliable but also good-quality electricity. Consumers usually face several dips or even blackouts in power supply when a large motor load is turned on. Moreover, numerous sensitive loads such as hospitals, industrial processing plants, financial organizations, data processing, and emergency resource providers demand quality power [2,19]. PQ problems can result from starting large motor loads, nonlinear loads, adjustable-speed drives, lightning, switching transients, and faults. Several PQ issues manifest in today's power system, and commonly listed problems are transients (impulsive and oscillatory), short-duration voltage variation (voltage sag, voltage swell, and interruption), long-duration voltage variations (undervoltage, overvoltage, and sustained interruptions), harmonics, low power factor, notching, DC offset, voltage imbalance, and voltage flicker [1]. Transients are PQ problems that involve an immensely high magnitude of voltage, current or both. It may reach thousands of volts and amperes even on a low-voltage system. Short-duration voltage problems relate to deviations in supply voltage for less than a minute. Examples of short-duration voltage variations are mainly voltage sags, voltage swells, and interruptions [3,20]. These are usually created by faults or energization of large loads; they also may be triggered by sporadic, loose fastening in power wiring. Long-duration voltage variations last longer than 1 min. These variations are further categorized as undervoltage, overvoltage, and sustained interruptions. These are mainly due to system load variations and switching behavior. Harmonics are sinusoidal voltage or currents having frequencies that are integer multiplies of the frequency (fundamental frequency; usually 50 or 60 Hz). Harmonic distortion is mainly contributed by nonlinear attribute loads in the power system. Sources of harmonic distortion in the system are nonlinear loads, inverters/converters used in grid integration of renewable energy sources, uninterruptable power supplies, adjustable-speed drives, etc. Notching is a periodic disturbance that causes current commutation from one phase to another even under normal operation of power electronic devices. Voltage imbalance results from differences in voltage magnitude or phase angle in a three-phase system. Sources of voltage imbalance can include large single-phase loads and line-to-ground faults. Voltage flicker is systematic random variation in the supply voltage; these are very rapid and caused by arc furnaces and oscillating loads [19].

3. Sources of poor power quality

PQ concerns in modern power systems can arise for several reasons. These can be natural or man-made and lead to distortions in voltage, current, and frequency. Lightning and bad weather conditions contribute to several faults in power systems and may severely affect voltage and current waveforms. Man-made sources of PQ problems are excessive use of nonlinear loads, adjustable-speed drives, startup of large motor loads, arc furnaces, traction loads, etc. Nonlinear load causes harmonics in the system produced by controlled/uncontrolled rectifiers, switched-mode power supply, inverters used in integration of renewable energy sources, converters used in electric vehicle charging, and electronic ballast for discharge lamps [1]. Sources of voltage sag/swell are faults in the power system, startup of large motor loads, and heavy load demand. Transients in the power system are due to both natural and man-made causes. Natural causes can be lightning strikes, bad weather, and animals, whereas man-made causes are switching activities, loose connections in the distribution system, and tap changing of the transformer. Long-duration voltage variations may also be produced by load variations in the system and switching behavior [19]. Voltage imbalance is due to large single-phase loads (such as inductance furnaces and traction loads) and asymmetrical distribution of three-phase power supply. PQ problems such as noise in the system are due to electromagnetic interferences, welding machine radiation, arc furnaces, and electronics equipments. Voltage flickers are produced by arc furnaces, any arcing condition in the power system, continuous starting/stopping of the motor, and oscillating loads.

4. Effects of poor power quality on end users

PQ issues affect end users in many ways; viz., transients may damage the power line insulator, and voltage sag can manifest in production losses in automated-process industries because of tripped motors or controller malfunctions. The failure of data processing and computer systems commonly results from voltage sags. The impact of long-duration voltage variation can damage household appliances. Voltage imbalance can affect the operations of large motors, leading to increases in motor temperature and causing the motor to trip. Notching and harmonics lead to distortion in current and voltage waveforms and cause additional losses in the distribution system—electromagnetic interference with communication systems as well as the faulty operation/misoperation of many electrical appliances [2,3]. The financial implications of production losses caused by PQ problems are so great that thorough research into the causes and remedies is of paramount importance.

5. Compensators for mitigation of power quality problems

PQ problems affect utilities, consumers, and manufacturers directly and indirectly in terms of financial loss, production wastage, equipment damage, loss of important data, etc. These problems can be mitigated by designing equipment that is less sensitive to power system disturbances and can operate under significant voltage disturbances with additional installed devices. Power system conditioning devices can be installed in power lines to overcome disturbances and distortions. These conditioners are categorized broadly into three types, viz. passive, active, and hybrid compensators, as discussed next.

5.1 Passive compensator

Passive compensators, namely tuned inductive capacitive (L-C) circuits, have been commonly employed in the past. The various configurations for such compensators include single-, double-, or triple-tuned filters, damped filters, and automatically tuned. The other configurations of passive compensators are passive-series filter, passive shunt filter, or a combination of both. Passive-series compensators are used to prevent a particular harmonic component from voltage-fed loads by offering large impedance to that frequency component [7]. Passive shunt compensators are used for twin purposes, viz. to mitigate harmonic current and meet the reactive power requirements of current-fed loads either partially or wholly. Combining both series and shunt passive compensators results in hybrid passive filters. These hybrid passive compensators can eliminate both voltage and current harmonics [1,3]. Additionally, they can be used to balance and regulate terminal voltages. Passive compensators have the lowest starting cost and are effectual, yet difficulties of large size, higher space demand, detuning with time, and the probability of resonance with system impedance are commonly encountered.

5.2 Active compensator

Active compensators are present-day solutions to PQ problems based on power converters. These solutions have evolved due to the high speed and capability of power semiconductor devices in power electronics. Converters are developed and employed for mitigating the PQ problems of poor power factor, harmonics, voltage sag, swell, flicker, and fluctuating loads. Such devices are particularly categorized as custom power devices or active power compensators. The classifications of devices are series, shunt, or hybrid [8,10].

An active-series compensator is connected in series with AC mains using an appropriate transformer to inject a voltage of required magnitude and frequency and restore the load voltage to rated value and fortify delicate loads in the system. It is also useful for damping out harmonic propagation due to resonance between the line impedance and passive shunt compensator. These compensators are commonly known as solid-state series compensators and DVRs [1,20]. Insulated-gate bipolar transistor (IGBT)-based VSCs are used in these compensators to inject equal and opposite voltage disturbance in series with AC mains to provide clean and regulated voltage.

Shunt active power compensators are effective for current-associated PQ problems and commonly recognized as DSTATCOMs. They are used to mitigate common current-associated PQ issues of the load such as PFC, harmonics, load unbalancing, and unregulated voltage [1,17]. Recent advances in shunt compensators are due to the introduction of fast solid-state devices. Initially, bipolar junction transistors, power metal-oxide-semiconductor field-effect transistors, and gate-turnoff thyristors were used to develop shunt compensators. The commencement of IGBTs has made fast control feasible, and IGBTs are now appraised as ideal solid-state devices for shunt compensators.

5.3 Hybrid compensator

Hybrid filters can mitigate both voltage- and current-related PQ problems. Combining active-series and active shunt compensators can compensate for both voltage- and current-related PQ problems, and these are identified as hybrid filters. A custom power device known as a Unified Power Quality Conditioner (UPQC) is an amalgamation of both series and shunt compensators [14]. This consists of two VSCs joined back-to-back with a common DC bus capacitor. The shunt device for UPQC is known as DSTATCOM and used to compensate for current-associated PQ issues, viz. harmonics elimination, immoderate reactive power, and load unbalancing. The series-connected device of UPQCs is known as a DVR and is used to compensate for voltage harmonics, sag, swell, flickers, notches, and voltage unbalancing.

6. Classification of active compensators

Active compensators are classified as series, shunt, and hybrid. Active-series compensators are used to recompense voltage-associated PQ issues viz. voltage spikes, voltage sag/swell, notches, flickers, voltage unbalancing and harmonics presented in voltage waveforms [1]. Active-series compensators topology can be half-bridge VSC and full-bridge VSC type. It can be single-phase two-wire, three-phase three-wire and three-phase four-wire system; and at the DC input side it can be rectifier supported, battery supported or capacitor supported.

Active shunt compensator can be based on VSCs and current-source converters. Topologically, it can be classified as two-wire (single-phase), three-phase three-wire, and three-phase four-wire with and without isolation. Three-wire active shunt compensators are based on three-leg VSC, three single-phase VSC, star/delta transformer with three-leg VSC, and two-leg VSC with split capacitor. Four-wire active shunt compensators are based on four-leg VSC, three-leg VSC with split capacitors, and three-leg VSC with zigzag transformer/star—delta transformer/T-connected transformers [1,20]. The hybrid active compensator is an amalgamation of active-series and shunt compensators that can overcome both voltage- and current-related PQ problems. It can be based on different supply systems and converter configurations. Supply systems for active hybrid compensators can be two-wire (single-phase), three-wire, and four-wire.

7. Selection of active compensators

Due to the availability of a variety of active power compensators nowadays, its selection for a specific requirement becomes an imperative task for the end user. Selection of an active compensator is based on several factors with the primary factor being the type of PQ problems, which can be voltage associated, viz. spikes, sag/swell, notches, flickers, voltage unbalancing, and voltage harmonics, or current-associated, viz. harmonics, excessive reactive power, and three-phase current unbalancing. Series active compensators are mainly utilized for voltage-related issues, and shunt active compensators are used for current-related issues, whereas hybrid compensators can be utilized for both voltage- and current-associated issues. The selection of the active compensator also depends on the nature of the load, viz. voltage-fed or current-fed, supply system, viz. two-wire, three-phase three-wire, and three-phase four-wire, pattern of the load, viz. fixed or variable, level of compensation required, cost, size, and reliability [20].

8. Design and modeling of active shunt compensator

In this section, design and modeling of an active shunt compensator is presented. The selected shunt compensator consists of a three-leg VSC with IGBTs using antiparallel diodes and a DC capacitor. Simulation work of the active shunt compensator is developed in MATLAB-Simulink using sim power system (SPS) toolboxes. The simulation study is carried out using components such as a VSC, which works as a current-controlled voltage converter. Interfacing inductors along with ripple filters are used to limit ripple currents. A set of three-phase linear/nonlinear loads is considered in this distribution system, whose compensation is carried out by the proposed shunt compensator.

Fig. 11.1 represents the circuit of a three-phase DSTATCOM. The diagram shows a three-phase AC supply with line impedance, shown by the L_s–R_s series branch, feeding a variety of loads. The DSTATCOM is developed by a VSC and is realized using a three-leg converter bridge having IGBTs with antiparallel diodes and high-value capacitors at the DC side. Interfacing inductors (L_f) are used at the AC side of the VSC and couple it to the grid. Series-connected capacitive (C_f) and resistive (R_f) elements at the PCC help reduce switching noise produced by the IGBTs. Three-phase linear and nonlinear loads are connected at the PCC to be compensated by the proposed DSTATCOM. Linear load is realized using three-phase star-connected R-L branches and nonlinear load by a three-phase uncontrolled bridge rectifier.

8.1 Operating principle of distribution static compensator

The single line diagram of a shunt compensator is shown in Fig. 11.2, which presents the basic operating concept of the shunt compensator. The compensator is appropriately controlled to inject leading/lagging current of

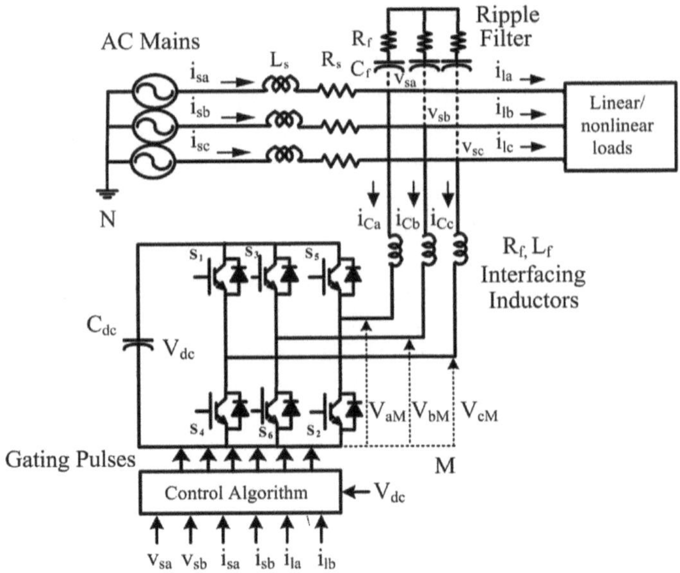

FIGURE 11.1 Schematic diagram of the DSTATCOM.

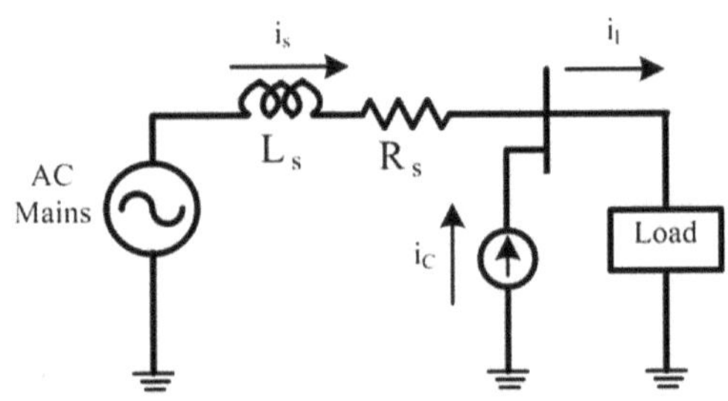

FIGURE 11.2 Basic principle of DSTATCOM.

a desired magnitude. The three-leg VSC is used in the system, where switches of each leg are switched on and off once in a complete cycle. The antiparallel diodes allow current flow in the reverse direction. VSC output RMS voltage is given as

$$V_{LL} = \frac{\sqrt{3} m V_{dc}}{2\sqrt{2}} \tag{11.1}$$

where V_{dc} is DC bus voltage and m is the modulation index. The shunt compensator is able to supply leading reactive power if V_{LL} is greater than the PCC line voltage (v_{ab}); otherwise, it draws lagging reactive power from the AC mains.

The load current (i_l) for the nonlinear load at the nth sampling instant is given as

$$i_l(t) = I_{lf}\sin(wt + \varnothing_f) + \sum_{h=5,7\ldots}^{\infty} I_{lh}\sin(h\omega t + \varnothing_h) \tag{11.2}$$

where I_{Lf}, I_{Lh}, $\varnothing_f$, and $\varnothing_h$ denote the fundamental constituents of load current, harmonic current, phase angle (fundamental constituent), and phase angle (harmonic constituent), respectively. It is clear from Eq. (11.2) that for achieving load compensation, the load current can be distributed into two parts, viz. fundamental or inphase constituent, or harmonic constituent. Fundamental and harmonic constituents can be further separated into two parts, viz. one part corresponding to active constituents (i_{lfp} and i_{lhp}) and another to reactive constituents (i_{lfq} and i_{lhq}), as depicted in Eqs. (11.3) and (11.4),

$$\begin{aligned} i_l(n) &= i_{lf}\sin\omega n\cos\varnothing_f + i_{lf}\cos\omega n\sin\varnothing_f \\ &+ \sum_{h=3,5,7\ldots}^{\infty}\left(i_{lh}\sin h\omega n\, cos\varnothing_f + i_{lh}\cos h\omega n\sin\varnothing_f\right) \end{aligned} \tag{11.3}$$

$$i_l(n) = i_{lfp} + i_{lfq} + i_{lhp} + i_{lhq} \tag{11.4}$$

Control of the shunt active compensator is designed such that it should compensate for harmonics and reactive current (i_{lfq}, i_{lhp} and i_{lhq}) so the supply provides only the fundamental active current (i_{lfp}) component of the load current.

8.2 Modeling of three-phase distribution static compensator

In Fig. 11.1, VSC as a shunt compensator is connected to a three-phase supply. Applying Kirchhoff's laws of voltage and current at the PCC leads to subsequent equations in the natural "abc" frame,

$$v_{sa} = R_f i_{Ca} + L_f\frac{di_{Ca}}{dt} + v_{aM} + v_{MN} \tag{11.5}$$

$$v_{sb} = R_f i_{Cb} + L_f\frac{di_{Cb}}{dt} + v_{bM} + v_{MN} \tag{11.6}$$

$$v_{sc} = R_f i_{Cc} + L_f\frac{di_{Cc}}{dt} + v_{cM} + v_{MN} \tag{11.7}$$

where v_{sa}, v_{sb}, and v_{sc} represent the measured line-to-ground voltages at the PCC of a three-phase balanced system. The following assumptions have been considered to obtain v_{MN}:

$$v_{sa} + v_{sb} + v_{sc} = 0,\ \ i_{Ca} + i_{Cb} + i_{Cc} = 0 \tag{11.8}$$

From these assumptions v_{MN} is obtained as

$$v_{MN} = -\frac{1}{3}\sum_{m=1}^{3} v_{mn}. \tag{11.9}$$

The switching function c_k of the of the k_{th} leg of the VSC is defined as

$$c_k = \begin{cases} 1, & \text{if } S_k \text{ is ON and } S'_k \text{ is OFF} \\ 0, & \text{if } S_k \text{ is OFF and } S'_k \text{ is ON} \end{cases} \tag{11.10}$$

Therefore, $v_{kM} = c_k v_{dc}$.
The phase-k dynamic equation of the shunt compensator model is given as

$$\frac{di_k}{dt} = -\frac{R_c}{L_c} i_k - \frac{1}{L_c}\left(c_k - \frac{1}{3}\sum_{m=1}^{3} c_m\right) v_{dc} + \frac{v_k}{L_c} \tag{11.11}$$

A switching state function (d_{nk}) is given as

$$d_{nk} = \left(c_k - \frac{1}{3}\sum_{m=1}^{3} c_m\right)_n \tag{11.12}$$

The value of the switching state function depends on the switching states (n) and phase (k). This function simultaneously depends on the switching functions of the three legs of the shunt compensator. The subsequent transformed model in the synchronous orthogonal rotating frame is expressed as [21]:

$$\frac{d}{dt}\begin{bmatrix} i_d \\ i_q \\ v_{dc} \end{bmatrix} = \begin{bmatrix} -\frac{R_f}{L_f} & \omega & -\frac{d_{nd}}{L_f} \\ -\omega & -\frac{R_f}{L_f} & -\frac{d_{nq}}{L_f} \\ \frac{d_{nd}}{C_{dc}} & \frac{d_{nq}}{C_{dc}} & 0 \end{bmatrix} \begin{bmatrix} i_d \\ i_q \\ v_{dc} \end{bmatrix} + \frac{1}{L_f}\begin{bmatrix} v_d \\ v_q \\ 0 \end{bmatrix} \tag{11.13}$$

where i_d, i_q and v_d, v_q are inphase and quadrature-transformed components of currents and voltages, respectively.

9. Development of active shunt compensator controller

Performance of the shunt compensator is predominantly decided by the control technique operated to acquire the reference currents. The control techniques can be broadly classified as conventional or modern control algorithms. Various signals, viz. PCC voltages, supply/load currents, and DC bus voltage

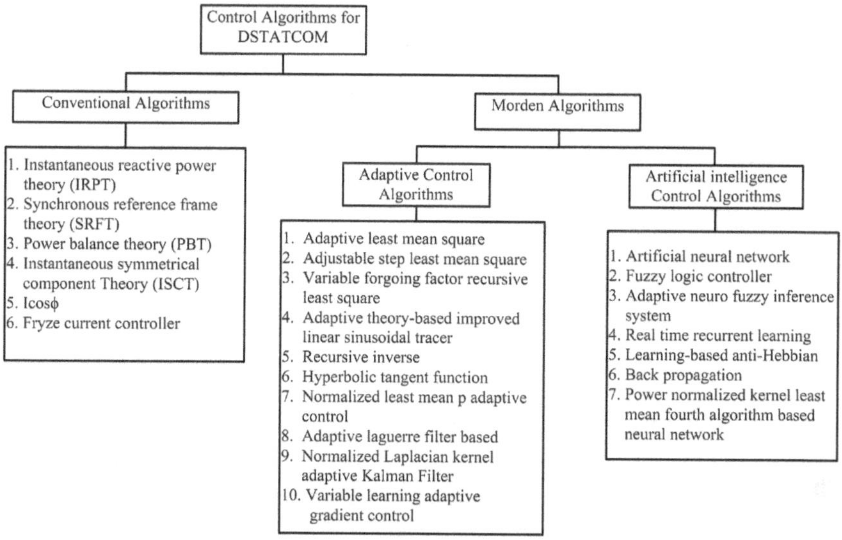

FIGURE 11.3 Classification of control algorithms for DSTATCOM.

of the VSC (used as DSTATCOM), are detected and used by different control algorithms. Various features of DSTATCOM, viz. harmonics eradication, reactive power recompense, and current balancing, are performed with these controllers. This section also deals with mathematical modeling and simulation of control algorithms for the DSTATCOM. Classification of control algorithms for the shunt compensator is given in Fig. 11.3.

9.1 Conventional control algorithms

The control algorithms under this category have been employed during the last decade for the control of DSTATCOM. A few well-known control algorithms employed are instantaneous reactive power theory, synchronous reference frame theory, power balance theory (PBT), and instantaneous symmetrical component theory. These algorithms may be designed and developed for controlling compensators in various modes. In this section, mathematical modeling and development of a PBT-based controller is explained.

9.1.1 Power balance theory–based control

The PBT control technique is used to estimate reference supply currents [22]. This controller action has been divided into three parts. The first presents the estimation of active constituents of loads and then reference currents. The second deals with estimation of reactive power components of reference supply currents and reference supply currents, and generation of switching pulses is presented in the third part.

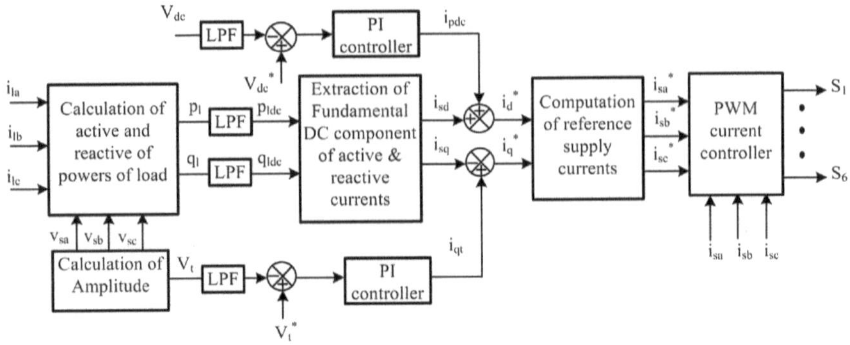

FIGURE 11.4 Block diagram of PBT control technique [22].

(i) Computation of fundamental active constituents of reference currents

Fig. 11.4 shows a block diagram of a PBT controller. The phase voltages (v_{sa}, v_{sb} and v_{sc}) at the PCC are sensed and their amplitude (V_t) computed as

$$V_t = \sqrt{\frac{2}{3}\left(v_{sa}^2 + v_{sb}^2 + v_{sc}^2\right)} \tag{11.14}$$

Using V_t, the unit inphase constituents are computed as

$$u_{pa} = v_{sa}/V_t; \quad u_{pb} = v_{sb}/V_t; \text{ and } u_{pc} = v_{sc}/V_t \tag{11.15}$$

The voltage V_{dc} of VSC used as a DSTATCOM is sensed and subtracted from the reference value (V_{dc}^*) for generation of the error signal (V_{de}). This signal is passed to a proportional integral (PI) controller. The output of PI (i_{pdc}) regulates the DC bus voltage at the reference value and also meets VSC losses. The PI controller output at the *nth* sampling time is given as

$$i_{pdc}(n) = i_{pdc}(n-1) + k_{pd}(v_{de}(n) - v_{de}(n-1)) + k_{id}(v_{de}(n)) \tag{11.16}$$

where k_{pd} and k_{id} are proportional and integral gains of the DC bus voltage PI controller.

The instantaneous active power (p_l) of load is calculated as

$$p_l = (v_{sa}i_{la} + v_{sb}i_{lb} + v_{sc}i_{lc}) = p_{ldc} + p_{lac} \tag{11.17}$$

The active power component of load contains DC as well as AC components. This is passed through a low-pass filter (LPF) to filter out AC ripple components, and the DC fundamental active power component of load (p_{ldc}) is extracted. The fundamental active power component of load current (i_{sd}) is calculated as

$$i_{sd} = \left(\frac{2}{3}\right)\left(\frac{p_{ldc}}{V_t}\right) \tag{11.18}$$

The reference active constituent of supply current (i_d^*) is now computed by adding i_{pdc} and i_{sd} as

$$i_d^* = i_{pdc} + i_{sd} \tag{11.19}$$

Active constituents of reference currents (i^*_{pa}, i^*_{pb}, and i^*_{sc}) are estimated from u_a, u_b, and u_c and the reference active power component (i^*_d) of load current as

$$i^*_{pa} = i^*_d u_a,\ \ i^*_{pb} = i^*_d u_b \text{ and } i^*_{pc} = i^*_d u_c \tag{11.20}$$

The components given in (11.20) are used to estimate the total reference supply currents given below.

(ii) Computation of reactive components

Using the unit inphase voltages given in Eq. (11.15), the unit quadrature vectors are computed as

$$u_{qa} = -u_{pb}/\sqrt{3} + u_{pc}/\sqrt{3} \tag{11.21}$$

$$u_{qb} = \sqrt{3}u_{pa}/2 + (u_{pb} - u_{pc})/2\sqrt{3} \tag{11.22}$$

$$u_{qc} = -\sqrt{3}u_{pa}/2 + (u_{pb} - u_{pc})/2\sqrt{3} \tag{11.23}$$

Voltage regulation using the shunt compensator can be performed using another PI controller over the terminal voltage. The error (v_{te}) between V_t and its set value (V^*_t) is calculated. The error is passed through PI, and output (i_{qt}) can be obtained at the *nth* sampling instance as

$$i_{qt}(n) = i_{qt}(n-1) + k'_{pq}(v_{te}(n) - v_{te}(n-1)) + k'_{iq}(v_{te}(n)) \tag{11.24}$$

where k'_{pq} and k'_{iq} are gains of PI.

The instantaneous reactive power constituent (q_l) of load is calculated as

$$q_l = \left(\frac{1}{3}\right)\{(v_{sa} - v_{sb})i_{la} + (v_{sb} - v_{sc})i_{lb} + (v_{sc} - v_{sa})i_{lc}\} = q_{ldc} + q_{lac} \tag{11.25}$$

The DC fundamental constituent of load reactive power (q_{ldc}) is extracted after the instantaneous reactive power component of load (q_l) is passed through an LPF. The magnitude of the fundamental active component of i_l is given as

$$i_{sq} = \left(\frac{2}{3}\right)\left(\frac{q_{ldc}}{V_t}\right) \tag{11.26}$$

The reference reactive constituent of load current (i^*_q) is computed by i_{qt} and i_{sq} as

$$i^*_q = i_{qt} - i_{sq} \tag{11.27}$$

The reference supply current constituents (i^*_{qa}, i^*_{qb}, and i^*_{qc}) are estimated from the output unit quadrature templates (w_{qa}, w_{qb} and w_{qc}) and reference reactive power components as

$$i^*_{qa} = i^*_q w_{qa},\ \ i^*_{qb} = i^*_q w_{qb} \text{ and } i^*_{qc} = i^*_q w_{qc} \tag{11.28}$$

These reactive constituents of reference currents are added with active reference currents to estimate total reference currents.

(iii) Gating pulses for voltage-source converter

Reference supply currents (i^*_{sa}, i^*_{sb}, and i^*_{sc}) for the three phases are determined by the addition of the respective active and reactive supply currents calculated in Eq. (11.20) and Eq. (11.28) as

$$i^*_{sa} = i^*_{pa} + i^*_{qa}, \; i^*_{sb} = i^*_{pb} + i^*_{qb} \text{ and } i^*_{sc} = i^*_{pc} + i^*_{qc} \tag{11.29}$$

The references currents i^*_{sa}, i^*_{sb}, and i^*_{sc} are compared with the supply currents i_{sa}, i_{sb}, and i_{sc}, which gives current errors (i_{sae}, i_{sbe}, and i_{sce}). These are passed through the pulse width modulation (PWM) current controller to produce gating pulses.

9.2 Modern control algorithms

The control algorithms under this category have been recently developed. These algorithms may be based on adaptive control, artificial intelligence techniques, and even nature-inspired genetic algorithms. The purpose of these new control algorithms is to find active and reactive power constituents from load currents, and with the help of PWM current control, generate gating pulses for the VSC. Since adaptive least mean square (LMS) techniques have become hugely popular in the last few years, one technique, viz. normalized least mean square (NLMS)-based adaptive control algorithm, is discussed here in detail for PQ improvement.

9.2.1 *Normalized least mean square control technique for shunt compensator*

LMS is a linear adaptive technique based on a simple approach. It involves automatic adjustment of system parameters in accordance with estimation error. Least mean square techniques have several variants, but the most popular include basic LMS, NLMS, and signed error least mean square (SLMS) [23]. These variants of LMS techniques are developed as controllers to estimate the weights associated with fundamental constituents of load currents and gating pulses for VSCs.

This section deals with the mathematical formulation of LMS algorithms for three-phase DSTATCOM control. The methods considered are basic LMS, NLMS, and SLMS designed and verified for DSTATCOM. A detailed block diagram of the controller is illustrated in Fig. 11.5.

(i) Development of normalized least mean square—based control algorithm

The voltages at PCC (v_{sa}, v_{sb} and v_{sc}) are sensed and magnitude V_t is calculated as

$$V_t = \sqrt{2\left(v^2_{sa} + v^2_{sb} + v^2_{sc}\right)/3} \tag{11.30}$$

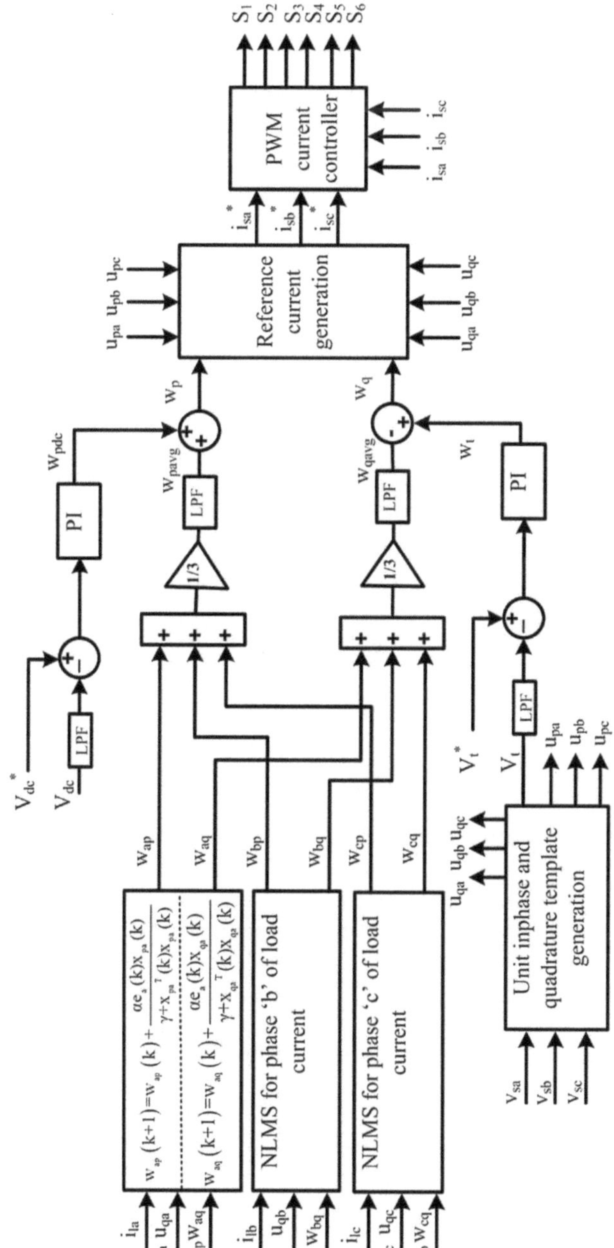

FIGURE 11.5 Block Diagram of the control algorithm for DSTATCOM using NLMS-based control algorithm.

Unit inphase templates are calculated from PCC voltages and their magnitude as given in Eq. (11.15). Voltage at the DC bus (V_{dc}) is detected and compared with reference DC bus voltage (V_{dc}^*), and thus the generated error is passed through PI. The output (w_{pdc}) of the DC bus PI controller is the same as given in Eq. (11.16). Fundamental weights associated with active and reactive power constituents of load currents are calculated from variants of LMS techniques (particularly NLMS) given below in Eq. (11.31).

A problem commonly encountered in the design of adaptive filters is selection of the step size parameter, μ. In the harmonic reduction problem, a step size parameter $\mu(n)$ is used. To evaluate the a *posteriori* estimation error $\acute{\varepsilon}$ at the nth instant,

$$\acute{\varepsilon}(n) = d(n) - w(n+1)^T x(n) \tag{11.31}$$

where d and x denote the desired and input signal, respectively.

Using $d(n) = w^T x(n)$ in Eq. (11.31), this can be modified as

$$\acute{\varepsilon}(n) = \varepsilon(n)[1 - \mu(n)x^T(n)x(n)] \tag{11.32}$$

If $\mu(n) = \frac{1}{x^T(n)x(n)}$ then $\acute{\varepsilon}(n)$ turns into a zero value. With this, the recursive weight equation as depicted in the basic LMS equation $w(n+1) = w(n) + \mu e(n)x(n)$ is modified as

$$w(n+1) = w(n) + \frac{e(n)x(n)}{x^T(n)x(n)} \tag{11.33}$$

A similar type of NLMS technique is given in Eq. (11.34) [23]:

$$w(n+1) = w(n) + \frac{\alpha e(n)x(n)}{\gamma + x^T(n)x(n)} \tag{11.34}$$

where α is the new normalized adaptation constant and γ is a small positive number.

The active load current weights (w_{ap}, w_{bp} and w_{cp}) are extracted from one of the above techniques and their average weight (w_{pavg}) given as

$$w_{pavg} = \frac{w_{ap} + w_{bp} + w_{cp}}{3} \tag{11.35}$$

The reference active weight (w_p) is determined by adding w_{pavg} and w_{pdc} as

$$w_p = w_{pavg} + w_{pdc} \tag{11.36}$$

The active reference supply currents (i_{pa}^*, i_{pb}^*, and i_{pc}^*) are calculated using w_p and u_{pa}, u_{pb}, and u_{pc} as

$$i_{pa}^* = w_p u_{pa}, \quad i_{pb}^* = w_p u_{pb} \text{ and } i_{pc}^* = w_p u_{pc} \tag{11.37}$$

(ii) Estimation of fundamental reactive constituents from normalized least mean square–based control algorithm

Unit quadrature templates (u_{qa}, u_{qb}, and u_{qc}) are calculated from u_{pa}, u_{pb}, and u_{pc} the same way as given in Eqs. (11.21)–(11.23). The magnitudes of PCC voltages are calculated as in Eq. (11.30) and compared with reference PCC voltage magnitude (V_t^*). Thus the generated error signal (v_{te}) is passed through PI with output (w_t) the same as given in Eq. (11.16).

The weights (w_{qa}, w_{qb}, and w_{qc}) corresponding to reactive constituents are extracted from variants of LMS techniques in the same way as active power weights are calculated, and their average weight is computed as

$$w_{qavg} = \frac{w_{aq} + w_{bq} + w_{cq}}{3} \tag{11.38}$$

Reference reactive weight (w_q) is obtained by subtracting w_{qavg} from the output of the AC bus PI controller (w_t) as

$$w_q = w_t - w_{qavg} \tag{11.39}$$

The reference reactive constituents of supply currents (i_{qa}^*, i_{qb}^*, and i_{qc}^*) are obtained from w_q and u_{qa}, u_{qb}, and u_{qc} as

$$i_{qa}^* = w_q u_{qa},\ i_{qb}^* = w_q u_{qb} \text{ and } i_{qc}^* = w_q u_{qc} \tag{11.40}$$

These fundamental reactive constituents are added with fundamental active constituents to estimate reference supply currents.

(iii) Generation gating pulses for voltage-source converter

Reference supply currents (i_{sa}^*, i_{sb}^*, and i_{sc}^*) for the three phases are determined by the addition of the respective active and reactive supply currents calculated in Eq. (11.37) and Eq. (11.40) as

$$i_{sa}^* = i_{pa}^* + i_{qa}^*,\ i_{sb}^* = i_{pb}^* + i_{qb}^* \text{ and } i_{sc}^* = i_{pc}^* + i_{qc}^* \tag{11.41}$$

The currents i_{sa}^*, i_{sb}^*, and i_{sc}^* are compared with actual supply currents (i_{sa}, i_{sb}, and i_{sc}) and produce i_{sae}, i_{sbe}, and i_{sce}, which are used to generate gating pulses for VSC using the PWM current controller.

10. Simulation study and performance of active shunt compensator

The simulated performances of a power balance theory (PBT)-based control algorithm are discussed here. The enactment of DSTATCOM to achieve harmonic elimination, reactive power recompense, and current balancing is presented for a PBT-based control algorithm.

10.1 Behavior of shunt compensator in power factor correction mode

Fig. 11.6 demonstrates behavior of a shunt compensator under nonlinear load in PFC mode. The waveforms shown are PCC voltages, supply, load,

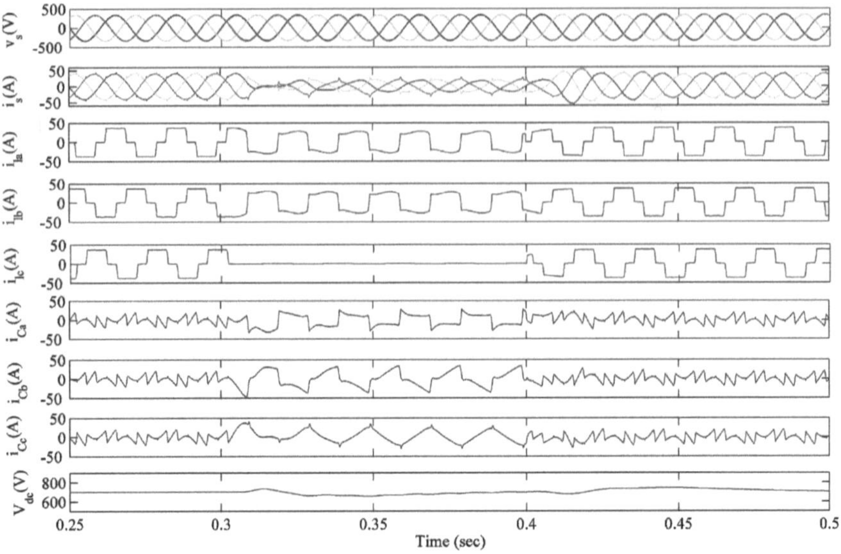

FIGURE 11.6 Performance of DSTATCOM in PFC mode.

compensator currents, and well-regulated DC bus voltage (v_s, i_s, i_l, i_C, and V_{dc}). Performance of the system is observed in steady-state (before $t = 0.3$ s) and unbalanced load ($t = 0.3$ to 0.4 s) conditions. Unbalancing is intentionally introduced in the load current (by removing phase c). Even after this, supply current is observed as balanced along with regulated DC bus voltage.

Fig. 11.7 shows spectra plots of phase "a" of PCC voltage (v_{sa}), supply current (i_{sa}), and load current (i_{la}). These results show THDs of 2.67%, 2.77%, and 26.61% in v_{sa}, i_{sa}, and i_{la}, respectively. After compensator action, the THD of supply current is observed within the limit specified by IEEE-519.

11. Application of active shunt compensator in grid integration of solar photovoltaic system

This section gives an application of DSTATCOM to grid integration of a solar photovoltaic (SPV) array. An SPV array is connected at the DC link of the VSC. The real power demand of linear and nonlinear loads is now met in addition to the features of the shunt compensator, viz. harmonics elimination, recompense of excessive reactive power, and current balancing. Results are demonstrated for power factor correction (PFC) in the presence of a grid-connected SPV array [24].

A two-stage SPV system is used to feed active power to the grid. The first stage consists of SPV power tracked using the maximum power point tracking (MPPT) technique realized through a DC–DC boost converter. In this stage, design of the SPV array, MPPT, and DC–DC boost converter is considered.

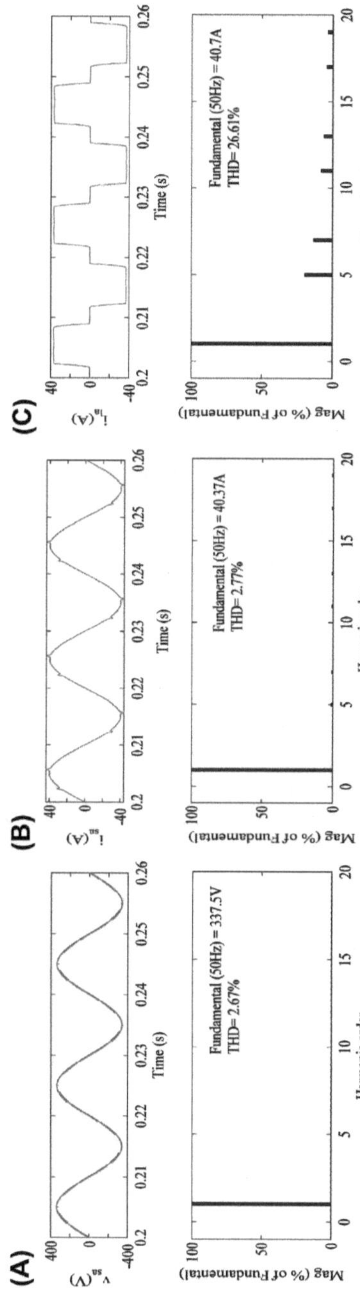

FIGURE 11.7 Harmonic spectra of (A) PCC voltage, v_{sa} (B) supply current, i_{sa} (C) load current, i_{la} in PFC mode.

The second stage consists of coupling the SPV generating system with the grid using VSC and a suitable control algorithm. This stage involves the design of DC bus voltage, DC bus capacitance, VSC rating, interfacing inductors, and ripple filters [25]. A suitable control algorithm for control of a three-phase VSC is designed to evacuate power generation from the SPV array during daytime and provides load compensation. This system is designed and modeled using the MATLAB environment with Simulink and SPS toolboxes. A detailed description of the system configuration and design of various components of a grid-connected SPV power generation system are presented in the subsequent sections.

12. Design, modeling, and control of solar photovoltaic-distribution static compensator system

The PV cell is the basic unit of the PV array. It is a semiconductor diode that converts sunlight into electricity by exposing its p–n junction to light. The semiconductor device is made of silicon material (fabrication process is economically feasible at large scale) and gives output voltage around 0.5–0.7 V under open-circuit condition. The light incident on the solar cell generates charge carriers that cause electric current to flow if the cell is short-circuited.

The SPV array is designed to provide 20 kW of peak power capacity. Each cell has 0.62 V of open-circuit voltage and 4 A of short-circuit current. One module has 30 series cells that produce 30 * 0.62 = 18.6 V of open-circuit voltage and 4 A of short-circuit current. Maximum power for each module occurs at the product of 85% of open-circuit voltage (18.6 * 0.85 = 15.81 V) and 85% of short-circuit current (4 * 0.85 = 3.4 A). The SPV array is formed by combining these modules appropriately in series and parallel. To achieve 20 kW of peak power capacity, 43 such modules are required in series, and 9 such modules are required in parallel.

Fig. 11.8 illustrates the schematic diagram of the grid-connected SPV system. A three-phase AC mains with grid impedance (series R_s-L_s branch) feeds three-phase linear and nonlinear loads. The proposed system consists of a 20 kW SPV array, DC–DC boost converter, and VSC. The DC–DC boost converter is used to boost output voltage of the SPV array using an inductor (L_b), a diode (D), and an IGBT as a switch. DC–AC conversion is performed with the help of a three-phase VSC, and it also functions as a DSTATCOM. Interfacing inductors (L_f) are used at the AC side of the VSC and couple the VSC to the grid. The high-frequency switching noise produced by the switching of IGBTs of the VSC is reduced with the help of series-connected capacitive (C_f) and resistive (R_f) elements at the PCC. The linear load is represented as a combination of three series R-L branches connected in star configuration. The nonlinear load is represented as an uncontrolled bridge rectifier with series R-L branch.

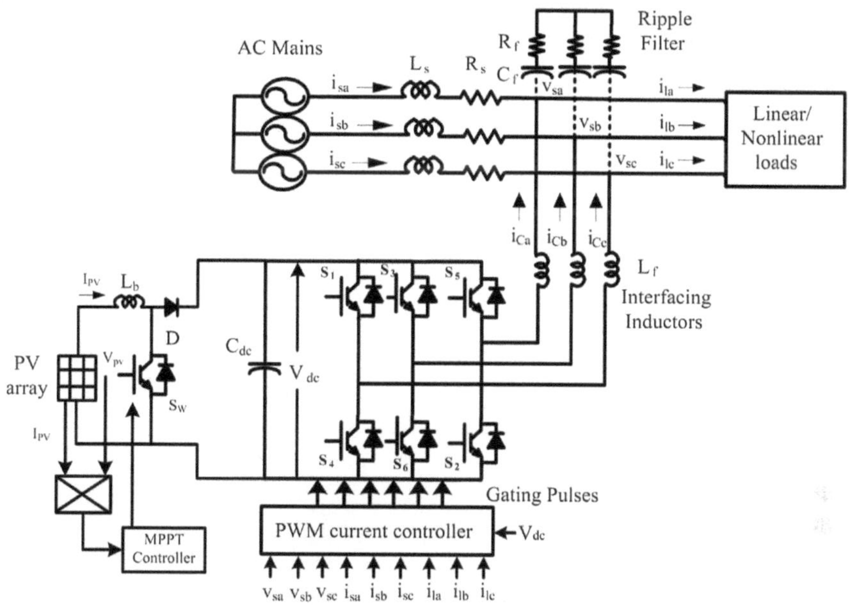

FIGURE 11.8 System configuration of grid-connected SPV system.

12.1 Control algorithm for grid-connected solar photovoltaic system

Different control algorithms have recently been reported for the grid-interfaced PV system. This chapter discusses the Wiener filter-based control algorithm to estimate the reference supply for control of the VSC. The Wiener filtering technique is known as the linear optimization discrete time filter [26]. This requires a priori information about the data to be processed.

12.1.1 Estimation of fundamental active power components of reference supply currents

Fig. 11.9 shows the block diagram of the control algorithm. The computations of inphase unit templates and PI controller were discussed in Eq.(11.15) and Eq.(11.16). The weights associated with the fundamental active constituent of load currents for the three phases are obtained using the Weiner filter approach as

$$w_{ap}(n+1) = [I - 2\mu R_{xa}]w_{ap}(n) + 2\mu P_{dxa} \tag{11.42}$$

$$w_{bp}(n+1) = [I - 2\mu R_{xb}]w_{bp}(n) + 2\mu P_{dxb} \tag{11.43}$$

$$w_{cp}(n+1) = [I - 2\mu R_{xc}]w_{cp}(n) + 2\mu P_{dxc} \tag{11.44}$$

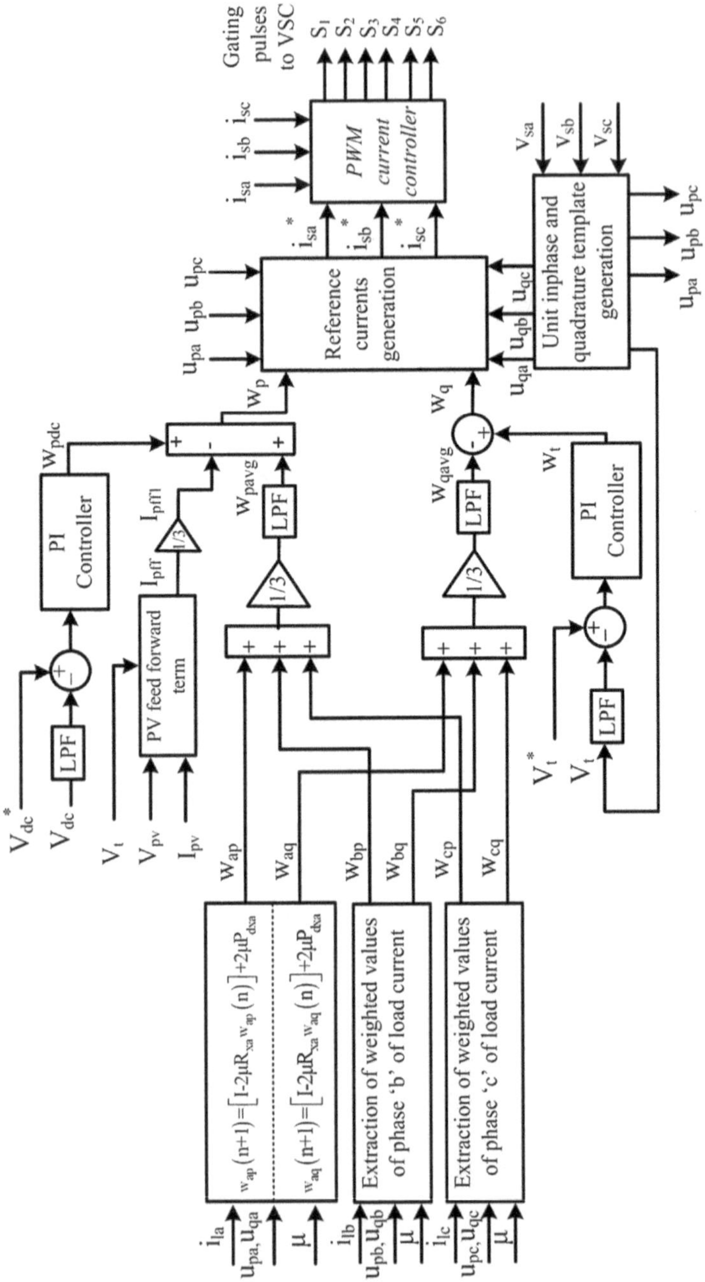

FIGURE 11.9 Block diagram of Wiener filter controller for grid-connected SPV system.

where R is autocorrelation matrix, P is cross-correlation vector, and μ is the step size. The average magnitude of weights (w_{pavg}) corresponding to active power components is given as

$$w_{pavg} = (w_{ap} + w_{bp} + w_{cp})/3 \tag{11.45}$$

The computation of the SPV instantaneous feed-forward compensation term (I_{pff}) is given as

$$I_{pff} = 2P_{pv}/V_t \tag{11.46}$$

The effective feed-forward compensation term (I_{pff1}) is computed by dividing I_{pff} by 3 for a three-phase system.

The total active reference weight (w_p) is obtained by the average weight (w_{pavg}), output of w_{pdc}, and I_{pff1} as

$$w_p = w_{pavg} + w_{pdc} - I_{pff1} \tag{11.47}$$

The reference currents $\left(i^*_{pa}, i^*_{pb} \text{ and } i^*_{pc}\right)$ associated with fundamental active components are estimated using u_{pa}, u_{pb}, and u_{pc} and w_p as

$$i^*_{pa} = w_p u_{pa},\; i^*_{pb} = w_p u_{pb} \text{ and } i^*_{pc} = w_p u_{pc} \tag{11.48}$$

These active reference supply currents $\left(i^*_{pa}, i^*_{pb} \text{ and } i^*_{pc}\right)$ are used to estimate reference supply currents.

12.1.2 Computation of fundamental reactive components

The unit quadrature vectors are acquired from unit inphase voltages (u_{pa}, u_{pb}, and u_{pc}) as in Eqs. (11.21)–(11.23), and the PI controller is realized on the PCC voltage as discussed in Eq. (11.16).

The reactive power weights are obtained as

$$w_{aq}(n+1) = [I - 2\mu R_{xa}]w_{aq}(n) + 2\mu P_{dxa} \tag{11.49}$$

$$w_{bq}(n+1) = [I - 2\mu R_{xb}]w_{bq}(n) + 2\mu P_{dxb} \tag{11.50}$$

$$w_{cq}(n+1) = [I - 2\mu R_{xc}]w_{cq}(n) + 2\mu P_{dxc} \tag{11.51}$$

The average weight (w_{qavg}) of reactive power component of load currents can be calculated from the amplitude of load-reactive weight components (w_{aq}, w_{bq}, and w_{cq}) as

$$w_{qavg} = (w_{aq} + w_{bq} + w_{cq})/3 \tag{11.52}$$

The amplitude of reactive power component of reference supply currents can be calculated by subtracting the output of PI controller (w_t), which is the leading reactive power component used to compensate for voltage drop due to supply impedance and loading on the system and average reactive weight (w_{qavg}), given as

$$w_q = w_t - w_{qavg} \tag{11.53}$$

The reactive components of reference current can be calculated as

$$i_{qa}^* = w_q u_{qa}, \ i_{qb}^* = w_q u_{qb} \text{ and } i_{qc}^* = w_q u_{qc} \tag{11.54}$$

These fundamental reactive constituents are used along with fundamental active constituents to generate total reference currents given as follows.

12.1.3 *Gating pulses for solar photovoltaic–voltage-source converter*

The reference supply currents (i_{sa}^*, i_{sb}^*, and i_{sc}^*) are computed using i_{pa}^*, i_{pb}^*, and i_{pc}^* and i_{qa}^*, i_{qb}^*, and i_{qc}^* as

$$i_{sa}^* = i_{pa}^* + i_{qa}^*, \ i_{sb}^* = i_{pb}^* + i_{qb}^* \text{ and } i_{sc}^* = i_{pc}^* + i_{qc}^* \tag{11.55}$$

The supply currents (i_{sa}, i_{sb}, and i_{sc}) are sensed and compared with i_{sa}^*, i_{sb}^*, and i_{sc}^*, and the current errors (i_{sae}, i_{sbe}, and i_{sce}) are generated, which are used to generate gating pulses for VSC using the PWM current controller.

13. Simulation study and performance of grid interfaced photovoltaic-distribution static compensator system

The performance of an SPV system using the Wiener filter-based control algorithm is presented here. The algorithm developed in MATLAB/Simulink and performance results depict power demand of the load fed from the SPV system. The algorithm is also tested for achieving PQ improvement features.

13.1 Behavior of solar photovoltaic system in power factor correction mode

The performance of SPV system with the Wiener filter-based control algorithm in PFC mode is presented here.

13.1.1 *Behavior of solar photovoltaic system in power factor correction mode under nonlinear load*

Fig. 11.10 shows the behavior of the SPV system with Wiener filter controller in PFC mode. The wave shapes of SPV voltage, current, output power, PCC voltages, supply currents, load currents, VSC currents, and DC bus voltage are shown in these results. During an unbalanced condition ($t = 0.3$ s to 0.4 s), supply currents balanced (with reduced magnitude) due to PQ features incorporated in the VSC controller. The SPV system feeds active power demand of the load connected, and excess power is fed to the grid. Solar irradiance is decreased to 500 W/m^2 at $t = 0.5$ s, which results in reduction of power generated up to approximately 10 kW. Since SPV output power is reduced, load demand is now moderately delivered by the grid.

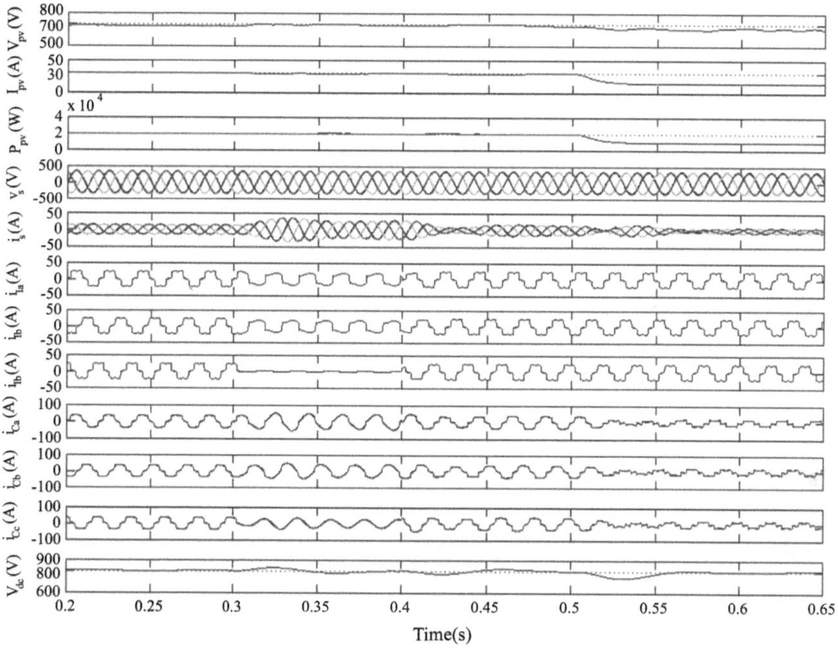

FIGURE 11.10 Behavior of the system under nonlinear load.

Fig. 11.11 shows the THD of phase "a" of PCC voltage (v_{sa}), supply current (i_{sa}), and load current (i_{la}) to be 1.18%, 3.38%, and 27.55%, respectively. THD in the supply current is drastically reduced to 3.38% when nonlinear load current has appreciable THD of 27.87%. These results show two aspects, viz. satisfactory performance of the SPV system as a load compensator and in reducing burden on the grid. Both the real and the reactive power burden on the grid are reduced by the SPV system.

13.1.2 Performance and power of solar photovoltaic system under nonlinear load

Fig. 11.12 illustrates behavior of the system with a Wiener filter-based control algorithm. These results present the plots of SPV output voltage (V_{pv}), current (I_{pv}), power (P_{pv}), load, supply and SPV real power (P_l, P_s and P_C), load, supply, and SPV reactive power (Q_l, Q_s and Q_C). The output voltage (V_{pv}), current (I_{pv}), and power (P_{pv}) of the designed SPV system are 680 V, 30.5 A, and 20 kW, respectively. Under steady-state circumstances (before $t = 0.3$ s), the active power requirement of the nonlinear load (P_L) is around 10 kW, which is fed by the solar system (P_C), and the additional power is moved to the grid (P_s). The reactive power requirement (Q_l) of the load is completely provided by the SPV system (Q_C), and there is almost an insignificant Q_l burden on the supply side.

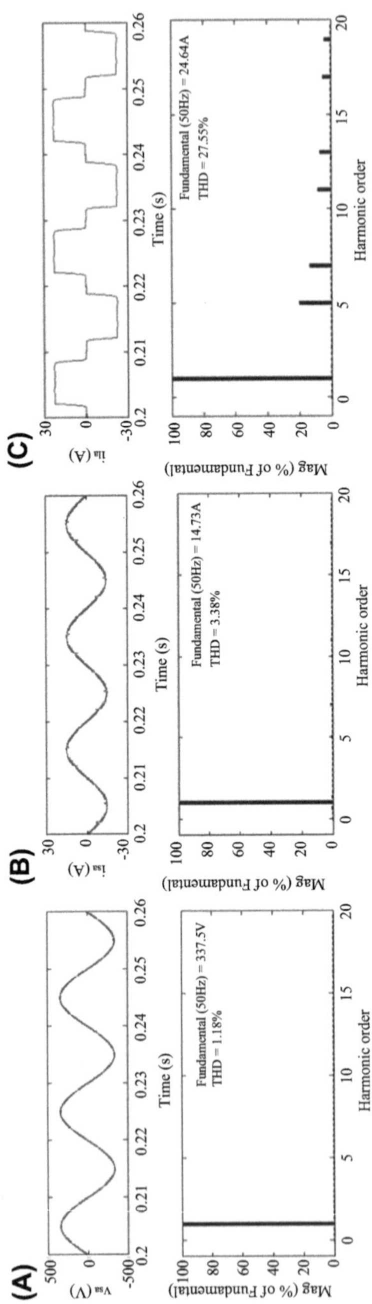

FIGURE 11.11 Harmonic spectra of phase 'a' of (A) v_{sa} (B) i_{sa} (C) i_{la} in PFC mode.

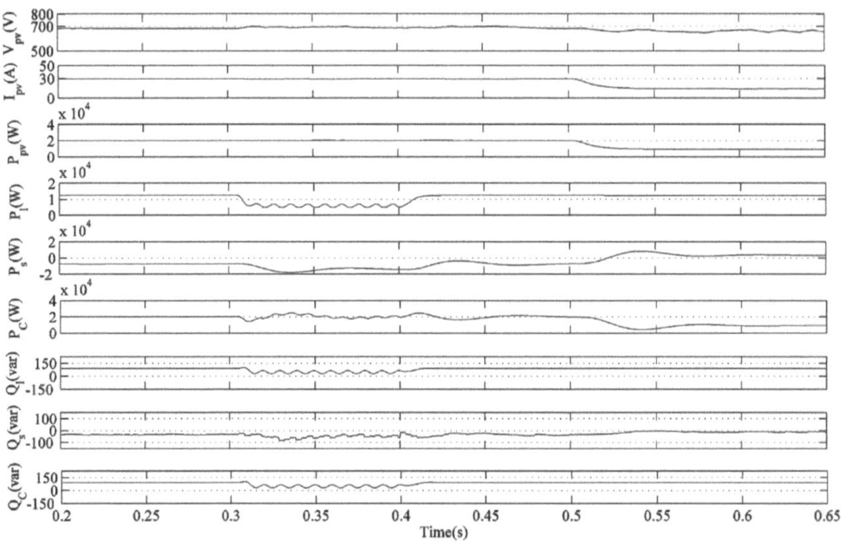

FIGURE 11.12 Performance and power of SPV system under nonlinear load.

13.2 Experimental validation

The above simulated control algorithm is verified on the implemented experimental setup (scaled down) in the laboratory. The experimental setup is developed using a real-time SPV array simulator, digital signal processor, Hall effect voltage/current sensors with appropriate buffer circuitry, interfacing inductor, and linear/nonlinear load. The Wiener filter-based control technique is implemented in the processor to estimate reference currents. With the use of a PWM current controller and gate driver circuitry, suitable gating pulses are generated for SPV-VSC.

Fig. 11.13 shows experimental results obtained from a scaled-down grid-integrated SPV setup. These results represent the power generated from an SPV system (P_{pv}), power transferred to the grid (P_g), load power demand (P_l), and harmonic spectra of supply current. The active power generated from the SPV system is 890 W, and the load demand is only 173 W; therefore, the rest of the power generated (718 W) from the SPV system is transferred to the grid. The THD in grid current is achieved 4.4% after compensator action, whereas load current is highly distorted. Fig. 11.14 presents performance of the SPV system with a Wiener filter-based controller in dynamic load condition. Fig. 11.14A shows performance of the system under unbalanced loading conditions. Due to PQ compensation features of the SPV system, the effect of unbalancing is not reflected in the supply current (i_{sa}) and PCC voltage (v_{sa}). Moreover, upon removing the SPV current (i_{pv}) in Fig. 11.14B, the supply current (i_{sa}) is still balanced with reduced amplitude.

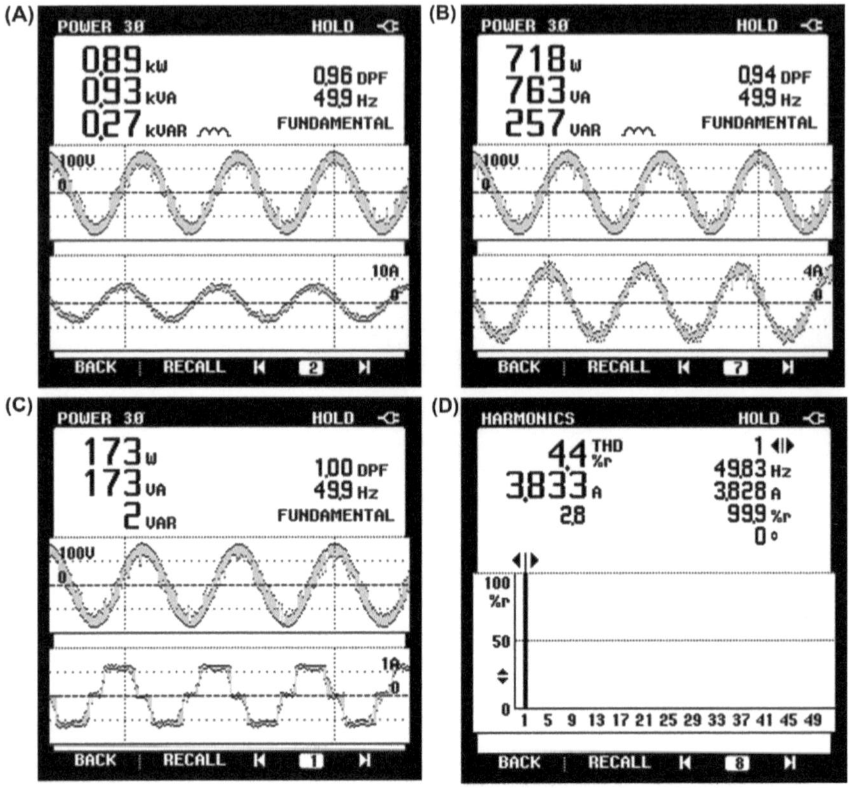

FIGURE 11.13 Experimental performance of grid integrated SPV system with Wiener filter-based control technique in steady state (A) SPV generated power (B) grid power (C) load power demand (D) harmonic spectra of grid current.

14. Conclusions

The importance of power quality (PQ) and several power quality problems have been discusses in this chapter. Some natural and humankind causes of poor PQ are discussed. Solutions to improve poor PQ using passive filters have been realized for decades; however, advances in power electronics and control technology have led to a renewed search for compensators. These compensators, based on fast switching, can be controlled to inject lagging or leading reactive power. This chapter discusses the detailed design of DSTATCOM in a three-phase, three-leg configuration used as a shunt compensator. The multifunctional capabilities of the shunt compensator are exemplified for PQ control. Thereafter, the compensator is integrated with SPV, and real power transfer capability is also exhibited. Simulation and experimental results are depicted for the PBT-based conventional algorithm and the Weiner filter-based modern control algorithm. Good performance results highlight the role of active compensators in overcoming PQ problems effectively. The role and importance of such devices are bound to play a significant role in the hands of the power engineers of tomorrow.

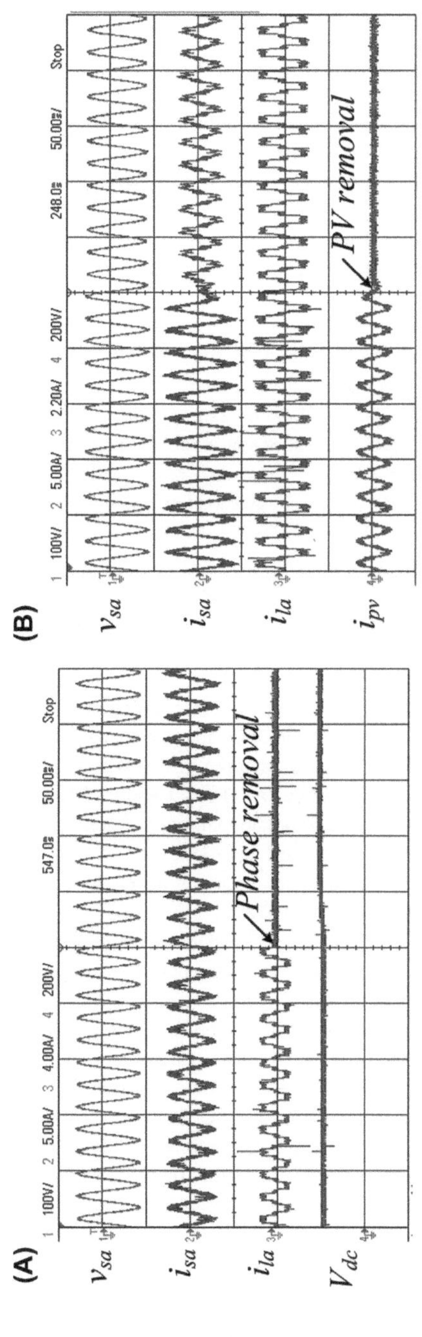

FIGURE 11.14 Experimental performance of grid integrated SPV system in dynamic condition (A) PCC voltage (v_{sa}), supply current (i_{sa}), load current (i_{la}) and dc bus voltage (V_{dc}.) (B) v_{sa}, i_{sa}, i_{la} and SPV-VSC current (i_{pv}).

References

[1] B. Singh, A. Chandra, K. Al-Haddad, Power Quality: Problems and Mitigation Techniques, John Wiley & Sons, 2014.

[2] M.H. Bollen, Understanding power quality problems, in: Voltage Sags and Interruptions, IEEE press, 2000.

[3] A. Ghosh, G. Ledwich, Power Quality Enhancement Using Custom Power Devices, Springer Science & Business Media, 2012.

[4] J. Schlabbach, D. Blume, T. Stephanblome, Voltage Quality in Electrical Power Systems (No. 36, IET, 2001.

[5] E.F. El-Saadany, M.M.A. Salama, A.Y. Chikhani, Passive filter design for harmonic reactive power compensation in single-phase circuits supplying nonlinear loads, IEE Proc. Generat. Transm. Distrib. 147 (6) (2000) 373–380.

[6] G. Benysek, Improvement in the Quality of Delivery of Electrical Energy Using Power Electronics Systems, Springer Science & Business Media, 2007.

[7] V. Verma, B. Singh, Genetic-algorithm-based design of passive filters for offshore applications, IEEE Trans. Ind. Appl. 46 (4) (2010) 1295–1303.

[8] H. Akagi, New trends in active filters for power conditioning, IEEE Trans. Ind. Appl. 32 (6) (1996) 1312–1322.

[9] P. Verdelho, G.D. Marques, An active power filter and unbalanced current compensator, IEEE Trans. Ind. Electron. 44 (3) (1997) 321–328.

[10] B. Singh, K. Al-Haddad, A. Chandra, A review of active filters for power quality improvement, IEEE Trans. Ind. Electron. 46 (5) (1999) 960–971.

[11] Y. Tang, P.C. Loh, P. Wang, F.H. Choo, F. Gao, F. Blaabjerg, Generalized design of high performance shunt active power filter with output LCL filter, IEEE Trans. Ind. Electron. 59 (3) (2011) 1443–1452.

[12] Y. Pal, A. Swarup, B. Singh, A review of compensating type custom power devices for power quality improvement, in: 2008 Joint International Conference on Power System Technology and IEEE Power India Conference, IEEE, October, 2008, pp. 1–8.

[13] M. Ochoa-Giménez, J. Roldán-Peréz, A. Garcia-Cerrada, J.L. Zamora-Macho, Efficient multiple-reference-frame controller for harmonic suppression in custom power devices, Int. J. Electr. Power Energy Syst. 69 (2015) 344–353.

[14] N. Gowtham, S. Shankar, UPQC: a custom power device for power quality improvement, Mater. Today Proc. 5 (1) (2018) 965–972.

[15] J. Ye, H.B. Gooi, B. Wang, Y. Li, Y. Liu, Elliptical restoration based single-phase dynamic voltage restorer for source power factor correction, Elec. Power Syst. Res. 166 (2019) 199–209.

[16] C. Xu, K. Dai, X. Chen, Y. Kang, Voltage droop control at point of common coupling with arm current and capacitor voltage analysis for distribution static synchronous compensator based on modular multilevel converter, IET Power Electron. 9 (8) (2016) 1643–1653.

[17] S.N. Duarte, F.T. Ghetti, P.M. de Almeida, P.G. Barbosa, Zero-sequence voltage compensation of a distribution network through a four-wire modular multilevel static synchronous compensator, Int. J. Electr. Power Energy Syst. 109 (2019) 57–72.

[18] S. Sannigrahi, S.R. Ghatak, P. Acharjee, Multi-objective optimisation-based active distribution system planning with reconfiguration, intermittent RES, and DSTATCOM, IET Renew. Power Gener. 13 (13) (2019) 2418–2429.

[19] R.C. Dugan, M.F. Mc-Granaghan, H.W. Beaty, Electrical Power Systems Quality, McGraw-Hill, New York, NY, 1996.

[20] S. Mikkili, A.K. Panda, Power Quality Issues: Current Harmonics, CRC press, 2015.
[21] A. Zouidi, F. Fnaiech, K. Al-Haddad, Voltage source inverter based three-phase shunt active power filter: topology, Modeling and control strategies, in: 2006 IEEE International Symposium on Industrial Electronics, vol. 2, IEEE, July, 2006, pp. 785–790.
[22] B.N. Singh, B. Singh, A. Chandra, K. Al-Haddad, Design and digital implementation of active filter with power balance theory, IEE Proc. Elec. Power Appl. 152 (5) (2005) 1149–1160.
[23] S.B.Q. Naqvi, S. Kumar, B. Singh, Implementation of a modified distributed normalized least mean square control for a multi-objective single stage SECS, in: 2018 2nd IEEE International Conference on Power Electronics, Intelligent Control and Energy Systems (ICPEICES), New Delhi, India, IEEE, 2018, pp. 706–711.
[24] B. Singh, C. Jain, S. Goel, ILST control algorithm of single-stage dual purpose grid connected solar PV system, IEEE Trans. Power Electron. 29 (10) (2013) 5347–5357.
[25] B. Singh, D.T. Shahani, A.K. Verma, Power balance theory based control of grid interfaced solar photovoltaic power generating system with improved power quality, in: 2012 IEEE International Conference on Power Electronics, Drives and Energy Systems (PEDES), IEEE, December, 2012, pp. 1–7.
[26] J. Chen, J. Benesty, Y. Huang, S. Doclo, New insights into the noise reduction Wiener filter, IEEE Trans. Audio Speech Lang. Process. 14 (4) (2006) 1218–1234.

Chapter 12

A novel adaptive fuzzy-based controller design using field programmable gate arrays for grid-connected photovoltaic systems

Narendra Babu P[1], Kaibalya Prasad Panda[1], Chittibabu B[2], Rangababu P[3], Gayadhar Panda[1]

[1]*Department of Electrical Engineering, National Institute of Technology Meghalaya, Shillong, Meghalaya, India;* [2]*Department of Electronics and Communication Engineering, Indian Institute of Information Technology Design and Manufacturing Kancheepuram, Chennai, Tamil Nadu, India;* [3]*Department of Electronics and Communication Engineering, National Institute of Technology Meghalaya, Shillong, Meghalaya, India*

1. Introduction

Clean energy production with quality-of-power enhancement has been growing daily to minimize dependence on existing energy sources (EES). EES—photovoltaic (PV) and wind—integration to the grid and off-grid for healthy energy production is increasing exponentially [1–3]. Apart from the above EES, PV power generation has been extensively utilized to reach requirements for energy, which boosts the quality of power and reliability while minimizing the overload on the central grid [4,5]. Maintaining the quality of power in power industries is of major significance for decreasing instability, losses, and perturbations in voltage. Hence, to preserve quality power during load and grid-transient conditions, there is a notable increase in novel topologies [6,7].

Reported literature [8] presents modeling of the PV system and the importance of maximum power point tracking (MPPT) techniques to yield maximum power from them. An outline of all existing, hybrid, and soft computing MPPT techniques recently published is rigorously examined in Refs. [9–13]. PV systems can be integrated with the grid in two ways. One is without a DC-DC conversion stage, known as a single-stage structure, and the other is dual-stage conversion, where a DC-DC converter is considered to raise

Advances in Smart Grid Power System. https://doi.org/10.1016/B978-0-12-824337-4.00012-6

PV voltage, which is further converted to AC through an inverter. The literature manifests that the flexibility and stability of the system can be enriched effectively by using a dual-stage PV system compared with a one-stage PV system.

The behavior of the system can be enhanced by injecting the active and reactive constituents into it through an appropriate adaptive control technique for the grid-forming power converter. Simple passive filters can be adopted for compensation of harmonics, but the major issue is tuning and series-parallel resonance in the passive filters. Thus, an active power filter (APF) is established to prevail over the above problems. Among the APF categories, the shunt APF (SAPF) has been extensively identified as a feasible solution. Many existing control scenarios [5—11] such as synchronous reference frame (SRF), proportional resonant, and instantaneous reactive power have been employed in this regard. The existing p-q control scheme is redeveloped in Ref. [5] to attain the compensation goal imposed on the system. A novel control strategy has been proposed based on the existing p-q theory in Ref. [9]. A multifunctional grid-integrated PV system employing a PI-based DC voltage controller has been analyzed and implemented in Refs. [4—7]. However, under load-unstable state, the system's dynamic performance degrades because of a very small cutoff frequency when utilizing the above-mentioned techniques. For enriching the dynamic performance of PV-fed utility grid-tied systems, various advanced controlled techniques have been proposed [12—15] that overcome the shortcomings associated with conventional control techniques. A few advanced robust control techniques are recursive least square [16], least mean square (LMS) [17], normalized LMS [18], least mean fourth (LMF) [19], etc. These control techniques have been implemented with the view of achieving good dynamic performance as well as mitigation of harmonics.

In distribution systems, estimation of parameters by the application of adaptive controllers has attained a lot of significance among researchers due to its merits. For quality-of-power advancement in grid-forming control systems, extensive research has been carried out using Kalman filter (KF) [13], deadbeat control [22], and modified LMS [20—22], echo state network (ESN) [23], etc.-based control techniques. In Ref. [24], the authors have proposed a control scheme based on fundamental current separation which gives better behavior than existing techniques. However, accountin for DC offset rejection, the control behavior is poor. Moreover, the steady-state performance of this control technique is also poor. An improved control technique combining the dual SOGI with fuzzy-tuned PID (FLPID) with adaptive frequency locked-loop control technique in Ref. [25] and the dual multilayer fifth-order generalized integrator-based control in Ref. [26] has been implemented in grid-forming inverter control. A new SOGI-based control scheme for dynamic behavior of a grid-friendly PV system has been validated by the authors in Refs. [25—27].

Different frequency- and time-domain techniques have been periodically utilized for frequency and specification estimation. Parallel computing functionality and faster complex function optimization have been tempted to utilize the fuzzy-tuned control topology [28] for precise estimation. Different categories of fuzzy-tuned control schemes have been employed for detection of harmonic applications. An adaptive PID-based control techniques have been used for DC voltage regulator in Ref. [29] and demonstrated its advantage over the existing PID control scheme. Compatibility of a fuzzy-tuned PID has been proved for precise detection of harmonic over the existing PI and PID control techniques. Detection of harmonic more precisely by employing the radial-basis-function-based ANN has been reported in Ref. [31], which gives better performance than the ADALINE, back-propagation network, and fast Fourier transform control techniques. In the literature, the recurrent-based ANN scheme for nonlinear load current harmonic estimation has also been studied.

Furthermore, a fixed-step-sized LMS iterative-based algorithm has been employed to separate the fundamental constituents of the load current as reported in Ref. [24]. For reforming the weight vector, a simple LMS algorithm has been employed by the researchers to solve complex problems, where the filtered weights are updated after each iteration. The LMS algorithm has several benefits, which include more straightforward and easy implementation, faster response, high-speed processing, and reduced computational burden, for which it finds applicability in various applications. The authors in Ref. [17] have discussed a detailed analysis of the LMS algorithm from an experimental implementation perspective. An in-depth analysis carried out in the literature on the LMS approach clears the misconception related to learning speed, learning accuracy, and nonconvergence problems. The normalized LMS technique is another control approach reported in Ref. [18] for power system frequency estimation, where a comparative study among various ANN-based control strategies with the LMS approach is disclosed. The main concept behind a variable normalized LMS algorithm is highlighted in Ref. [18], and using this stability analysis in a PV-interfaced grid-tied voltage source converter system has been detailed in Refs. [24,25].

It is clear from the above discussion that selecting a suitable control scheme for handling power quality issues is quite essential. With this motivation, this chapter aims to improve quality of power and balancing of DC-bus voltage using a novel fuzzy logic proportional integral derivative based on echo state network (FLPID-ESN). The SAPF is controlled by the proposed ESN technique, which is used to detach the fundamental weight element (FWE) from the load currents and evaluate the calculated currents and accordingly generate switching signals for the SAPF system. Additionally, an FLPID adaptive feedback controller is employed to stabilize the fixed DC-bus voltage in a dual-stage grid-friendly PV system. This chapter is organized by different sections: Section 2 describes the description and modeling of the complete system, Section 3 reports the proposed methodology and numerical analysis, Section 4 reports the simulation and experimental analysis during various operating conditions and also describes the comparative analysis of the proposed system with recent cited works, and Section 5 describes the conclusion of the chapter, references, nomenclature, and acronyms.

2. System description

Fig. 12.1 shows the grid-friendly PV system with FLPID-ESN control technique. The complete system incorporates a PV system, DC–DC (boost) converter, and an IC-based MPPT that is employed to extract the maximum power point (MPP) from the PV. Different types of DC loads are tied to the terminals of the DC bus, which is linked to the utility grid through the grid-interfacing converter and the coupling inductor. Different types of AC loads are tied to the PCC terminals. The FLPID controller is considered ripple free DC-bus terminal voltage, which is merged with the FWE separation of the irregular load current produced by employing an adaptive ESN reference current-generation scheme to produce the calculated currents for controlling the grid-tied PV inverter. A hysteresis current controller (HCC) is utilized to generate the desired pulse width modulation (PWM) pulses for the grid-connected inverter. A coupling inductor is tied between the grid-interfacing inverter and the AC bus to mitigate the high-order harmonics. The proposed FLPID-ESN control scheme is validated in real-time on a virtex-7 VC-707 field programmable gate array (FPGA) evaluation kit via a Matlab/Simulink-integrated XILINX system generator.

2.1 Designing the photovoltaic system

A current source and parallel inverted diode have been used to model the PV cell, and the PV cell has its own series as well as parallel resistances. Series resistance is used to move electrons from n to p junctions, the parallel resistance is used for leakage current, and PV cells process light energy from sunlight to DC power. The solar panel is an integration of solar cells that works on the principle of photoelectric effect. Generally, a solar cell has a normal P-N junction diode, the junction absorbs photon energy from the light, and the charge carriers are excited and separated at the junction. This creates an electric field across the junction causing electric current to flow, thereby transforming solar energy to electricity. A PV cell-equivalent circuit is displayed in Fig. 12.2. The PV cell is represented by a current source representing the photoproduced current from a PV cell. The output current (i_{pv}) produced by the PV cell is given as follows:

$$i_{PV} = i_{ph} - i_o - i_{sh} \tag{12.1}$$

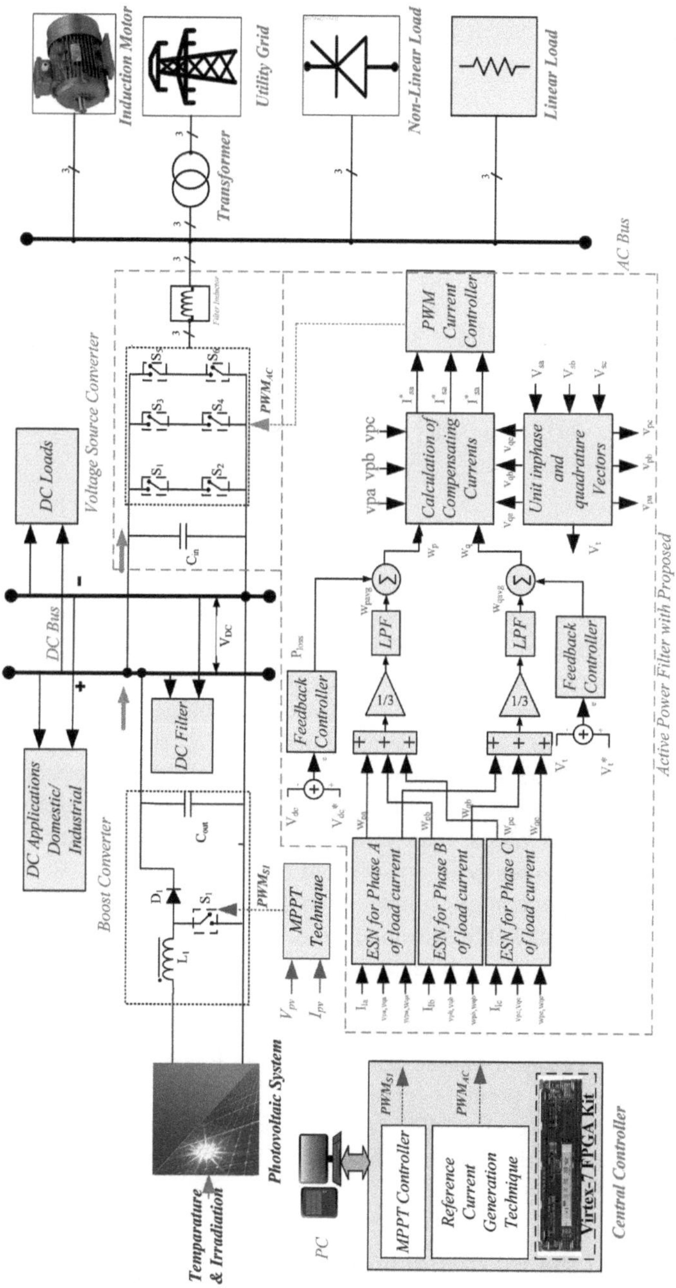

FIGURE 12.1 Grid-friendly photovoltaic system with the fuzzy logic proportional integral derivative based on echo state network scheme.

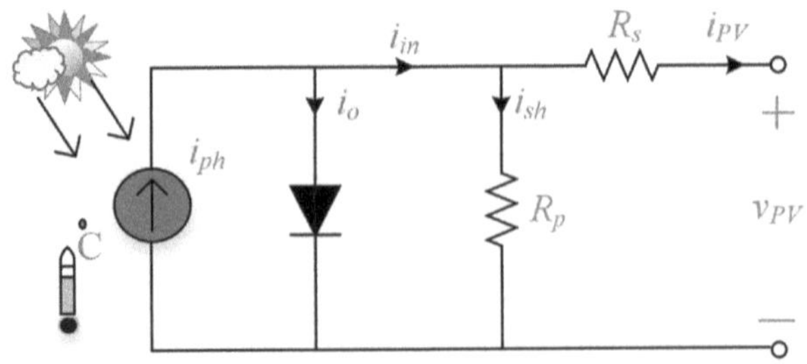

FIGURE 12.2 Equivalent diagram of photovoltaic system.

$$i_{ph} = [i_{sc} + k_i(T - 298)] \times \frac{Irr}{1000} \tag{12.2}$$

$$i_o = i_{o,sat}\left[\exp\left(\frac{v_{PV} + i_{PV} \times R_s}{n \times Kb \times \frac{T}{q}}\right) - 1\right] \tag{12.3}$$

$$i_{sh} = \left(\frac{v_{PV} + i_{PV} \times R_s}{R_p}\right) \tag{12.4}$$

where i_{PV}-current produced by the PV cell, v_{PV}-terminal voltage across the PV cell, i_{ph}- photogenerated current of the PV cell, i_{sc}-short circuit current, k_i temperature coefficient, Irr-solar irradiance incident on the PV, $i_{0,sat}$-diode reverse saturation current, n-ideality factor, q-charge of an electron, K_b-Boltzmann constant, T-cell temperature in Kelvin, R_p-shunt resistance, and R_s-series resistance. The PV system output power is heavily dependent on temperature and irradiation variations. Due to atmospherically unstable conditions, output power varies. The utilized incremental-conductance MPPT technique [26–32] ensures peak power from the PV system.

The PV power generation source is more dominant renewable power generation over nonrenewable (gasoline, coal, etc.) energy sources because it is pollution-free, clean, and inexhaustible. Stand-alone grid-connected applications are the major applications of the PV system. In addition, the PV cell has curves during variations in temperature and irradiation. Each P – V and I-V curve has a unique point where the PV system works at peak efficiency. That unique point is represented as MPP. MPP techniques have been employed to trace the MPP from the PV system by matching the impedance level from the PV and load sides. MPP techniques are used to enhance system efficiency and cost.

2.2 Modeling of the maximum power point tracking technique

In general, the PV panel converts solar to electrical energy from 30% to 40% based on temperature and irradiation levels. MPPTs have been utilized to improve the PV system's efficiency. The peak power of a circuit is extracted when the load impedance equals the source impedance. Therefore, the MPPT is acting as an impedance matching device in the grid integrated with the PV system. To enhance the voltage from the PV system, a boost converter is interconnected between the DC bus and PV system so that several DC applications can be utilized. By varying the duty cycle of the MPPT, the boost converter can extract the peak power from the PV system. The PV system's power mostly depends on temperature, radiation, and the position of the panel. Fig. 12.3 shows the varying irradiation levels of the PV system, which are also the product of current and voltage. Varying the V and I, power can be maximized. To extract peak power from the renewable energy sources, MPPTs have been implemented. Common MPPT techniques exist in the literature as follows:

(1) incremental conductance method
(2) perturb and observation method
(3) voltage-based peak power tracking method
(4) parasitic capacitance method
(5) current-based peak power tracking method

In the above-mentioned existing MPPT techniques, an incremental conductance (IC)-based MPPT method is one of the most useful and best MPPT techniques compared with other MPPT techniques. The limitations of the perturb and observation method are that performance is poor during changes in atmosphere condition. In order to defeat this drawback, an IC MPPT technique has been introduced. The IC technique contains a slope of the current derivative with respect to the derivative voltage to get the MPP. The condition to acquire the MPP is that $\left(\frac{di}{dv}\right)$ should be equal to $\left(\frac{-i}{v}\right)$. For changes in voltage, either biggest or smallest value, the MPP varies. If the power increases and decreases, the curves should be the same and reverse directions respectively. Fig. 12.4 shows the IC-based MPPT method flowchart.

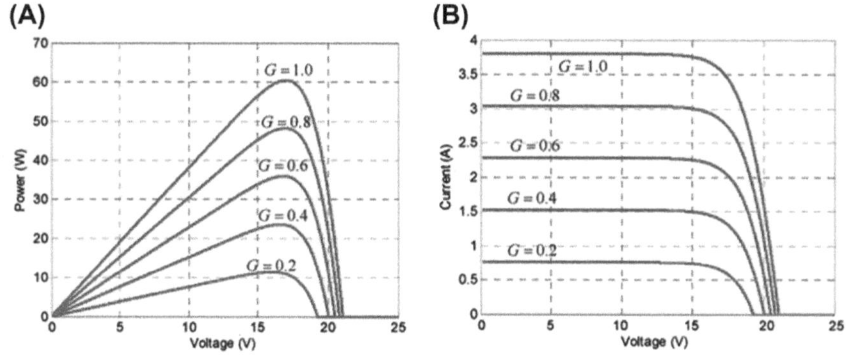

FIGURE 12.3 Output characteristics of photovoltaic system.

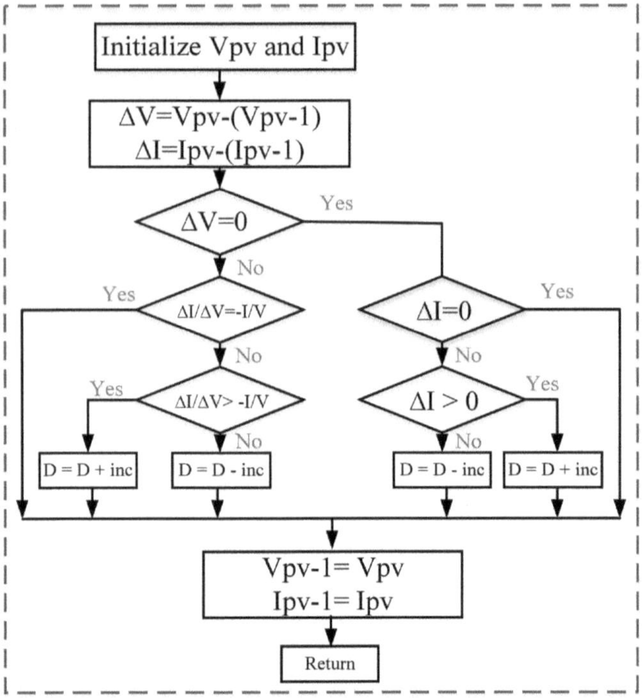

FIGURE 12.4 Incremental-conductance-based maximum power point tracking technique flowchart.

If the $\frac{dP}{dV}$ is negative, the MPPT lies on the right side of the present location, and if the $\frac{dP}{dV}$ is positive, the MPPT is on left side. Mathematical analysis of the IC technique is as follows:

$$\frac{dP}{dV} = \frac{d(V \cdot I)}{dV} = I\frac{dV}{dV} + V\frac{dI}{dV} = I + V\frac{dI}{dV} \tag{12.5}$$

The MPP is reached when $\left(\frac{dP}{dV}\right) = 0$ and

$$\frac{dI}{dV} = -\frac{I}{V} \tag{12.6}$$

$$\frac{dP}{dV} > 0 \quad \text{then } V_P < V_{mpp} \tag{12.7}$$

$$\frac{dP}{dV} = 0 \quad \text{then } V_P = V_{mpp} \tag{12.8}$$

$$\frac{dI}{dV} < 0 \quad \text{then } V_P > V_{mpp} \tag{12.9}$$

2.3 Operation of boost converter

The voltage increase converter (DC–DC) is applied to improve PV system voltage at a required output voltage for several DC applications by using the MPPT technique in the absence of a transformer. A high-frequency switch, an inductor, and a diode are the major components of the boost converter. These components are connected in a coordinated manner to improve the input voltage of the PV system and deliver peak power to the grid and load. The manipulation of the duty cycle in the MPPT control technique occurs when the boost converter switch is working, which raises the voltage. The internal structure of the DC–DC converter is shown in Fig. 12.5. The DC–DC converter works on dual modes: 1. switch-ON mode and 2. switch-OFF mode. In mode 1, the switch is closed; this is represented as the charging mode function. In mode 2, the switch is open; this is called the discharging mode function.

2.3.1 Switch-on mode operation

In this operation, the inductor is charged from the PV source. In this mode, the current of the inductor increases aggressively. However, for simplicity we have assumed that inductor behavior is linear in discharging and charging states. In the circuit, the diode blocks the opposite flow of current. However, the load current carry is stable and supplies energy to the capacitor when it is discharging.

2.3.2 Switch-off mode operation

In mode 2, the diode becomes short-circuited when the switch is in ON state. In this state, the capacitor is charged from the inductor with opposite polarities. However, the load current carry is stable throughout the operation.

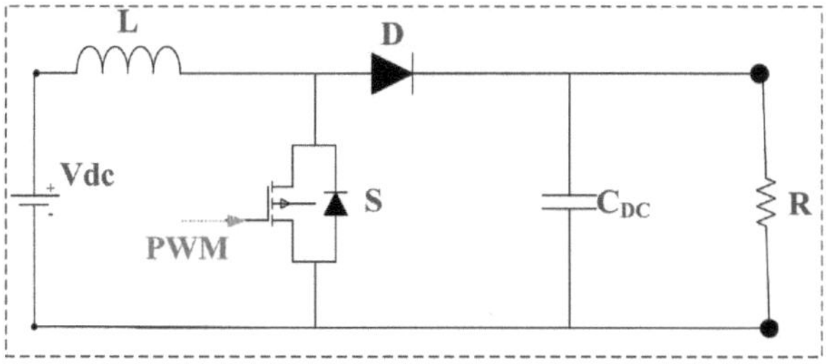

FIGURE 12.5 Circuit diagram of boost converter.

3. Proposed fuzzy logic proportional integral derivative based on echo state network reference current generator

The ESN reference current-generation technique is implemented to calculate the three-phase currents by separating the FWC from the load current and consequently generate antiharmonic currents. The mathematical analysis of active current constituents of the load currents is presented in this section. By sensing the three-phase grid voltages (v_{sa}, v_{sb} and v_{sc}) at the point of common coupling (PCC), the peak magnitude value of the PCC voltage is calculated by using the equation below:

$$V_t = \sqrt{\frac{2\left(v_{sa}^2 + v_{sb}^2 + v_{sc}^2\right)}{3}} \tag{12.10}$$

where V_t is the peak value of magnitude at the PCC.

By using peak value of magnitude at the PCC, the unit in-phase values are computed as follows:

$$v_{pa} = \frac{v_{sa}}{V_t}, v_{pb} = \frac{v_{sb}}{V_t}, v_{pc} = \frac{v_{sc}}{V_t} \tag{12.11}$$

The FLPID DC voltage controller is applied to compute the switching losses of the APF system and to stabilize the constant DC-bus voltage. The output of FLPID at the nth sampling moment is

$$\{w_{pdc}(n) = w_{pdc}(n-1) + K_p(V_{de}(n) - V_{de}(n-1)) + K_i(V_{de}(n)) + K_d(V_{de}(n)) \tag{12.12}$$

The fundamental equation representing calculation of real power weights of load currents is expressed as

$$w_p(n+1) = w_p(n) + \eta x(n+1)^T \times e(n+1) + \gamma x(n)^T \times e(n) \tag{12.13}$$

where e is the error between estimated load currents and sensed load currents, η is the learning gain, γ is the momentum gain, and $x(n)$ is the input vector.

The weights are updated and built in such a way that the mean square error (MSE) is reduced. The MSE at the nth sampling point is given as follows:

$$MSE = \frac{1}{r}\sum_{n=1}^{r}(i_l(n) - i_{est}(n)) = \frac{1}{r}\sum_{n=1}^{r}(i_l(n) - x(n) \times w_p) \tag{12.14}$$

where $i_l(n)$-load current, $i_{est}(n)$ represents the estimated load current, and r represents the length of input and output for $[2 \times 1]$ scheme sequences. To compute the weight component of the fundamental real powers (w_{pa}, w_{pb}, w_{pc}) corresponding to load currents, Eq. (12.8) is used as follows:

$$w_{pa}(n+1) = w_{pa}(n) + \eta v_{pa} e_a(n+1) + \gamma v_{pa} e_a(n+1) \tag{12.15}$$

$$w_{pb}(n+1) = w_{pb}(n) + \eta v_{pb} e_b(n+1) + \gamma v_{pb} e_b(n+1) \tag{12.16}$$

$$w_{pc}(n+1) = w_{pc}(n) + \eta v_{pc} e_c(n+1) + \gamma v_{pc} e_c(n+1) \tag{12.17}$$

The average weight magnitude (w_{pavg}) corresponding to real power constituents ($w_{pa}, w_{pb},$ and w_{pc}) is expressed as

$$w_{pavg} = \frac{w_{pa} + w_{pb} + w_{pc}}{3} \tag{12.18}$$

The total *reference* weight (w_p) corresponding to average weight magnitude (w_{pavg}) and the feedback controller weight component (w_{pdc}) is expressed as

$$w_p = w_{pavg} + w_{pdc} \tag{12.19}$$

The SAPF currents $\left(i^*_{pa}, i^*_{pb}, \text{ and } i^*_{pc}\right)$ corresponding to the fundamenal active weight element are evaluated by utilizing Eqs. (12.6) and (12.14), as

$$\begin{cases} i^*_{pa} = w_p \times v_{pa}, \\ i^*_{pb} = w_p \times v_{pb}, \\ i^*_{pc} = w_p \times v_{pc} \end{cases} \tag{12.20}$$

3.1 Calculation of fundamental reactive constituents of reference currents using echo state network

The unit quadrature voltage vectors at PCC are evaluated from unit in-phase templates ($v_{pa}, v_{pb},$ and v_{pc}) as

$$v_{qa} = \frac{-v_{pb}}{\sqrt{3}} + \frac{v_{pc}}{\sqrt{3}} \tag{12.21}$$

$$v_{qb} = \frac{\sqrt{3}v_{pa}}{2} + \frac{(v_{pb} - v_{pc})}{2\sqrt{3}} \tag{12.22}$$

$$v_{qc} = -\frac{\sqrt{3}v_{pa}}{2} + \frac{(v_{pb} - v_{pc})}{2\sqrt{3}} \tag{12.23}$$

To regulate the voltage, an APF is implemented through a secondary FLPID controller realized over the peak magnitude of PCC voltage. The error in the middle of the actual and reference voltages is processed through the FLPID controller, and the output (w_t) is calculated at the nth sampling moment as

$$\{w_t(n) = w_t(n-1) + k_p(V_t(n) - V_t(n-1)) \\ + k_i(V_t(n)) + k_d(V_t(n)) \tag{12.24}$$

where k_p, k_i, and k_d are defined as the gain parameters of the controller.

Similarly, to compute the fundamental weights of reactive powers (w_{qa}, w_{qb}, and w_{qc}) corresponding to load currents, Eq. (12.8) is used as follows:

$$w_{qa}(n+1) = w_{qa}(n) + \eta v_{qa} e_a(n+1) + \gamma v_{qa} e_a(n+1) \tag{12.25}$$

$$w_{qb}(n+1) = w_{qb}(n) + \eta v_{qb} e_b(n+1) + \gamma v_{qb} e_b(n+1) \tag{12.26}$$

$$w_{qc}(n+1) = w_{qc}(n) + \eta v_{qc} e_c(n+1) + \gamma v_{qc} e_c(n+1) \tag{12.27}$$

The average weight magnitude (w_{qavg}) can be expressed as follows:

$$w_{qavg} = \frac{w_{qa} + w_{qb} + w_{qc}}{3} \tag{12.28}$$

The reactive reference component (w_q) is calculated as

$$w_q = w_t - w_{qavg} \tag{12.29}$$

The fundamental reference currents of reactive constituents using Eqs. (12.16)–(12.18) and (12.24) are expressed as follows:

$$\begin{cases} i_{qa}^* = w_q \times v_{qa}, \\ i_{qb}^* = w_q \times v_{qb}, \\ i_{qc}^* = w_q \times v_{qc} \end{cases} \tag{12.30}$$

Finally, the reference currents for PWM generation can be calculated using Eqs. (12.15) and (12.25) as follows:

$$\begin{cases} i_{sa}^* = i_{pa}^* + i_{qa}^*, \\ i_{sb}^* = i_{pb}^* + i_{qb}^*, \text{ and} \\ i_{sc}^* = i_{pc}^* + i_{qc}^* \end{cases} \tag{12.31}$$

The source currents are compared with reference currents, which results in the error currents (i_{sae}, i_{sbe}, and i_{sce}). The errors of the currents are processed through the PWM controller, which generates a PWM signal to the APF. Three-phase compensating currents from the APF then compensate the load harmonics and thereby maintain power quality.

3.2 Adaptive DC-link voltage controller

The FLPID controller is utilized for implementing a feedback mechanism that maintains DC-bus voltage constant as shown in Fig. 12.6. It is easy to

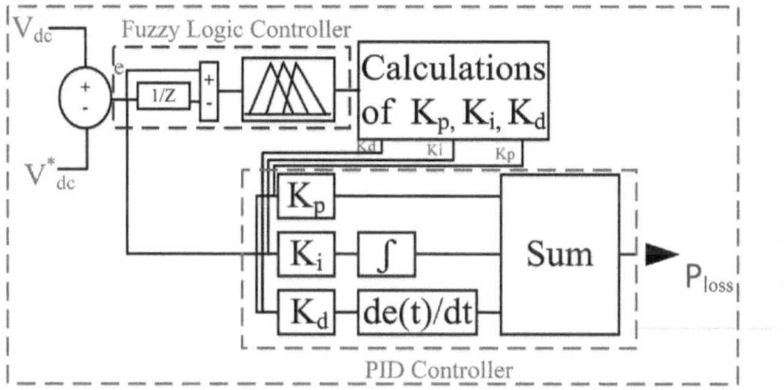

FIGURE 12.6 Control diagram of fuzzy logic proportional integral derivative technique.

implement and gives robust performance, but the tuning of controller gains is difficult. So far, many techniques have been researched for tuning PID controller gains. The fuzzy logic control for tuning of PID controller gains is one of the popular techniques based on the error and time difference.

The mathematical analysis for tuning PID parameters is expressed as follows:

$$H_c(s) = K_p + \frac{K_i}{s} + K_d(s) \tag{12.32}$$

$$H_c(s) = K_p\left(1 + \frac{1}{\frac{K_p}{K_i}(s)} + \frac{K_d}{K_p}(s)\right) \tag{12.33}$$

The discrete-time similar formula for PID is given by

$$x_k = K_p{}^*e(k) + K_i \cdot T_s \sum_{i=1}^{n} e(i) + \frac{K_d}{T_s} \cdot e(k) - e(k-1) \tag{12.34}$$

where x_k denotes the control signal, $e(k)$ denotes the error between the reference and actual voltages, and T_s denotes sampling time.

The mathematical analysis of fuzzy gain scheduling for the PID controller is expressed as follows:

$$\begin{cases} K'_p = \dfrac{K_p - K_{p,\min}}{K_{p,\max} - K_{p,\min}} \\[2ex] K'_d = \dfrac{K_d - K_{d,\min}}{K_{d,\max} - K_{d,\min}} \end{cases} \tag{12.35}$$

The defuzzification yields are denoted as follows:

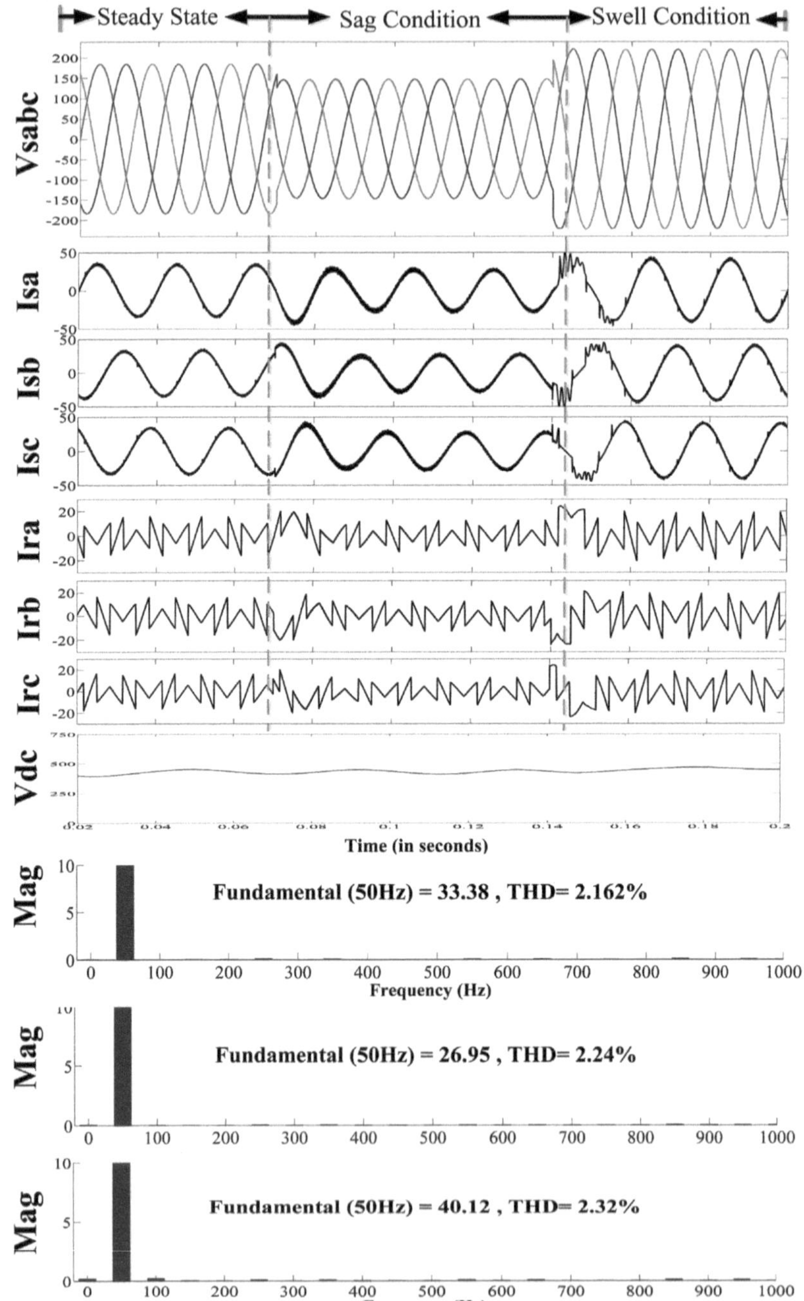

FIGURE 12.7 Simulation results under test case 1 and 2.

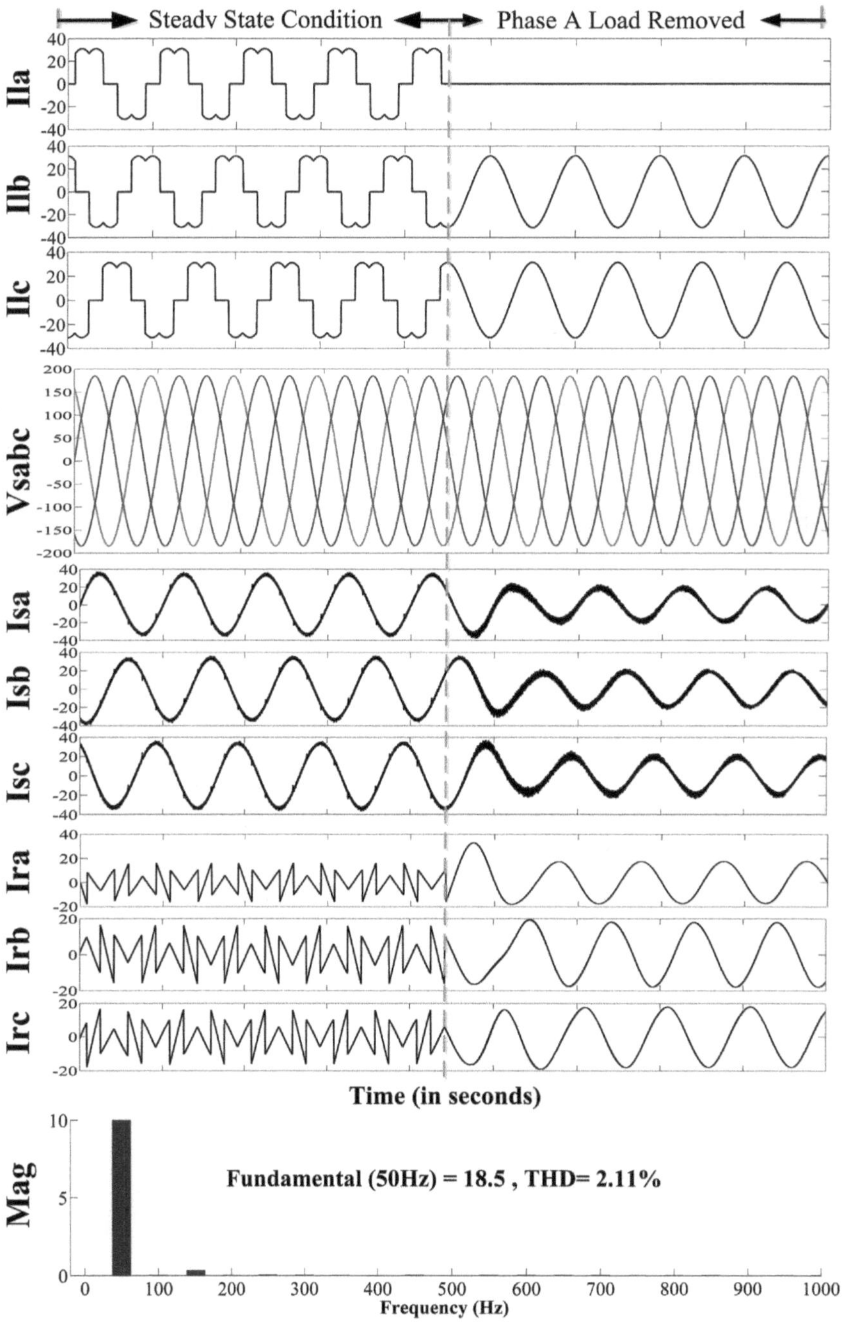

FIGURE 12.8 Simulation results under test case 3.

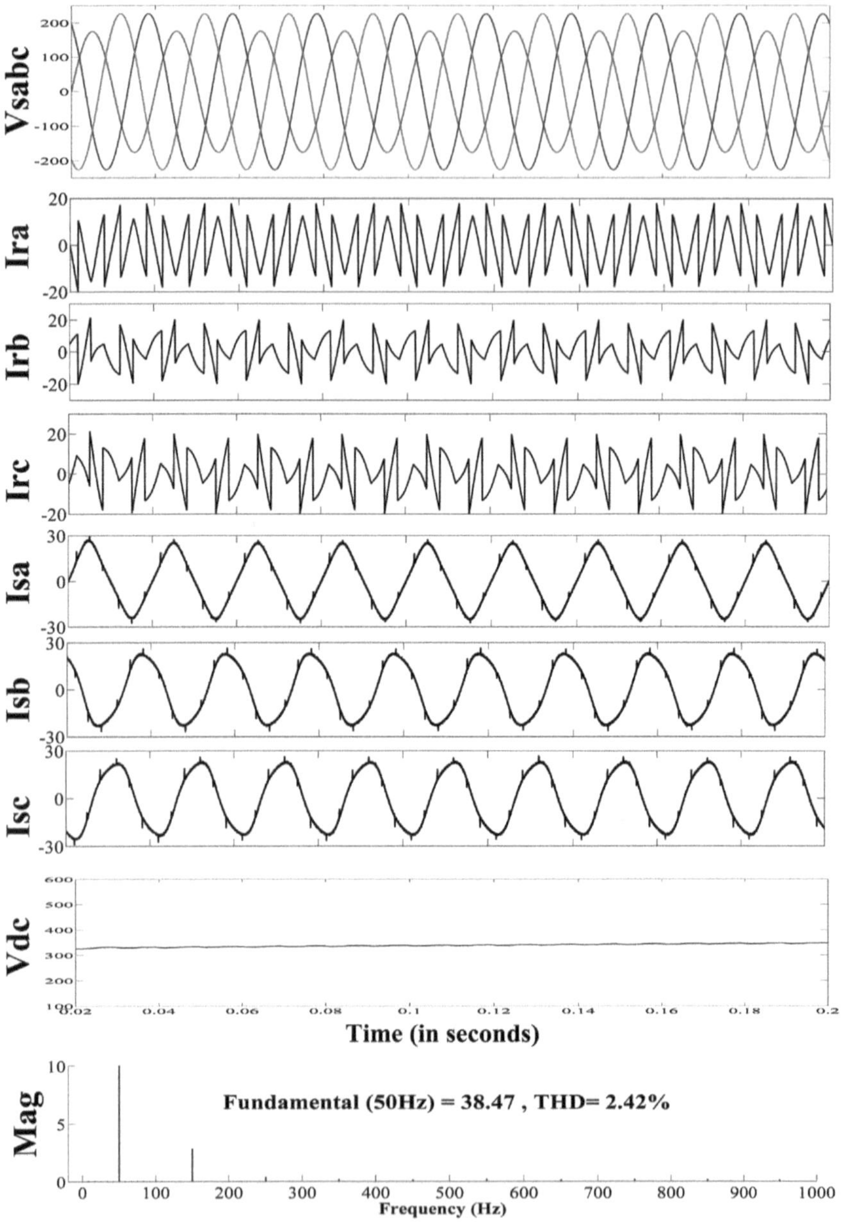

FIGURE 12.9 Simulation results under test case 4.

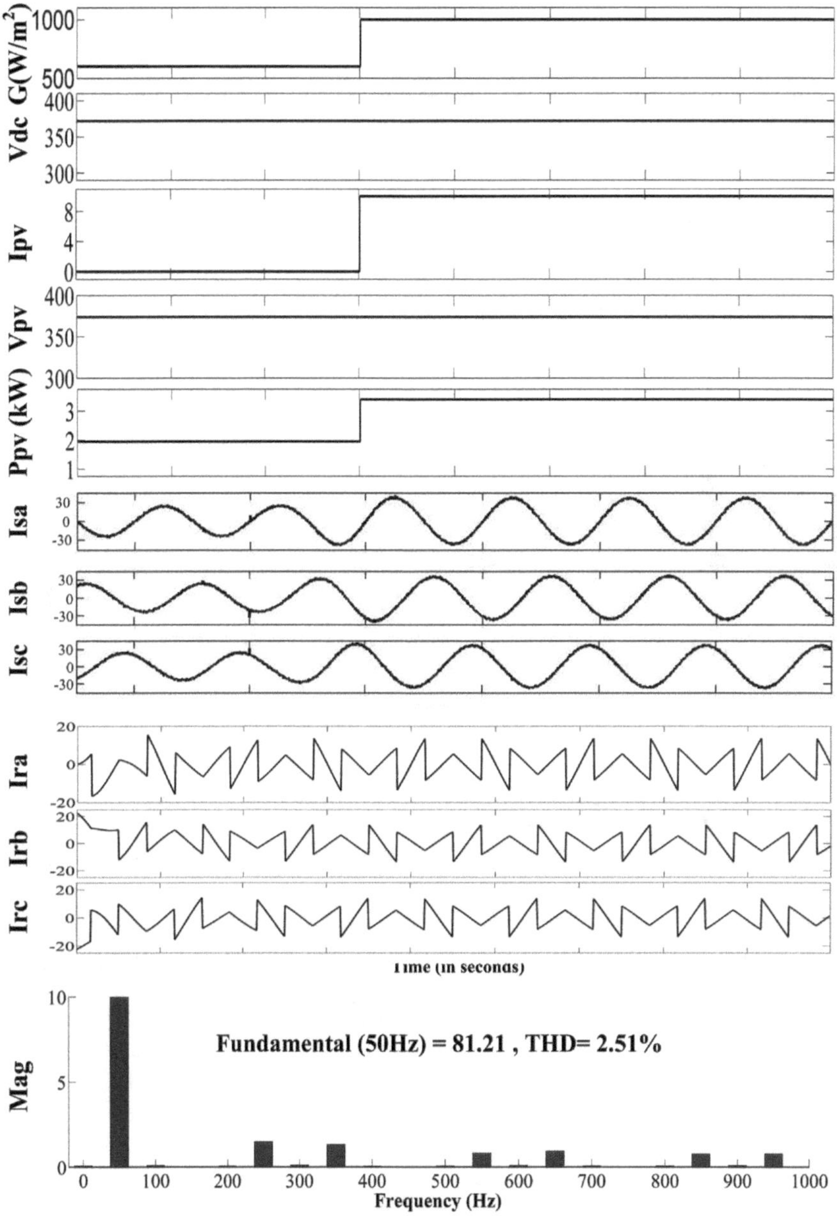

FIGURE 12.10 Simulation results under test case 5.

$$\begin{cases} K_p' = \sum_{i=1}^{m} \mu_i \cdot K_p' \cdot i \\ K_d' = \sum_{i=1}^{m} \mu_i \cdot K_d' \cdot i \\ \alpha = \sum_{i=1}^{m} \mu_i \cdot \alpha_i \end{cases} \tag{12.36}$$

Tuning of PID gains by using the fuzzy logic controller are expressed as

$$\begin{cases} K_p = \left(K_{p,\max} - K_{p,\min}\right) \cdot \left(K_p' + K_{p,\min}\right) \\ K_d = \left(K_{d,\max} - K_{d,\min}\right) \cdot \left(K_d' + K_{d,\min}\right) \\ K_i = \dfrac{K_p^2}{\alpha \cdot K_d} \end{cases} \tag{12.37}$$

4. Simulation and experimental results under different scenarios

The proposed FLPID-ESN control scheme has been executed on a Matlab/Simulink. System and PV parameters of the proposed system are shown in Tables 12.1 and 12.2 respectively. A PV integrated with grid system is investigated for quality-of-power enrichment of the distribution system using the proposed FLPID-ESN technique under steady-state, grid-voltage dynamic, load removed, unbalanced supply voltage, and change in irradiation state conditions. A Acomparative analysis of the FLPID-ESN with traditional and a few adaptive control techniques examines the advantage of the proposed scheme.

To evaluate the behavior and efficiency of the FLPID-ESN proposed technique for the PV-integrated grid system, an FPGA-based experimental setup is developed with a 60 V prototype laboratory setup. Fig. 12.11 shows the FPGA-based prototype laboratory setup that incorporates the grid-interfacing converter, uncontrolled diode rectifier load, 1.5kW PV panel setup, PV simulator, power analyzer, power quality analyzer, FPGA interface circuit, coupling inductor, three-phase variac, three-phase load, and master computer. To examine the advantage of the proposed scheme, the system is operated under several transient conditions. Table 12.3 shows the total harmonic distortion (THD) content of the grid currents under various operating conditions. The suggested FLPID-ESN technique is based on a reference current-generation technique of the VSI and the feedback controller of the DC

TABLE 12.1 System specifications.

Variables	Simulation	Experimental
Grid voltage (ph-ph) (V_s)	230 V	60 V
Frequency (f)	50 Hz	50 Hz
Grid-interfacing inductor (L_{apf})	2.2 mH	1.5 mH
Irregular load (RL)	60Ω, 50 mH	40Ω, 30 mH
DC bus capacitor (C_{DC})	1000 μF	2200 μF
Reference DC bus voltage	375 V	100 V

TABLE 12.2 Parameters of the photovoltaic (PV) system.

Parameters	Simulation	Experimental
Power of PV (P_{pv})	3.5 kW	300 W
Irradiation (G)	1000 W/m^2	1000 W/m^2
Open-circuit voltage (V_{oc})	21.5 V	21.5 V
Short-circuit current (I_{sc})	6.25 A	4.55 A
Max. PV voltage (V_{mp})	17.65 V	17.65 V
Max. PV current (I_{mp})	5.65 A	4.35 A
Series modules (N_S)	11	2
Parallel modules (N_P)	3	2

voltage regulator. An HCC is employed to produce the desired PWM signals to the APF. The APF creates antiharmonic currents that feed into the grid for harmonic rejection.

4.1 Test case 1: steady-state with RL load and sudden switching of the shunt active power filter

The execution of the suggested SAPF scheme with FLPID-ESN is performed under balanced voltage conditions. Fig. 12.7 displays the simulation results under balanced grid-voltage conditions. From 0 to 0.07 s, the system operates at steady-state case. Before compensation, the THD content in the grid currents is 29.96%. After compensation, the source current becomes more like a sinusoidal wave, and the THD in the grid current decreases to 2.162%. In this

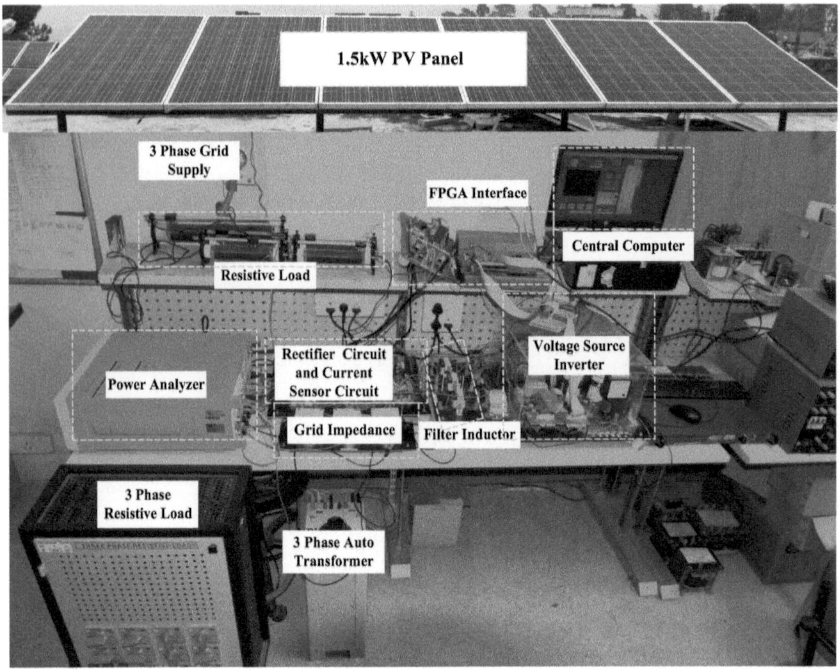

FIGURE 12.11 Field programmable gate array-based active power filter prototype laboratory setup.

TABLE 12.3 Total harmonic distortion values under several test cases.

Scenario's	Parameters	Simulation (%)	Experimental (%)
Case 1	Load current	29.96	26.452
	Source current	2.162	2.261
Case 2	Load current	29.96	26.583
	Source current	2.24	2.555
Case 3	Load current	29.96	26.275
	Source current	2.11	2.289
Case 4	Load current	29.96	26.399
	Source current	2.42	2.652
Case 5	Load current	29.96	26.130
	Source current	2.51	2.586

case, the PV supply is maintained constant, and better performance of the APF is achieved using the proposed control technique. Fig. 12.7 represents the grid voltages (V_{sabc}), grid currents (I_{sa}), (I_{sb}), (I_{sc}), reference currents in all three phases (I_{ra}), (I_{rb}), (I_{rc}), and the DC-bus voltage (V_{dc}). The corresponding THD analysis of the phase-a current of the grid after the filter is also shown in Fig. 12.7. To validate the simulation results, the PV-integrated APF system experimental setup is implemented with the FLPID-ESN, which is displayed in Fig. 12.11. The real-time PV characteristics under different irradiation levels are shown in Figs. 12.12 and 12.13 respectively. Phase-a voltage of the grid (V_{sa}), and the phase-a current of the grid (I_{sa}), filter current (I_{fa}), and load current (I_{la}) for phase-a are shown in Fig. 12.14. The phase-a current of grid THD without the SAPF is 26.452% and is reduced to 2.261% with SAPF. The current of the grid is clearly sinusoidal, and both the current and voltage are in-phase with unity power factor.

4.2 Test case 2: voltage of the grid with sag–swell condition

Proposed grid-friendly PV system behavior subjected to sag and swell disturbances in PCC voltage is also evaluated. From 0.07 to 0.14 s, the system operates under a sag condition; from 0.14 to 0.2 s, the system operates under a swell condition. Fig. 12.7 represents the grid voltages, currents, calculated currents of the SAPF, and DC link bus voltage. During sag and swell test cases, the phase-a grid current's THD is 2.24% and 2.32%, respectively. The simulation findings under this condition are also validated by experimental results. Fig. 12.15 displays the phase-a voltage of the grid with current, current of the

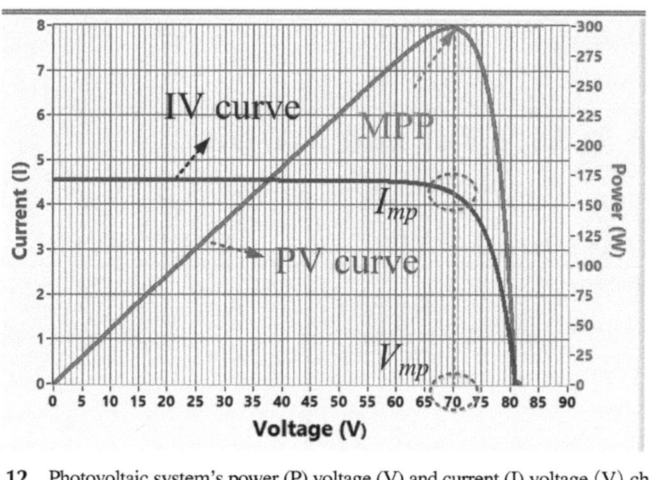

FIGURE 12.12 Photovoltaic system's power (P) voltage (V) and current (I) voltage (V) characteristics during 1000W/m^2.

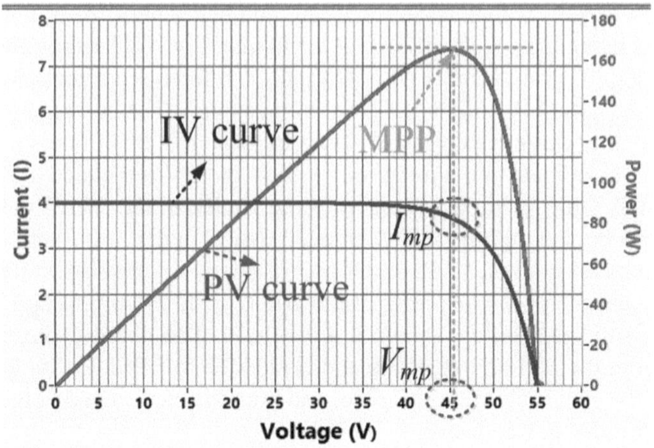

FIGURE 12.13 Photovoltaic system's power (P) voltage (V) and current (I) voltage (V) characteristics during $600W/m^2$.

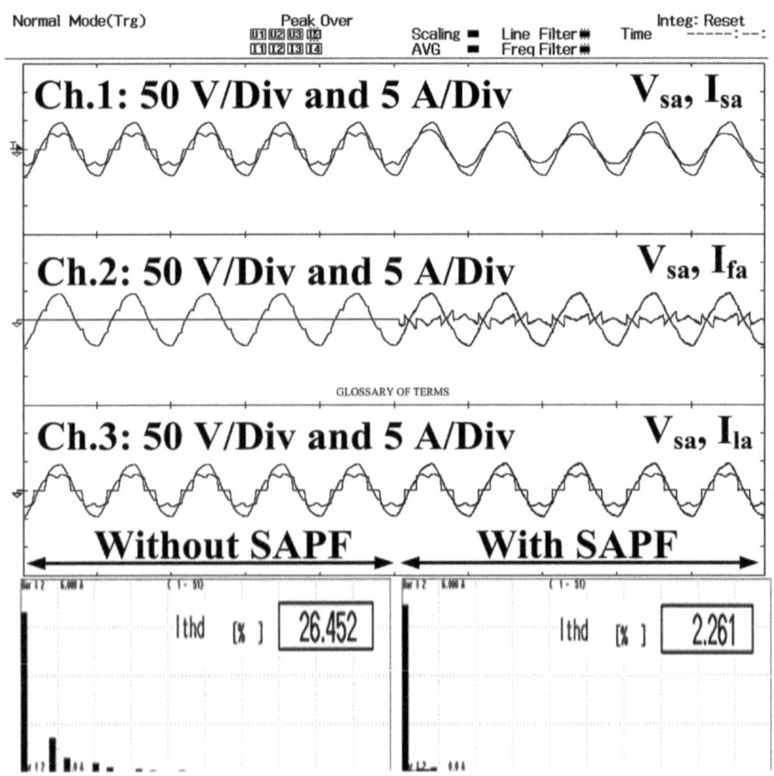

FIGURE 12.14 Real-time test results under test case 1.

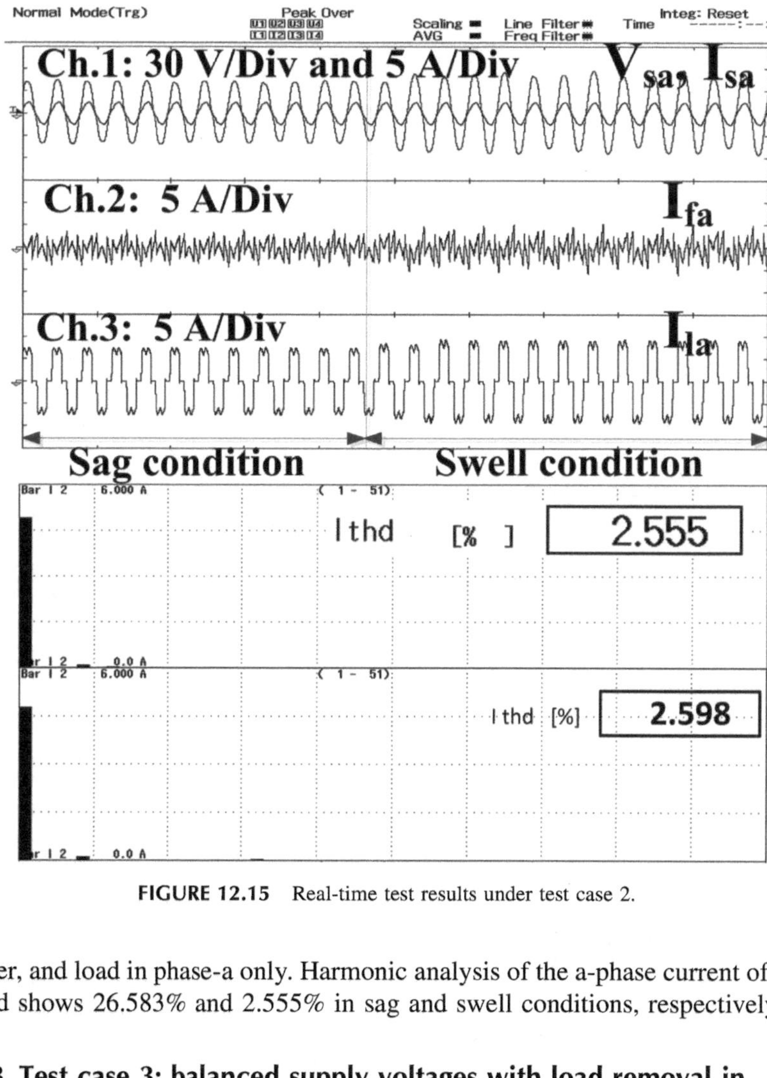

FIGURE 12.15 Real-time test results under test case 2.

filter, and load in phase-a only. Harmonic analysis of the a-phase current of the grid shows 26.583% and 2.555% in sag and swell conditions, respectively.

4.3 Test case 3: balanced supply voltages with load removal in phase-a

Practically, systems are heavily subjected to sudden load connections and disconnections. So the proposed SAPF scheme has been tested under such a performance condition. Up to 0.11 s, the system operates under normal condition, and at 0.11 s, the phase-a load current is removed. Fig. 12.8 shows the a-, b-, c-phase load currents, voltages of the grid, a-, b-, c-phase currents of the grid, and a-, b-, c-phase reference currents. Fig. 12.8 also displays the harmonic analysis of the a-phase grid current with 2.11% THD. The currents of the grid system continue to be sinusoidal. In this test condition, the power of the PV continues to be stable, and the DC-link voltage is stable. Similar to the simulation, the proposed control scheme under this condition is validated with experimental results. Current of the grid, filter, and load in phase-a, and THD analysis of a-phase current of the grid, are shown in Fig. 12.16. The harmonic analysis of load current before the load removal in phase-a is 26.275%, and the current of the grid THD after the SAPF is 2.289%.

4.4 Test case 4: unbalanced supply voltage condition

The execution of the suggested SAPF scheme with FLPID-ESN is performed under an unbalanced supply voltage case. This is an infrequent test case in practical systems. In this test case, the effectiveness of SAPF with FLPID-ESN is validated in simulation and real-time. Fig. 12.9 shows the unbalanced a-, b-, c-phase voltages of the grid, a-, b-, c-reference currents, a-, b-, c-phase current of the grid, and the DC-bus voltage. In phase-a current of the filter, a small swell occurs to compensate the source unbalancing—i.e., due to the drop in

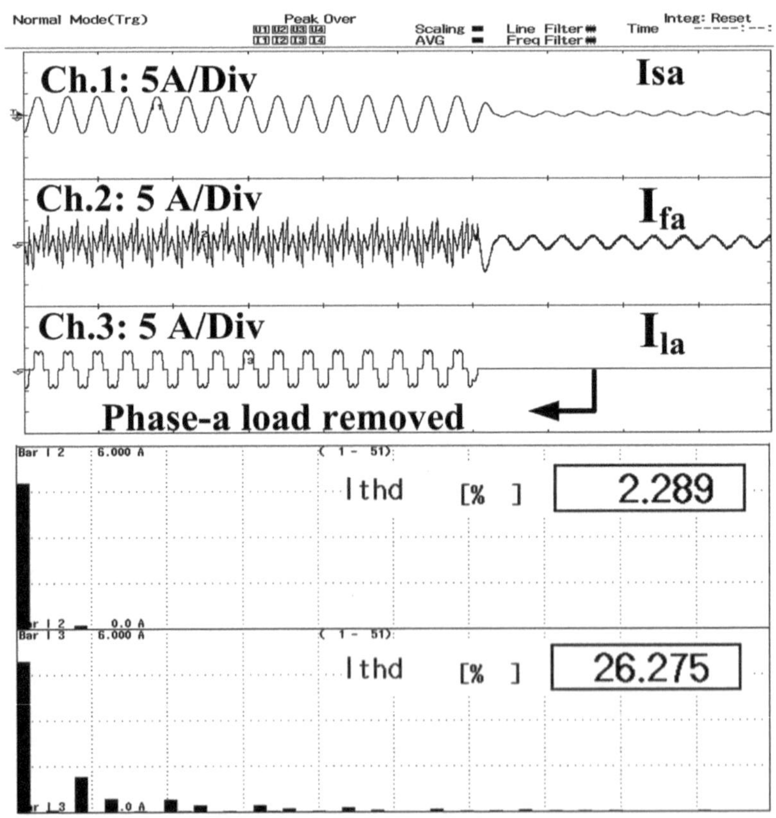

FIGURE 12.16 Real-time test results under test case 3.

voltage of phase-a. The source and load currents in phase-a decrease proportionally as the grid voltage decreases. The phase-a grid current's harmonic content with the FLPID-ESN technique in this scenario has a THD of 2.42% as shown in the figure. Fig. 12.17 shows the experimental results obtained under this condition. The first subfigure in this figure shows the unbalanced source voltages and currents of all three phases. The second subfigure shows the filter currents in a-, b-phases, and the a-, b-phase currents of the load are shown in the second subfigure, which closely match with simulation findings. The phase-a filter current compensates the unbalancing in phase-a supply voltage. The current harmonics of load before the SAPF is 26.399%, and the filter suitably reduces the source current THD to 2.652% using the proposed adaptive control approach, which is also observed in Fig. 12.17.

4.5 Test case 5: change in photovoltaic operating conditions

Unlike the test cases discussed earlier, in this test case, system performance has been verified with a rapid change in solar irradiation. As the irradiation enhances rapidly from 600 to 1000 W/m^2 at 0. 09s (in Fig. 12.10), the PV current also increases proportionally; however, the voltage of the DC link is kept stable by using the incremental conductance MPPT control technique. The PV variables with a-, b-, c-phase currents of the grid and reference currents appear in Fig. 12.10. The source current in phase-a has a lower THD of 2.51%, which satisfies the IEEE-519 standard. A similar test case is validated experimentally by increasing and decreasing the irradiation. As appears in Fig. 12.18, the source, filter, and load current in phase-a vary proportionally as the irradiation changes. The PV current also changes accordingly with changes in environmental conditions. The source current THD in phase-a drops to about 2.586% by using the proposed controller even under a rapid change in irradiation, which can be observed in Fig. 12.18. However, the load current THD of phase-a is still 26.130% and needs to be compensated.

4.6 Comparative performance of the fuzzy logic proportional integral derivative based on echo state network technique with existing techniques

Table 12.4 displays comparative performance under steady-state condition with IRP, SRF technique, Fryze compensation theory, and ADALINE technique based on their characteristics. Table 12.5 summarizes the comparative analysis of weight convergence with LMF and LMS with the FLPID-ESN technique. Under the steady-state condition, the performance of the FLPID-ESN technique is found to be the best among all these techniques. The harmonic analysis of grid currents of phase-a with IRP, SRF, ADALINE, and Fryze techniques is shown in Fig. 12.19A–D. Harmonic spectra of phase-a source currents result in THD of 3.15%, 3.58%, 4.28%, and 3.02%,

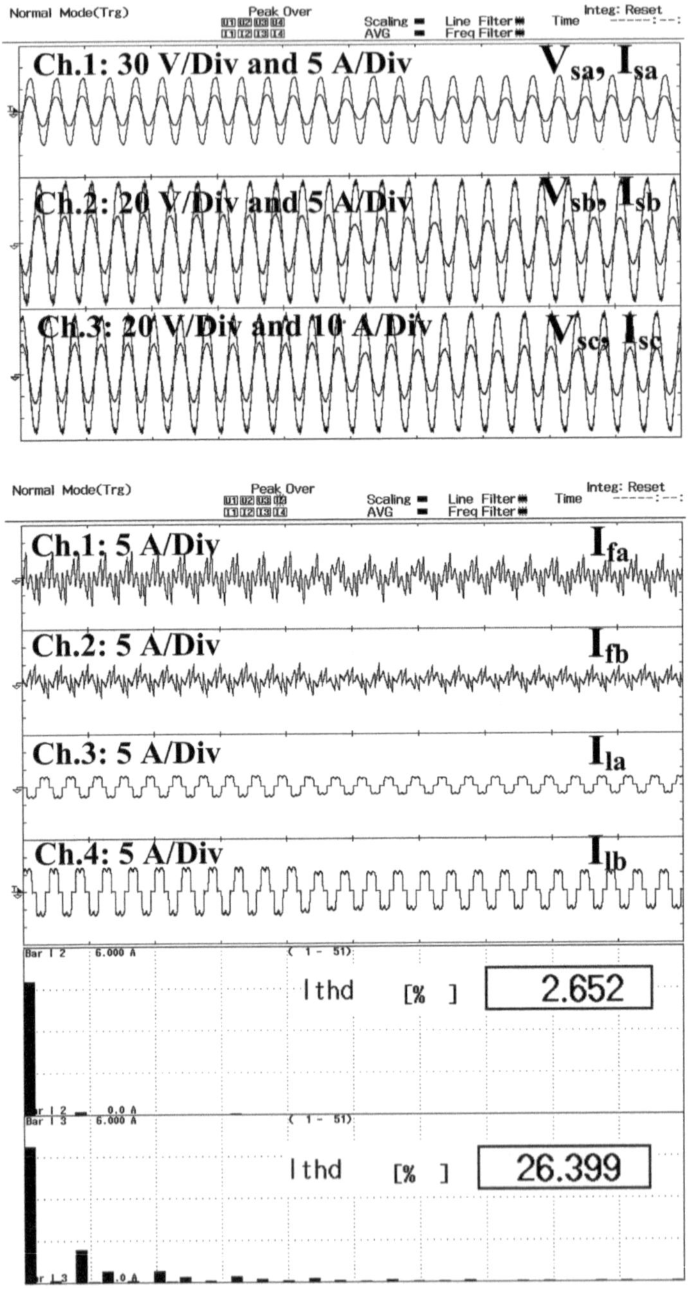

FIGURE 12.17 Real-time test results under test case 4.

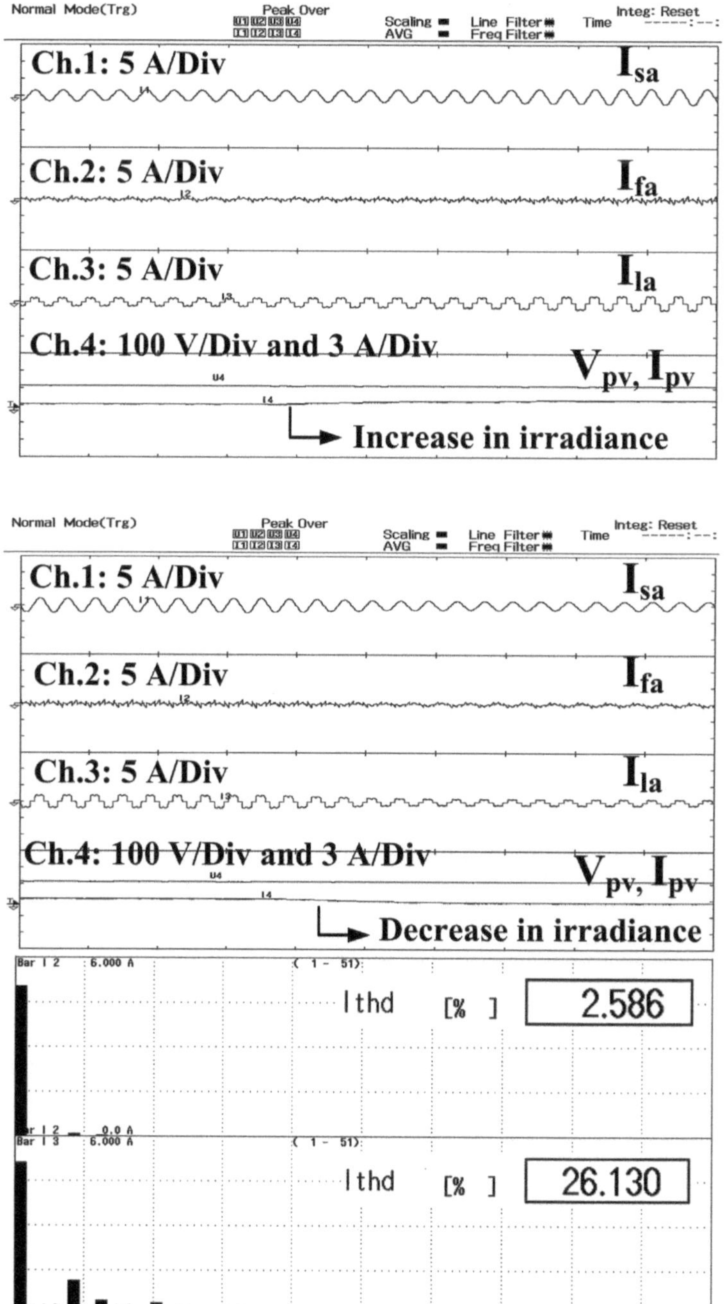

FIGURE 12.18 Real-time test results under test case 5.

TABLE 12.4 Comparative analysis of fuzzy logic proportional integral derivative based on echo state network with existing techniques.

S.no.	Characteristics	FLPID-ESN	IRP	SRF	Fryze	Adaline
1.	PLL needed	No	No	Yes	No	No
2.	Clark's/Park's transformation needed	No	Yes	Yes	No	No
3.	Steady-state convergence (1 cycle)	0.021 s	0.016 s	0.022 s	0.015 s	0.028 s
4.	THD of grid current (I_{sa})	2.162%	3.15%	3.58%	3.02%	4.28%
5.	Weight convergence	Fastest	NA	NA	NA	Slow
6.	Sampling time	55 μs	65 μs	70 μs	60 μs	55 μs

TABLE 12.5 Comparative analysis of fuzzy logic proportional integral derivative based on echo state network with adaptive techniques.

S.no.	Characteristics	FLPID-ESN	LMF	LMS
1.	Type of filter	Adaptive	Adaptive	Adaptive
2.	Oscillations	Less	More	More
3.	Complexity	Moderate	Low	Low
4.	Amplitude tracking capability	Better	Good	Good
5.	Accuracy	Better	Moderate	Moderate
6.	THD of grid currents	Low	High	High

respectively using these techniques. However, source-current THD using the proposed technique is lower than it is with these techniques. Comparative analyses of fundamental weight convergence (W_{pa}) with ESN, LMS, and LMF techniques are shown in Fig. 12.19E. The fastest weight convergence of the proposed techniques can be inferred from the figure.

5. Conclusion

A two-stage grid-friendly photovoltaic system employing an adaptive FLPID-ESN control scheme for quality-of-power enrichment has been highlighted in this chapter. Extensive real-time analysis is used to validate simulation outcomes under various load and grid-transient conditions. The grid-interfacing inverter is controlled by the proposed ESN technique, which is used to separate the FWC from the currents of the load and evaluate the calculated SAPF currents and accordingly generate switching signals for the SAPF system. Additionally, an FLPID adaptive voltage regulator is implemented to continue the reference voltage of the DC-bus in a two-stage PV power conversion system. The proposed FLPID-ESN adaptive control scenario uncovers several features such as better FWE separation, adaptive behavior at the DC and AC sides, harmonic rejection ability, and steady and transient behavior. The dynamic behavior of the proposed FLPID-ESN is analyzed and compared with a few recently developed control scenarios and shows better performance than those recently developed control schemes. The currents of the distributed terminal point that were obtained are well within the IEEE-519 standard limits.

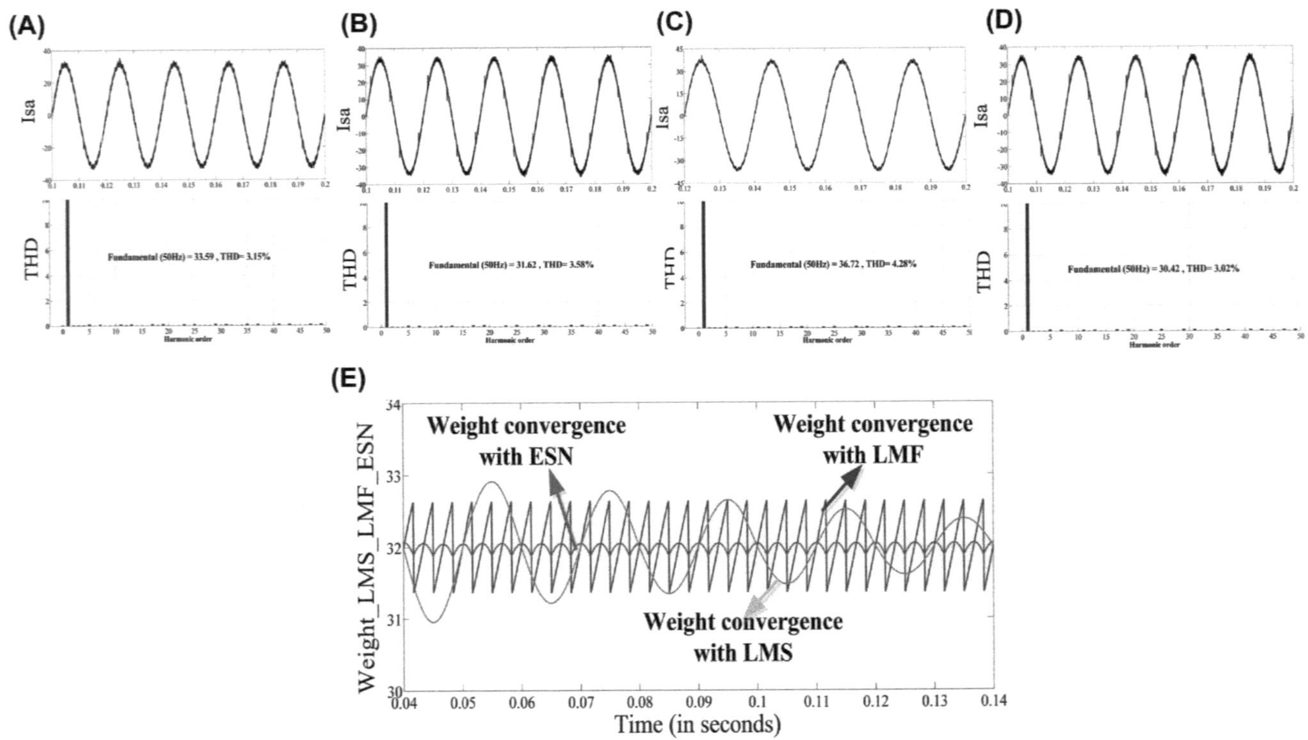

FIGURE 12.19 Comparative analysis of proposed control scheme with conventional control techniques: (A) grid current and THD spectra of IRP theory, (B) grid current and THD spectra of SRF theory, (C) grid current and THD spectra of ADALINE algorithm, (D) grid current and THD spectra of Fryze theory, and (E) comparative analysis of weight convergence of proposed control scheme with LMF and LMS techniques.

Nomenclature

I_{pv} Current of the PV
Δ I Change in current of the PV
V_{pv} Voltage of the PV
ΔV Change in voltage of the PV
P_{pv} Power of the PV
D Duty cycle
G Irradiation level
D_1 DC–DC boost converter's diode
L_1 DC–DC boost converter's inductor
S_b DC–DC boost converter's switch
V_{dc} Voltage of the DC-link bus
V_{dc}^* Reference voltage of the DC-link bus
C_{out} Capacitor of the DC-link
I_{labc} Nonlinear load currents
I_{sabc} Currents of the utility grid
V_{labc} Currents of the nonlinear load
V_{sabc} Voltages of the utility grid
V_{pa}, V_{pb}, V_{pc} Voltages of the PCC point
V_{qa}, V_{qb}, V_{qc} Unit vector quadrature voltages
W_{pa}, W_{pb}, W_{pc} FAWC of nonlinear currents
W_{qa}, W_{qb}, W_{qc} FQWC of nonlinear currents
V_t Voltage of the unit-vector template
V_t^* Reference voltage of the unit-vector template
W_{pdc} Active DC weight magnitude
W_{pavg} Average active weight magnitude
W_{qavg} Average reactive weight magnitude
W_p real power weights of load currents
W_q Reactive power weights of load currents
W_{net} Net weight component of the load current
S_1 to S_6 Switches of the grid-interfacing inverter
$I_{sa}^*, I_{sb}^*, I_{sc}^*$ Proposed controller calculated currents
P_{loss} Current loss components of the interfacing inverter
e DC-bus error voltage
$K_{p,iandd}$ PID gains
η Leaming gain
γ Momentum gain
$x(n)$ Input vector

Acronyms

PV Photovoltaic
MPPT Maximum power point tracking
PCC Point of common coupling
VSI Voltage source inverter
IC Incremental conductance
THD Total harmonic distortion
RES Renewable energy system
LPF Low pass filter

PID Proportional integral derivative
FLPID Fuzzy logic tuned PID voltage regulator
FWE Fundamental weight element
FAWC Fundamental active weight component
FQWC Fundamental quadrature weight component
ESN Echo state network
PWM Pulse width modulation
MF Membership function

Acknowledgment

This work is supported by the RECTPCL-CSR Funded Projects Company Limited Grant RECTPCL/CSR/2016–17/693.

References

[1] X. Liang, C. Andalib-Bin-Karim, Harmonics and mitigation techniques through advanced control in grid-connected renewable energy sources: a review, IEEE Trans. Ind. Appl. 54 (4) (2018) 3100–3111.

[2] N. Babu, J.M. Guerrero, P. Siano, R. Peesapati, G. Panda, An improved adaptive control strategy in grid-tied PV system with active power filter for power quality enhancement, IEEE Syst. J. (2020), https://doi.org/10.1109/JSYST.2020.2985164.

[3] S.K. Khadem, M. Basu, M.F. Conlon, Harmonic power compensation capacity of shunt active power filter and its relationship with design parameters, IET Power Electron. 7 (2014) 418–430.

[4] S. Mikkili, A. Panda, Real-time implementation of PI and fuzzy logic controllers based shunt active filter control strategies for power quality improvement, Int. J. Electr. Power Energy Syst. 43 (2012) 1114–1126.

[5] P.N. Babu, B. Kar, B. Halder, Modelling and analysis of a hybrid active power filter for power quality improvement using hysteresis current control technique, in: 2016 7th India International Conference on Power Electronics (IICPE), 2016, pp. 1–6.

[6] P.N. Babu, B. Kar, B. Halder, Comparative analysis of a Hybrid active power filter for power quality improvement using different compensation techniques, in: 2016 International Conference on Recent Advances and Innovations in Engineering (ICRAIE), 2016, pp. 1–6.

[7] Z. Wang, F. Zeng, P. Li, C. Wang, X. Fu, J. Wu, Kernel solver design of FPGA-based real-time simulator for active distribution networks, IEEE Access 6 (2018) 29146–29157.

[8] P. NB, B. CB, R.B. Peesapati, G. Panda, An optimal current control scheme in grid-tied hybrid energy system with active power filter for harmonic mitigation, in: International Transactions on Electrical Energy Systems 30, Wiley, 2020, p. e12183, https://doi.org/10.1002/2050-7038.12183.

[9] S. Leszek, Czarnecki, Constraints of instantaneous reactive power pq theory, IET Power Electron. 7 (9) (2014) 2201–2208.

[10] S. Zeliang, G. Yuhua, L. Jisan, Steady-state and dynamic study of active power filter with efficient FPGA-based control algorithm, IEEE Trans. Ind. Electron. 55 (4) (2008) 1527–1536.

[11] P.N. Babu, B.K. Choudhury, B. Kar, B. Halder, Modelling of a hybrid active power filter for power quality improvement using synchronous reference frame theory, Int. J. Eng. Res. Technol. 6 (3) (2017) 369–374.

[12] M. Qasim, V. Khadkikar, Application of artificial neural networks for shunt active power filter control, IEEE Trans. lnd. Inform. 10 (3) (2014) 1765–1774.

[13] J. Kanieski, R. Cardoso, H. Pinheiro, et al., Kalman filter based control system for power quality conditioning devices, IEEE Trans. lnd. Electron. 60 (11) (2013) 5214–5227.

[14] M.A.S. Masoum, S. Jamali, N. Ghaffarzadeh, Detection and classification of power quality disturbances using discrete wavelet transform and wavelet networks, IET Sci. Meas. Technol. 4 (4) (2010) 193–205.

[15] M.A. Platas-Garza, J.A. de la O Serna, Polynomial implementation of the Taylor-Fourier transform for harmonic analysis, IEEE Trans. Instrumentation and Measurement 63 (12) (2014) 2846–2854.

[16] M. Fallah, M. Imani, H.M. Kojabadi, et al., Novel structure for unbalance, reactive power and harmonic compensation based on VFF-RLS and SOGI-FLL in three-phase four wire power system, in: IEEE Energy Conversion Congress and Exposition (ECCE), Montreal, QC, 2015, pp. 6254–6260.

[17] R.K. Agarwal, I. Hussain, B. Singh, Application of LMS-based NN structure for power quality enhancement in a distribution network under abnormal conditions, IEEE Trans. Neural Netw. Learn. Syst. 29 (5) (2017) 1598–1607.

[18] P.N. Babu, B.C. Babu, P.R. Babu, G. Panda, Three-phase grid-tied photovoltaic system with an adaptive control scheme in active power filter, in: IEEE International Conference on Energy, Systems and Information Processing (IEEE ICESIP 2019), 2019, pp. 1–6.

[19] R.K. Agarwal, I. Hussain, B. Singh, LMF-based control algorithm for single stage three-phase grid integrated solar PV system, IEEE Trans. Sustain. Energy 7 (2016) 1379–1387.

[20] B. Singh, C. Jain, S. Goel, ILST control algorithm of single-stage dual purpose grid connected solar PV system, IEEE Trans. Power Electron. 29 (10) (2014) 5347–5357.

[21] S. Pradhan, I. Hussain, B. Singh, et al., Modified VSS-LMS-based adaptive control for improving the performance of a single-stage PV integrated grid system, IET Sci. Meas. Technol. 11 (4) (2017) 388–399.

[22] A.B. Shitole, H.M. Suryawanshi, et al., Grid interfaced distributed generation system with modified current control loop using adaptive synchronization technique, IEEE Trans. Ind. Inform. 13 (5) (2017) 2634–2644.

[23] D. Ali, R. Abolfazl, Application of echo state network for harmonic detection in distribution networks, IET Gener. Transm. Distrib. 11 (5) (2017) 1094–1101.

[24] P.N. Babu, P.R. Babu, G. Panda, An adaptive current control technique in grid-tied PV system with active power filter for power quality improvement, in: TENCON 2019 – 2019 IEEE Region 10 Conference (TENCON), 2019, pp. 187–191.

[25] P.N. Babu, P.R. Babu, G. Panda, A pre-filtering based current control strategy in grid-tied photovoltaic systems with active power filter for harmonic mitigation, in: TENCON 2019 – 2019 IEEE Region 10 Conference (TENCON), 2019, pp. 1003–1008.

[26] P.N. Babu, P.R. Babu, G. Panda, An adaptive differentiation frequency based advanced reference current generator in grid-tied PV applications, IEEE J. Emerg. Sel. Top. Power Electron. (2019), https://doi.org/10.1109/JESTPE.2019.2933140. Early Access.

[27] V.N. Kumar, P.N. Babu, R. Kiranmayi, P. Siano, G. Panda, Improved power quality in a solar PV plant integrated utility grid by employing a novel adaptive current regulator, IEEE Syst. J. (2019), https://doi.org/10.1109/JSYST.2019.2958819. Early Access.

[28] P. Gil, C. Lucena, A. Cardoso, L.B. Palma, Gain tuning of fuzzy PID controllers for MIMO systems: a performance-driven approach, IEEE Trans. Fuzzy Syst. 23 (4) (2015) 757–768.

[29] P.N. Babu, P.R. Bana, P.R. Babu, G. Panda, An interleaved buck converter based active power filter for photovoltaic energy application, in: 2019 First International Conference on Power Electronics Applications and Technology in Present Energy Scenario (PETPES-2019), 2019, pp. 1–6.

[30] T.K. Soon, S. Mekhilef, A fast-converging MPPT technique for photovoltaic system under fast-varying solar irradiation and load resistance, IEEE Trans. Ind. Inform. 11 (1) (2015) 176–186.

[31] P.N. Babu, B. Chittibabu, P.R. Babu, G. Panda, Three-phase grid-tied photovoltaic system with an adaptive current control scheme in active power filter, in: Energy Sources, Part A: Recovery, Utilization, and Environmental Effects, 2020, https://doi.org/10.1080/15567036.2020.1762807. Early View.

[32] S.J. Pinto, P.N. Babu, P.R. Babu, G. Panda, Monitoring and control of multibus microgrid system using FPGA platform, in: IEEE TENSYMP 2019, 2019.

Chapter 13

Renewable energy integration in modern deregulated power system: challenges, driving forces, and lessons for future road map

Divya Asija, Rajkumar Viral
Department of Electrical & Electronics Engineering, Amity School of Engineering and Technology, Amity University Uttar Pradesh, Noida, India

1. Introduction

1.1 Renewable energy

The estimated share of renewables in global electricity generation was more than 26% by the end of 2019. Renewable energy systems (RESs) such as small hydropower, biomass, wind, and solar photovoltaic have been effectively harnessed in past years for various uses. Presently, the overall stake of all renewable energy (RE) in electricity generation stands at about 28%, with hydropower making up an enormous majority (83%) [1]. Electricity production using biomass delivers feasible and continuing options for grid and off-grid electrification, though with its inept use; biomass resources currently offer about 26% of electricity generation share by the use of advanced, efficient, modern technologies [2] and has been rising globally. Wind and solar power have developed as the world's fastest-growing renewable sources due to local availability; they are mainly attractive for grid electrification at a large scale. Fig. 13.1 demonstrates the worldwide growth of RES electricity generation by source [2]. The main factor motivating the growth of distributed energy resources (DERs) is cost savings and the enhanced reliability of the energy system. DERs can make economic sense in situations where both generated electrical energy and processed heat are used by the facility. The equipment used for DERs are available in a wide range from 20 kW to over 200 MW. Steam, natural gas, fuel oil, and diesel oil can be used as generator

Advances in Smart Grid Power System. https://doi.org/10.1016/B978-0-12-824337-4.00013-8

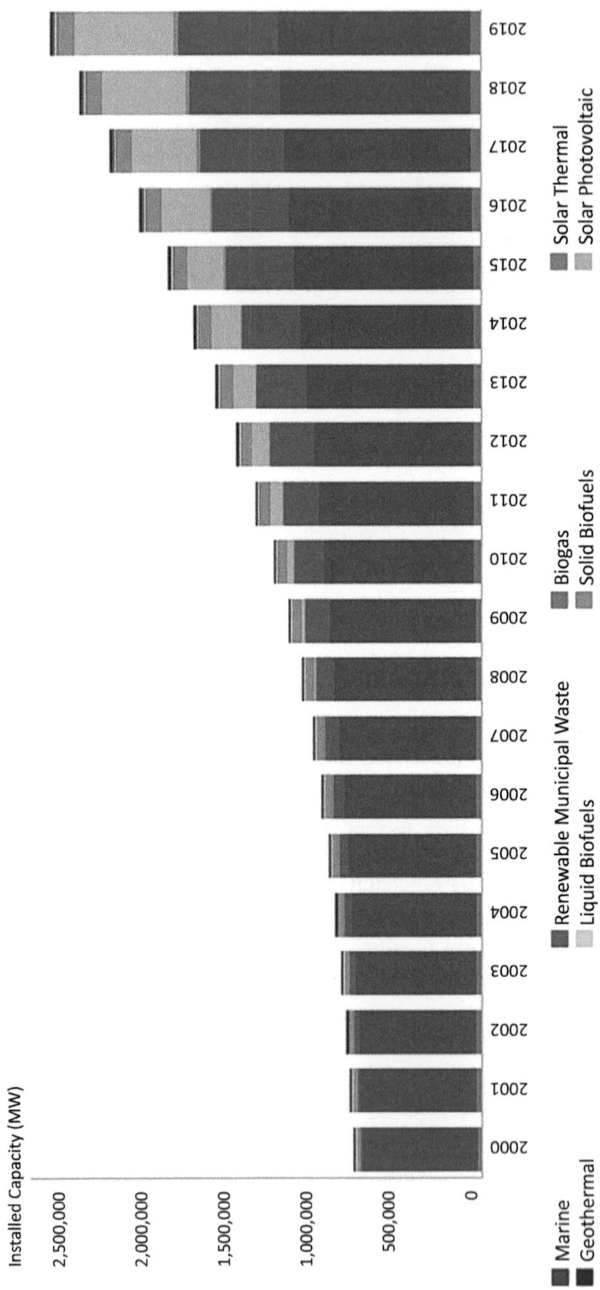

FIGURE 13.1 Growth of renewable energy systems in electricity generation by source worldwide.

fuel, and many prime movers can operate on dual fuels, thereby capitalizing on fuel sources that are more economical.

Prompt utilization of RE and energy efficiency, as well as technical divergence of energy resources, would bring about substantial energy security and economical profits. Recent years have seen growth in the exploitation of RESs in power grid systems. Numerous programs have been employed in different parts of the world, mostly in developed and developing nations. Many studies reveal that RESs can deliver reliable and reasonably low-cost electricity. To meet global energy demand, RE offers alternate solutions that allow for concerns about the environment as well as other socioeconomic factors [3]. RE-based technologies have cost structures that tend to disrupt traditional wholesale electricity prices in a liberalized market because they have either zero or near-zero marginal costs. The overall returns of RE are higher because of its lower marginal costs relative to the fixed marginal costs of fossil fuel—based electricity generation. With the displacement of fossil fuels in the present liberalized market, these returns would not be accepted, and certainly new rules need to be set according to the new market architecture. The major advantage of deregulation in the context of RESs comprises inexpensive electricity, efficient expansion of preset capacity, minimization of cost, diverse choice, and obviously reliable service to consumers.

1.2 Indian power scenario

The Indian economy has been growing very rapidly, accelerating the need to meet energy demands in a sustainable manner, though the economy faces many challenges in meeting required energy demand in coming decades. Cumulative energy requirements combined with a more sluggish than predicted upsurge in local fuel production indicate that ingress of RE into the energy mix is rising speedily. The population of India was estimated to be 1387 million as of January 1, 2020, with 70% of that in rural areas. Among them, nearly 44.7% are rural residents who have limited/erratic or no access of electricity [3]. The national grid of India had an estimated 370.106 GW of installed capacity as of March 29, 2020, whereas renewable power generation, which includes large and small hydroelectric plants and other sources, constituted 35.86% of total installed capacity [3]. India's total installed capacity (MW) regionwise as of March 31, 2020, is shown in Fig. 13.2.

India has extraordinary possibilities for harnessing RE because it is imparted with substantial natural and locally available resources and favorable geographic and climatic surroundings that sustain the promotion of RESs like solar, wind, biomass, small hydro, and others. Total power generation from renewables in India was estimated as 11,245.19 MU in February 2020 compared with 9342.95 MU in February 2019. The percentage share of total generating capacity by renewables sourcewise in India is shown in Fig. 13.3. It can be observed from this figure that solar generation has achieved the highest

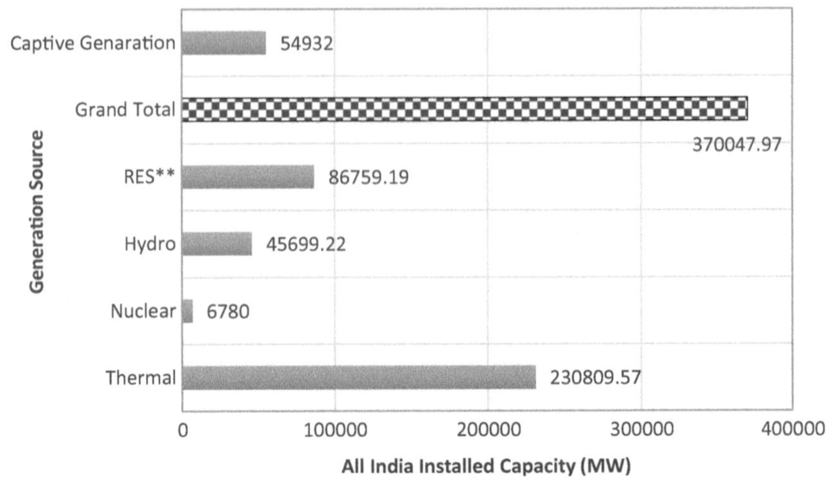

FIGURE 13.2 Toal installed capacity (MW) in India by region, March 31, 2020.

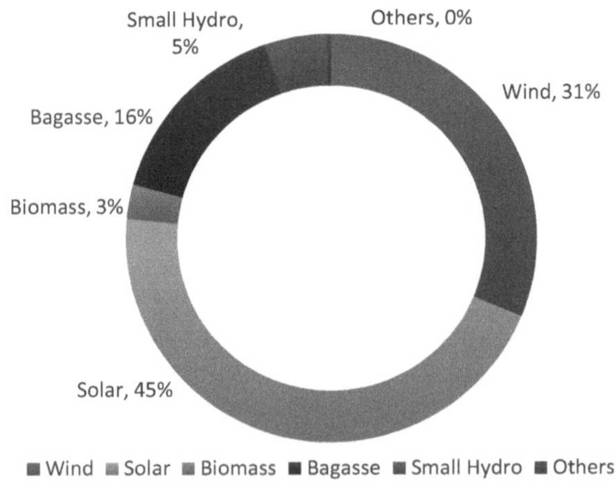

FIGURE 13.3 Percentage share of total generating capacity in India for all renewables by source, February 2020.

penetration, 45%, followed by wind at 31%, with the remainder from other RE resources [4].

Additionally, there is substantial potential for decentralized energy resources to meet various application needs, such as hot water requirements for residential, commercial, and industrial segments by solar energy and cooking energy demand in rural areas by biogas and biomass. RE also has significant potential for providing worldwide energy access. In decentralized or off-grid environments, RE is an alternative, scalable, and a feasible solution for supplying power to unelectrified or power-scarce rural, remote villages and hamlets.

India is fully determined in its goals for promoting RE. In India, RE is playing a gradually more vital role in the expansion of grid power, providing clean energy, minimizing the use of fossil fuels, and serving India in achieving its low-carbon goals. India's Intended Nationally Determined Contribution aim is to install 175 GW of renewable generating capacity by 2022 with a recent goal to increase the country's stake in nonconventionally based installed capacity to 40% by 2030 [4]. To achieve these manifold targets, the Indian government has implemented many schemes, policies, and dedicated central/state agencies. They are working to use and promote RE-based power generation for all applications countrywide. The government is conscious of the cutting-edge development that India can deliver in the crucial RE sector and has pushed the essential impetus for funds and incentives for innovative technology. The accessibility of assets at modest prices for additional progress of the segment is an added and significant challenge.

2. Structure of the modern power system

During the 1990s, the operational structure of electric utilities was completely transformed worldwide. Power transmission companies and electric utilities have been compelled to discriminate their business and operations from those of traditional vertically integrated mechanisms—i.e., bundled power systems to open market systems or unbundled power systems. This can be explicitly noticed in countries like Norway, the United States, Sweden, Finland, the United Kingdom, and some parts of South America. The numerous reasons for structural variations in power systems vary from region to region and by parts of countries. High demand growth combined with incompetent system administration and nonrational tariff policies were the prevailing issues in developing countries. Thus, the financial support required to develop improved generating and transmission capacity was reduced to a large extent. In such situations, many electrical utilities were compelled to reorganize their power sectors due to undue pressure created from external forces such as international funding agencies [1].

On the other hand, developed countries aimed to provide electricity at lower prices and with greater choices for purchasing electricity from cheaper sources. The objective of altering the mode of operation—i.e., deregulation or reformation—as we mentioned, is to enrich competition and attract consumers with new energy-selling options and accompanying financial benefits. When the power system is deregulated, the earlier vertically integrated utility performing the different power-related functions in sequence—i.e., generation to transmission, transmission to distribution, and distribution to retail sales—is dismantled into separate companies dedicated to each function. The electricity bill for the end consumer now involves at least two components: one from the company that provides generated electrical units at the distribution end and the other from the transmission network operator liable for providing network services. Deregulation mainly refers to reformulating the policies and economic incentives that governments set up to drive and control the electricity supply industry. It is known by different names such as reregulated market, competitive power market, open power market, vertically unbundled power system, open access, and power system restructuring (privatization and deregulation) and reform [5]. Several forces are behind deregulation:

- high tariffs
- global economic crunch
- overstaffing in each phase of the power sector
- regulatory failure
- inefficient managerial systems
- political pressure
- scarcity of resources required by the public for anticipated growth
- rise of environmentalism
- technological advancements
- rise in public awareness
- burden of financial bodies

The configuration of a vertically integrated electric utility is shown in Fig. 13.4 [6]. It shows that money flow is not bidirectional but unidirectional—i.e., from the customer or user to the electric company. Likewise, information flow is present only between generator companies and

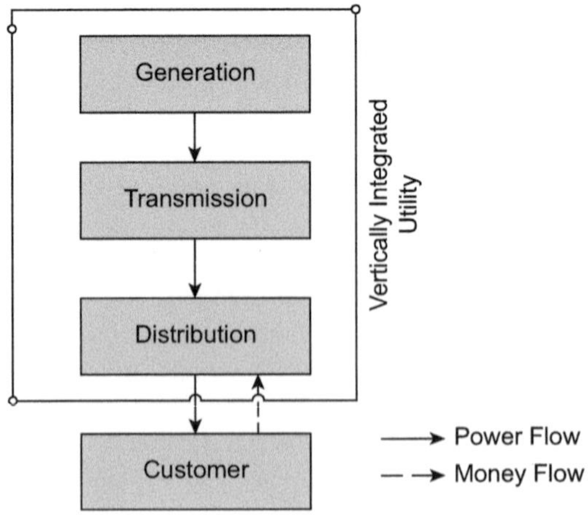

FIGURE 13.4 Vertically integrated electric utility.

the transmission systems involved. Since utilities were vertically integrated, it was often difficult to separate out the costs of individual factors related to generation, transmission, or distribution. Thus, utilities have been charging based on an average tariff plan that considers the average payment trend within a specific period [6].

The deregulated system configuration shown in Fig. 13.5 is not a universal configuration. This configuration has several variations with respect to different electrical systems and the geographic area. A system operator is selected for the whole system and has the duty of maintaining whole-system balance by keeping records of import and export of the electricity supply so that it matches generation and consumption without creating ambiguity in the system. An independent authority or system operator not involved in any type of market competition and not owning any generating bodies is known as an independent system operator (ISO). Comparing the components of Fig. 13.4 with those of Fig. 13.5, no alteration is identified as long as power flow is considered [6–8]. The customer has the ability to conduct power transactions directly with either a generating company or a retailer depending on the system category. A number of power vendors will transport their generated energy units to customers with the help of retail dealers on a shared platform of transmission and distribution (T&D) lines driven by the ISO. Generator companies, T&D utilities, and energy retailers communicate with the ISO. Most times, the customer communicates with the retailer for required energy. The retailer associates with generator companies and buys power from them. After this, they transfer it to the customer through regulated transmission and

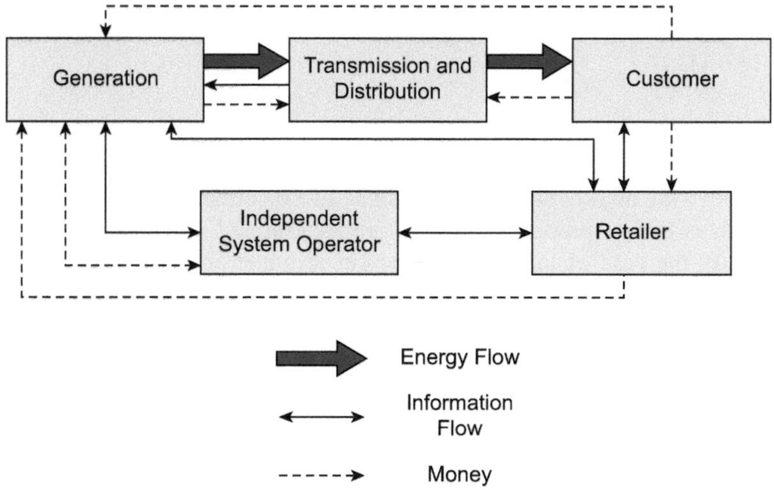

FIGURE 13.5 Typical arrangement of a deregulated power system.

distribution corridors. All these transactions taking place between various entities are tracked by a single responsible authority—i.e., the ISO [6–11].

2.1 Different markets

In the current scenario of a deregulated power system, electricity market players determine electricity prices by considering the demand and supply forces for trading or exchanging electrical units. In the bilateral mode, the combined auction and acquisition of electricity is done at a price decided by a competitive bidding process. Regulatory tariffs are imposed for central and state generating plants to transact electricity through specified network paths or routes. The market structure is decided based on certain significant parameters, such as the social welfare factor, that provide benefit to all entities—i.e., generation, T&D, low power loss, maintenance of power quality, voltage support, frequency control, and system reliability [9].

For each parameter specified, we have a service provider that requires a distinct market, and some parameters require the amalgamation of several markets.

Services can be provided by unrestricted markets, state electricity bodies, regulated electricity providers, or a hybrid of these providers. The most critical task is performed by the system operator who handles functionality in a regulated or deregulated market. In a deregulated market environment, the system operator provides the balancing and scheduling functions for operating in day-ahead and real-time markets [10].

Not all services at the two ends of demand and supply are entirely regulated; some are still under the roof of the regulated power system. For example, demand- and supply-side markets are competitive while certain ancillary services and the sourcing of transmission rights are mainly determined by the system operator in accordance with hourly demand as shown in Fig. 13.6 [11].

2.1.1 Day-ahead market

This type of market is based on auctioning electricity by an exchange or a pool technique. This market structure is designed to be either centralized or bilateral with two-way power flow. Bids are determined for maximizing total surplus. In the case of injecting free power at numerous locations, the locational energy price is set equivalent to the marginal change in total surplus. Three sets of conditions—guidelines to bid, directions for acceptance of bid, and settlement condition—are implanted before every auction. The multipart bid mechanism has been used by power pooling. An efficient auction is predominantly executed using a single-price day-ahead auction [12].

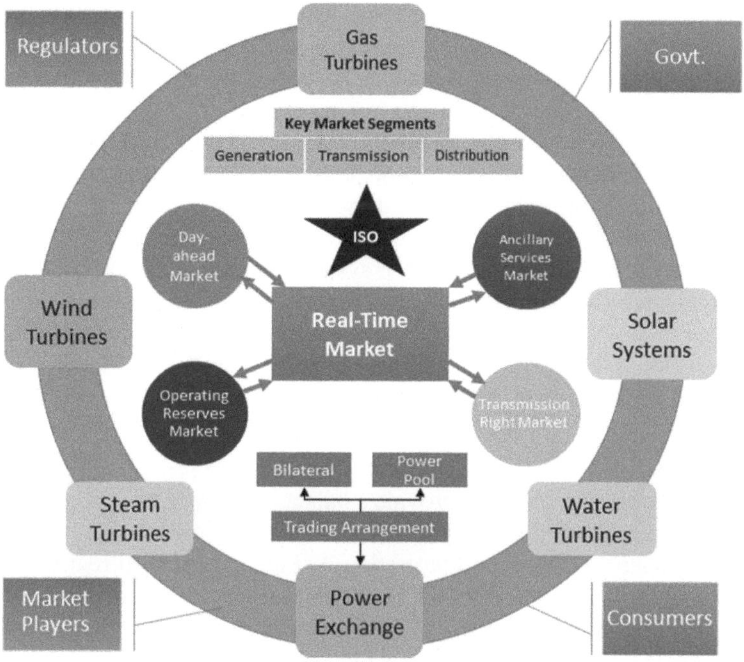

FIGURE 13.6 Electricity market arrangement.

2.1.1.1 Day-ahead market pricing structure

Day-ahead markets follow two main methodologies, the centralized nodal and bilateral pricing methods. Price is determined by examining all supply and demand bids, and then the supply is selected from the highest supply bid, and demand is selected as the lowest demand bid for market transactions. Price determination considers marginal surplus, which is the variation in total surplus when one unit of costless supply is augmented.

Unconstrained markets have certain evident features such as the availability of cost-free supply units, which would move the whole supply curve to the left as the quantities of power produced and consumed remain unaltered. Thus, if the marginal surplus is P, then the ultimate production cost savings would be P*1 kW. We must limit power consumption if the accessible generator has a high cost. But with the availability of an extra unit of 1 kW, consumption could be incremented. The cost of production is not altered in this case; it remains constant. Only consumption would be incremented to P*1 kW, thus further incrementing the total surplus by the equivalent value. For both considered cases, marginal surplus is equivalent to price.

Features of constrained markets are entirely different. In this market structure, both market players, the supplier and the consumer, lie on opposite sides of the market platform. Suppliers lie on the side of transmission constraints, while buyers are on the other. The free Kw of supply can be introduced on either side of the platform. Marginal surplus depends on the consumer and supplier sides for power consumption and availability. A kW increment on the consumer side would increment power demand, while a kW increment on the supply end would decrement the overall cost of supply. Due to price alterations by market type, the system operator seizes a portion of the total surplus, which is further equivalent to gross marginal surplus. Now at this point we can say that the competitive market price is set by considering only the marginal surplus, thereby inducing effectual performance. Market clearing is done and trade conducted at these available prices [11−15].

2.1.2 Real-time market

This type of market is meant to provide a balance of power and accomplish the retrieval of transmission security. It takes on the spot price and entails trades beyond the contract power demonstration that are available in real time [12]. Trading is done round the clock without any pause or delay until an accurate price is revealed. It is a centralized market based on approaches such as pool, exchange, or intermediaries, thereby achieving a balance between demand and supply. Thus, it is different from the bilateral approach, which is extremely slow to handle real-time power balancing and retrieval of transmission security, as trades depend on other partners. Trading is done in real time and at real-time prices without considering the contractual price. Real-time market price determination occurs on the basis of actual demand and supply conditions, whether forward contracts or real-time market scenarios are being considered [13].

2.1.3 Operating reserves market

This is a condition-based market where in some instances cheap generators are not allowed to produce more while expensive generators are allowed more production to generate spinning reserves for reliability and to control price hikes by maintaining reserves at optimum levels. Operating reserves are available in various forms. These markets behave differently from the conventional energy market by virtue of paying generators under a different price structure. Efficient and optimal capacity bid scoring is used for spinning reserves with the right information [14].

Without deliberation to maximize profit, suppliers are requested to bid for variable costing parameters such as capacity, opportunity, and energy. Bidders with the lowest supply bid are allocated spinning reserves, and those having the lowest energy bid are designated to fulfill the energy requirement. These bids would in turn intensify competition in determining the market clearing levels for energy and capacity prices [16]. The main causes of high costs come from utilizing the spin from running nonspinning generators [17–21]:

Generators have a certain operating limit below which they have great difficulty working and generating power efficiently. No load costs are borne by a generator at startup and shutdown, as it is not linked with output from the inframarginal generators. It is expensive at times when marginal costs are below the market price. In such a case, cheap generators are used less, thereby producing less quantity due to the availability of less spin from extramarginal and marginal generators, ultimately leading to an increased spin price. In this condition, the market price is above the margin cost for all generators that have variable costs for capacity provisioning.

2.1.4 Ancillary services market

Power is the main service in the electricity market, but to run the whole deregulated system smoothly, certain ancillary services are required. Of the six types of ancillary services, economic dispatch, transmission security, and trade enforcement are directly provided by the service operator. Leftover services such as black start capability, voltage stabilization, and balancing can be purchased from the competitive market. The system operator has the responsibility of balancing supply and demand by making them equivalent as shown in Fig. 13.7. Five markets are developed to operate in the deregulated environment. One market only serves to meet regulatory requirements and operates continuously, whereas the other four tackle deviations and emergency situations. Ancillary services benefit the whole market by providing extra services in the form of public goods or line items.

The demand and supply sides can be regulated jointly or individually operated according to market type. A nonprofit or profit-making entity provides the usual monopoly service—i.e., a system operator whose functionality is to provide regulated demand for ancillary service and coordinate the different markets. System operators also provide balancing services they have purchased and share the cost of services among generators or loads on a pro rata basis, or they assign a fraction of the cost to each supplier depending on usage of the physical infrastructure of transmission. The last approach is self-providing and is the least regulated.

Frequency correction is done by controlling the real power flow of each area system operator in a control area that individually do not provide frequency stability. Thus, the whole system is accountable for achieving frequency stability for the entire area. Generators and capacitor banks are exploited to provide voltage support. This provides reactive power support that cannot be transmitted easily. In a competitive environment it is difficult to purchase this service, but long-term contracts are there to assist.

The locational marginal pricing mechanism provides transmission security by controlling operation of the day-ahead market and transmission rights.

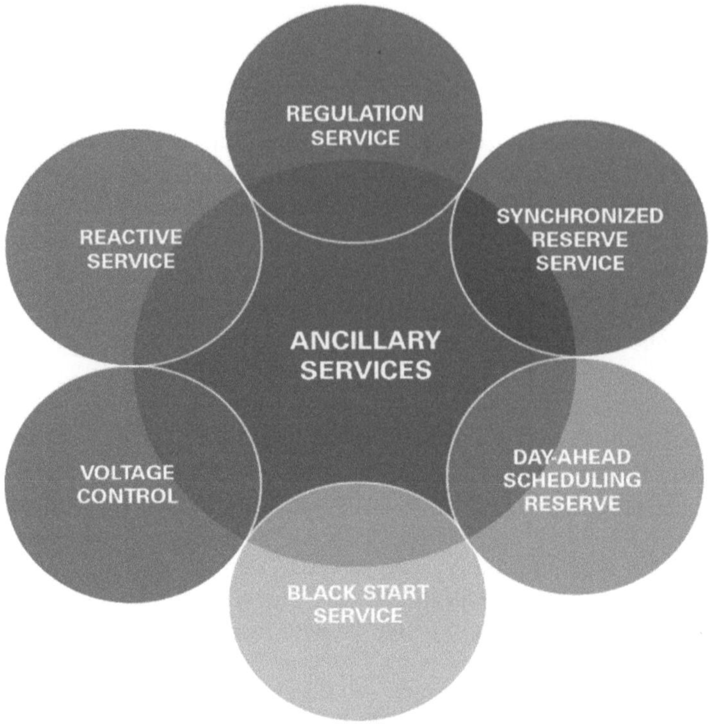

FIGURE 13.7 Types of ancillary services.

It considers optimal power flow through the entire network [17,18]. Economic dispatch of generators provides the best usage of generators with reference to selectivity and capacity. Dispatch and scheduling are used as a means to provide this service [19–21].

2.1.5 Transmissions rights market

The tools used to lower the influence of transmission price are physical transmission rights and financial transmission rights that protect the market from adverse situations. Under financial transmission rights, the transmission line owner receives the real-time price for transmitting power from that particular transmission path or line. On the other side, the physical transmission right provides the owner a right or priority to use its particular path or line. Thus, these rights provide a good structure for financial transactions. Consumers benefit from this market structure by getting power at the most competitive price because of the market's transparent and reliable methodology. It unbundles the vertically integrated system to a deregulated system with separate entities or market players for each phase of generation, transmission, and distribution. It has three main trading arrangements: pool, exchange, and bilateral. These arrangements discover the best price for trading and provide healthy competition [21–23].

2.2 Challenges faced by power markets

In the deregulated electricity market structure, power markets face certain major challenges in coping with newly reformed regulations [4,15,21,23,24]:

- Regulation that fails to keep pace due to structural factors such as limited access to low-cost fuels and variable electricity usage
- Higher industry risks due to desired investor returns
- Changing daily load curves
- Nonavailability of transmission network resulting in congestion risk
- Lack of government intervention in central policy matters
- Uncertainty of future environmental regulations
- Integration issues amid distributed energy sources and renewables
- Indeterminate demand and price levels
- Indeterminate fuel price

3. Renewable energy integration

3.1 Integration of renewable energy systems in the power market

The integration of RESs with conventional electricity grids is a very complex task. Numerous issues like the sporadic nature and inconsistency of RESs contribute to severe frequency variations and may expose the system to reverse power flow. The incorporation of RESs to the grid governs its impact on power system operation and its effect on the different upward contingencies of energy resources. This basic nature of RESs has many impacts on the power system. Consequently, the accumulation of these resources into the power system imposes a different set of system operating conditions like reverse power flow, rising system fault levels, harmonic distortion, instability, and voltage variation problems [25].

Large-scale integration of these resources may also adversely affect frequency and voltage control of the entire power system. For reliable and steady operation, certain protective measures must be taken on both sides of the system [26].

3.2 Major integration challenges

Variability in RE generating sources creates another concern for balancing the power system. Larger flexibility in the power network may be required to withstand supply-side inconsistencies and the connection to the generating side and loads. Based on the literature available, the major challenges of integrating RESs with existing power systems are categorized as technical and economic:

3.2.1 Technical challenges

Comprehensive types of technical challenges are associated with the system and disturb its performance [27]:

3.2.1.1 Performance of power system challenges

- Overvoltage issues due to remote generation of RES
- Increases in power quality disturbances
- Uncoordinated security of protecting systems

3.2.1.2 Line overloading and losses

The danger of overload is interrelated with the maximum standard value of the current, contributing to more losses. RESs are coupled closer to the load than conventional large-scale generation, causing less losses. Also, power flow from large voltage levels in the direction of the load is reduced and results in reduced risk of overloading.

3.2.1.3 Voltage variation challenges

The integration of RESs may change overvoltages and is a more serious concern. Especially for locations some distance from a main bus, rising voltage is the foremost restrictive factor for the integration of RESs [28].

3.2.1.4 Power quality challenges

Power quality disquiets the electrical interaction between the power network and its consumers. It mainly consists of two components, voltage quality subject to connected equipment and devices, and current quality issues for the mode in which equipment current influences the system [29].

3.2.1.5 Miscellaneous challenges

In addition to all these issues, a few other concerning factors reflect the impact of RESs—protection schemes, stability margins, voltage regulation, frequency distortions, etc.

3.2.1.6 Economic challenges

The huge investment deployed in establishing RE facilities for electricity generation, along with the aforementioned technical challenges, sometimes directly affects investment losses and returns. The various economic issues related to integration of RES into the existing power network are as follows [20–26,30]:

- Cost-effectiveness of techniques/methods and characteristics of the existing power grid network
- Appropriate grid infrastructure viability
- Best operational practices to incorporate RESs
- The generating fleets
- Regulatory frameworks that affect which solutions are the most economic and viable
- Risk hazing capacity
- Demand-response mechanisms
- Wide-scale development of storage
- Flexible expansion planning
- Reliable and affordable protection scheme

3.3 Challenges with power market deregulation

3.3.1 Impact on generating companies

Deregulation of electricity is very good for a developing nation, but the multiparty trust of generation companies (GENCOs) will be affected. This is because it is very difficult to store electricity, so it is disbursed when it is generated. Hence, ISOs that control user transmission access to the transmission line certainly affect the market players linked to generation [18].

3.3.2 Monetary conflicts of interest

Electricity prices are variable in a deregulated power system. Participants or market players are well versed in this fact. In some instances it happens that when the market price per unit of electricity slows, it has different valuations for generators and consumers. Generating companies generate profits at this phase due to bulk generation, while consumers are affected in terms of payment for this period. Thus, deregulation creates a monetary conflict of interest among different entities that leads to a loss of reliance among them and ultimately affects electricity transactions from one node to others in the network. This was experienced by Wales and England after privatization of their electrical entities. The Nigerian Electricity Regulatory Commission (NERC) has done a great job by acting as an impartial adjudicator in these power affairs [28].

3.3.3 Elevated demand in the transmission network

After implementation of a deregulated system, all entities become independent single entities moving from generation to transmission and transmission to distribution. In order to maximize profit, every generating company wants to transact its power from the path offering the least hindrance and highest profit. This results in an increment in the total number of transactions through the particular path, moving toward a complex situation for which the system is not designed. It sometimes leads to blackouts, which are not at all desirable for power system reliability. Elevated demand in the transmission network also hampers accountability for environmental and social constraints due to consideration of a comprehensive planning strategy that integrates generation (at the load end) or cogeneration with energy-efficiency arrangements [29].

3.3.4 Burden on the transmission grid

The transmission network has a complexity of transactions in a deregulated power system. Complexity in transactions through different paths of the transmission network are the root cause of accidents, and due to the volatile nature of electricity as a commodity, it will be further enhanced. Real-time control of power transactions along the transmission path is predominantly required due to its inability to store electricity and disposition to accidents that are difficult to fix. Balancing this abrupt change in the flow of power often requires power reserves and high transaction capacity in the transmission network for moving abundant power from one node to another or for immediate load shedding. But it is difficult to accommodate such changes, creating a burden on the transmission grid that is not at all desirable [27].

3.3.5 Augmented difficulty for coordinating operation of the electricity grid

The main task of the central entity in a bundled electricity system is to balance the load to segregate customers having different load usability patterns at different time slots of the day or for different seasons of the year. This helps allot the groups of consumers during peak and off-peak loads. In a deregulated electricity market, market players do not get any incentives for balancing the load relative to the load curve. Sellers and buyers of electricity are self-concerned, considering benefit or profit for themselves. They are not bothered by balance and coordination of the load. They may reserve generation capacity and misrepresent information. Thus, it is difficult to have reliable sources of information, which is a major concern for the new market structure. With this market structure, service contributors and brokers who bid for higher profits are the information providers. They have a tendency to skew information. Thus, it is quite difficult to coordinate the transmission grid of this new deregulated market structure [30].

3.3.6 Inhibition to invest

The transition of an electricity market structure from a bundled to an unbundled arrangement provides individual parameters of interest in gaining benefit. Individual entities are not ready to maintain and expand the existing transmission infrastructure. Thus, they are not willing to invest in increased capacity of the transmission network, as it would create more competition among electric utility merchants. In this era of deregulation, we do not rely completely on private players. In spite of that, we must have a mix of public and private utilities to serve the interests of all in a positive direction [28].

4. Driving forces and the future road map

Based on the discussion and aforementioned challenges, major driving forces and future directions can be engaged to promote and encourage the use of large integration of RESs into modern deregulated power markets:

- Market architecture should be well-defined according to the physical operating capacity of the grid and available grid capacity.
- Due to restructuring of the power market, GENCOs are free to generate and can incorporate significant integration of renewables, but due to a lack of large storage facilities and immature storage technology, new large storage systems must be developed so that flexible operation of RE is ensured.
- Due to deregulation, all RE base systems are considered at marginal price in the wholesale electricity market; this conflict of monetary interests among different entities should be avoided as in the case of the NERC market.
- Elevated demand in the transmission system is highly influential due to the large penetration of REs, and it closely depends on market type, so planning strategies should be constituted to consider these demands in advance to ensure system reliability and network resilience.
- RE base generation seems an opposite option for eliminating congestion within the transmission network, but suitable design mechanisms and well-incorporated implementation methods encourage generators and distribution companies to promote the increased use of RE-based generation.
- Further, an RES-based generation system provides opportunities for individual utilities/consumers to generate more renewable power and earn profits as well.
- RE-based generation could also help to create more incentives for sellers and buyers while considering balance and coordination of the load.
- RES options should evaluate the explicit requirements and be based on diverse services requirements related to suitable procurement processes as well as offer good rationalization to regulators.
- RESs are so integrated that they are much more flexible in designing favorable tariff mechanisms that are easily adoptable for system operators, utilities, and local consumers to achieve better incentives.
- Action plans and policy upgrades should be revisited from time to time to meet the most recent changes and advancements in the field of RESs.

- Transparency and implementation of the policies and regulatory framework are extremely important. All strategies and rules should be effortlessly available and self-explanatory so that no misunderstanding is caused.
- New persuasive schemes and plans should be employed to promote attentiveness and the development of RESs.

5. Conclusion

RESs lead to some distinct technical and economic complexities in power grids and electricity markets. Since RESs have little to no marginal costs, they can provide reduced prices in the day-ahead and real-time power markets, though this creates undesirable price variations in some of the cases. This produces issues for system operators in deregulated scenarios specifically, while meanwhile they stimulate significant generating capabilities for the power grid to meet dependability goals. Nonetheless, separate generating vendors stay in the market based only on trade criteria—whether they receive sufficient compensation to carry on meaningful, sustained operations. Energy output from RESs can also vary more quickly than system operators can control for, and thus a large addition of RESs could nearly upsurge the demand for present ancillary services (mostly frequency control) and may require new kinds of ancillary services, which could be delivered by demand or supply side sources. Consequently, this chapter resolves to deliberate the growth in RES-based generation, its addition to the modern liberalized power market, the building of the modern restructured power sector, key issues with RE integration in deregulated power markets, and motivating factors. Based on convincing upcoming trends, this should provide assistance for policymakers, power system planners and operators, managers/regulators, and RE stockholders to develop a more flexible framework for the future.

References

[1] BP-The Energy Outlook, 2019. https://www.bp.com/content/dam/bp/business-sites/en/global/corporate/pdfs/energy-economics/energy-outlook/bp-energy-outlook-2019.pdf.

[2] International Renewable Energy Agency (IRENA). https://www.irena.org/Statistics/View-Data-by-Topic/Capacity-and-Generation/Statistics-Time-Series. (Accessed May 2020).

[3] CEA. http://cea.nic.in/reports/monthly/installedcapacity/2020/installed_capacity-04.pdf. (Accessed May 2020).

[4] CEA Executive Summary Report March 2020, http://cea.nic.in/monthlyexesummary.html. (Accessed May 2020).

[5] S.C. Pirrong, The Economics, Law and Public Policy of Market Power Manipulation, vol. 24, Kluwer, Boston, 1996, p. 10, 59, 70.

[6] D. Phillipovic, Energy Risk: Valuing and Managing Energy Derivates, McGraw-Hill, New York, 1998, p. 3.

[7] D. Bohi, Analyzing Demand Behavior: A Study of Energy Elasticities, Resources for the Future/Johns Hopkins, Baltimore, 1981.

[8] P. Robert, The Structure of World Energy Demand, MIT Press, Cambridge, 1979.

[9] P. Joskow, R. Schmalensee, Markets for Power: An Analysis of Electric Utility Deregulation, MIT Press, Cambridge, 1984.

[10] K. Rose, Electric Restructuring Issue for Residential and Small Business Customers, National Regulatory Research Institute, Columbus, OH, June 2000.

[11] E. Hirst, B. Kirby, Retail-Load Participation in Competitive Wholesale Electricity Markets (Prepared for the Edison Electric Institute and the Project for Sustainable FERC Energy Policy, December 2000.

[12] S. Braithwaite, in: "Customer Response to Market Prices – How Much Can You Get when You Need it Most?" EPRI International Energy Conference, Washington, D. C, July 2000.

[13] Schwarz, et al., Industrial Response to RealTime Prices for Electricity: Short-Run and Long-Run, University of North Carolina, December 2000.

[14] Letter from Rederick H. Ritts, Attorney for Steel Dynamics Inc., a Large Industrial Consumer of Electricity to Alan Richardson of the American Public Power Association and Roy Thilly of Wisconsin Public Power, Inc., FERC Staff Report, August 19, 1998, pp. 5–6, 5-7.

[15] Outage Report, Identifies Numerous Problem with Information Including a General Lack of Data (Finding 6), Poor Load Projections and Forecasts (Findings 8, 17, 28), Unit Ratings (Finding 11), Cable Conditions and Incipient Failure (Finding 5, 14), Inadequate Notice (Finding 20) and Failure to Preserve Records (Finding 33).

[16] "Initial Comments of the Consumer Federation of America," Remedying Undue Discrimination through Open Access Transmission Service and Standard Electricity Market Design, Federal Energy Regulatory Commission, Docket No.RM01-12-000.

[17] "Initial Comments of the Consumer Federation of America," Remedying Undue Discrimination through Open Access Transmission Service and Standard Electricity Market Design, Federal Energy Regulatory Commission, Docket No. RM-01-12-000, October 15, 2002 (West and South).

[18] FERC Staff Report, pp. 3-20.

[19] The Information Problem Received the Most Attention in the Outage Report. A Number of Information and Management Weaknesses Are Noted Including Inadequate Forecasting Tools (Finding 8, 13, 17, 18 20), a Lack of Monitoring Instruments (Findings 5, 11, 14), and Little Real Time Information to Respond to Problems (Findings 6, 27).

[20] FERC Staff Report, 43 Cambridge energy research associates (CERA), electric power trends: 2001. High Tension: The Future of Power Transmission in North America (August 2000) (Hereafter, CERA, High Tension), 2000, 3-2, 4-3, 4-4, 4-16.

[21] D. Stipp, The real threat to America's power, Fortune (March 5, 2001).

[22] K.,E. Harris, Thoughts on Wild Prices, The Economic Theory of Auction (Nuffield College, July 2000), July 1998, p. 1, 44 Id. 45 Klemperer, Paul.

[23] R.C. McDiarmid, L.G. Dowden, D.I. Davidson, A modest proposal: revoke the nobel prize? Recognize the limitations of theory? Or grant a license to steal? Electr. J. (January/February 2001).

[24] IOSR Journal of Electrical and Electronics Engineering (IOSR-JEEE), May–June 2014, pp. 19–22, e-ISSN: 2278-1676, p-ISSN: 2320-3331, Volume 9, Issue 3 Ver. III, www.iosrjournals.org.

[25] O. Anyaka, Challenges of Deregulation of Electric Power Sector in a Third World Economy Boniface.

[26] Contribution to Bulk System Control and Stability by Distributed Energy Resources Connected at Distribution Network, January 15, 2017. IEEE PES Technical Report.

[27] B. Tomoiagă, M. Chindriş, A. Sumper, A. Sudria-Andreu, R. Villafafila-Robles, Pareto optimal reconfiguration of power distribution systems using a genetic algorithm based on NSGA-II, Energies 6 (2013) 1439–1455.
[28] R.E. Brown, L.A.A. Freeman, Analyzing the reliability impact on distributed generation, IEEE — Power Eng. Soc. 2 (2001) 1013–1018. Summer Meeting.
[29] M. bollen, F. Hssan, Integration of Distributed Generation in the Power System, IEEE Press, 2011.
[30] R.C. Dugan, T.E. McDermott, Operating conflicts for distributed generation on distribution systems, in: Rural Electric Power Conference, April 2001.

Index

Note: 'Page numbers followed by "f" indicate figures and "t" indicates tables.'

G

H

I

P

Q

R

FSC
www.fsc.org
MIX
Papier | Fördert gute Waldnutzung
FSC® C083411

Kolibri 360 GmbH
Ferdinand-Jühlke-Straße 7
99095 Erfurt, Deutschland
produktsicherheit@kolibri360.de